Experimental Brain Research Series 15

Generation and Modulation of Action Patterns

Edited by H. Heuer and C. Fromm

In Collaboration with
C.H.M. Brunia, J.A.S. Kelso, and R.A. Schmidt

With 90 Figures

Springer-Verlag
Berlin Heidelberg New York
London Paris Tokyo

Dr. Herbert Heuer
Abteilung für Psychologie
Fakultät für Psychologie und Sportwissenschaft
Postfach 8640, 4800 Bielefeld 1, FRG

Professor Dr. Christoph Fromm
Max-Planck-Institut für Biophysikalische Chemie
Postfach 2841, 3400 Göttingen, FRG

ISBN-13:978-3-642-71478-8 e-ISBN-13:978-3-642-71476-4
DOI: 10.1007/978-3-642-71476-4

Library of Congress Cataloging-in-Publication Data. Generation and modulation of action patterns. (Experimental brain research series; 15) Bibliography: p. Includes index. 1. Neuropsychology. 2. Motor ability. 3. Movement, Psychology of. 4. Efferent pathways. I. Heuer, H. (Herbert), 1948-. II. Fromm, C. (Christoph) III. Series.
QP360.G46 1986 599'.01852 86-20299

2125/3140-543210

Preface

During the academic year 1984/85 a study group on "Perception and Action" was established at the Center for Interdisciplinary Research of the University of Bielefeld. Among the various activities of this group was a series of symposia. This volume covers the proceedings of one of the symposia, held in October 1984 and entitled "Generation and Modulation of Action Patterns".

Bringing together scientists from diverse disciplines in general is challenging, but also hazardous. Aware of the notorious communication problems, it was tried to keep the number of representatives of at least the major areas of Neurobiology and Psychology about equal and to elicit presentations that combine experimental findings with theoretical statements on major issues in motor control. Despite the different viewpoints, concepts, research strategies, and methods as they are reflected in the present volume, there was the general impression among the participants that such a multidisciplinary effort was profitable and one step on the way to the realization of common goals and common conceptual grounds.

The presentations and discussions during the meeting were centered around the following issues: (1) characteristic properties of action patterns: do they exist and what is their origin? (2) formation and initiation of action patterns: what happens on the route from a general plan until the movement emerges on the surface? (3) spatio-temporal sequencing: what are the kinematic principles and the underlying mechanisms? (4) adjustment of on-going movements to environmental demands and maintenance of action patterns in the face of an unpredictable environment. This, of course, represents a "soft" structure with a lot of overlap. In the present volume it has been reduced to just two sections, i.e. "Formation and Initiation of Action Patterns" and "Modulation and Maintenance of Action Patterns".

As most of the study group's activities the planning of this symposium was a joint endeavor of several people. During a preparatory meeting in 1983 a committee of five persons – the present editors – was formed to realize the plans in cooperation with the organizers of the study year (W. Prinz, O. Neumann, P. Wolff, A.F. Sanders, and H. Heuer) and the staff of the Center for Interdisciplinary Research. We gratefully acknowledge the financial and organizational support of the Center. The resources and the special atmosphere of this institution made this symposium not only possible but also shaped the spirit of open-mindedness that is a necessary prerequisite for communication across the borders of established disciplines. We are indebted to the contributors to this volume and to those who helped to edit it by reviewing the contributions. Finally, we thank Wilfried Staemmler who devoted his skills of word processing to refashion the various formats of the manuscripts.

Bielefeld and Göttingen
August 1986

Herbert Heuer
Christoph Fromm

Contents

List of Contributors*

Annett, J. *187*[1]
Arbib, M.A. *158*
Atkeson, C.G. *41*
Bingham, G. *158*
Brunia, C.H.M. *28*
Chapman, C.E. *14*
Cruse, H. *263*
Dean, J. *263*
Georgopoulos, A.P. *242*
Graham, D. *252*
Haagh, S.A.V.M. *28*
Heuer, H. *87*
Hindorff, V. *64*
Hollerbach, J.M. *41*
Iberall, T. *158*
Kelso, J.A.S. *105*
Lamarre, Y. *14*
Larish, D.D. *55*
Lee, D.N. *217*
MacKay, D.G. *174*
Massey, J.T. *242*
Monsell, S. *72*
Morasso, P. *145*
Munro, E.M. *64*
Rosenbaum, D.A. *64*
Saltzman, E. *129*
Schwartz, A.B. *242*
Shapiro, D.C. *231*
Viviani, P. *201*
Wiesendanger, M. *3*
Young, D.S. *217*
Zaccaria, R. *145*
Zelaznik, H.N. *55*

---

* The address of the authors is given on the first page of each contribution
[1] Page, on which contribution commences

Formation and Initiation of Action Patterns

Initiation of Voluntary Movements and the Supplementary Motor Area

M. Wiesendanger*

Institut de Physiologie, Université de Fribourg, 1700 Fribourg, Switzerland

THE SEARCH FOR NEURAL EVENTS IN THE BRAIN WHICH PRECEDE VOLUNTARY MOVEMENTS

By introspection it seems evident that the execution of a voluntary movement is preceded by mental planning of that movement. Several contributors to this volume have addressed this general question in terms of instructions given to the subject before movement execution. Using the reaction–time paradigm, some indirect conclusions can be drawn about how this information may be used and how it affects the timing of the motor reaction. Besides this psychophysical approach to the problem, a major interest in the planning of movements can be traced back to clinical studies. It was mainly in the nineteenth century that movement disorders were correlated with brain pathology. The general concept emerged that destruction of the motor cortex and its output pathways affect movement execution; whereas, lesions in some nonprimary cortical areas lead to disorders which appear to concern movement planning rather than execution. Thus, the clinical picture of "apraxia" may intuitively be interpreted as a mental disorder in the construction of the correct spatiotemporal "kinematic melody" ("Bewegungsentwurf") of the intended action. In his classical studies, Liepmann (1920) made for the first time a phenomenological classification of apractic manifestations. Ever since, attempts to attribute the various forms of apraxia to lesions of defined cortical areas have been in the foreground of pathological studies (e.g., von Monakow 1914; Kleist 1934; Luria 1980; Geschwind 1974). These studies revealed a remarkably rich variety of different "higher–order" motor disorders which, of course, also aroused the interest of psychologists studying higher brain functions.

It is not the place here to review these different categories of apraxias, of motor neglect syndromes and of general reductions in motor drive (for recent accounts, see Heilmann and Valenstein 1979; Kertesz 1983). However, it may be noted in the present context that the cortical field between the agranular motor cortex and the granular frontal cortex has been considered to constitute a hierarchically superior motor cortical area, corresponding approximately to Brodmann's area 6 or Economo and Koskina's area FC (cf. Wiesendanger 1981). Interestingly, Campbell (1905) proposed (on grounds which are not clear) that this area, which he termed "intermediate" precentral cortex, may represent Hughling Jackson's highest level of motor control. This idea was remarkably resistant to some vehement criticism (see Fulton 1951; Walshe 1935; Woolsey et al. 1952). In more

* My work and that of my collaborators has been supported by the Swiss National Science Foundation (grant no. 3.752.80 and 3.522–0.83).

recent years, the writings of Luria (1980) and Geschwind (1974) as well as results of animal lesion experiments again revived interest in area 6.

The premotor cortex, as the lateral portion of area 6 is commonly named following Fulton's proposal, is now generally considered to be a higher–order motor field (cf. also Wiesendanger 1981, for review). Unfortunately, these lesion studies in neurological patients and in animal experiments do not lend themselves for a deeper understanding of the normal function of this area and they tell us nothing about the temporal implications in the planning of movements.

This then begs the question of whether there are other ways to gain further insights into the brain processes which precede movements, a question which was highly motivated by the clinical observations alluded to above. One first step in this direction was the introduction of electromyography (EMG) in the analysis of voluntary movements which was so brilliantly pioneered by Wachholder (1928). The EMG reflects the last–order control signal leaving the central nervous system (CNS). This method thus provided a "window" which allows one to glance into the actions of the brain and the spinal cord.

A major step forward was achieved with the technical innovation of recordings from single cortical neurones in the awake, performing monkey (Jasper et al. 1958). It thus became possible to analyze the activity of cell populations in various motor structures in a reaction time paradigm. The minimum reaction time for a hand movement in the monkey is about 120 ms. Recordings in the motor cortex by Evarts (1966, 1968) brought us back in time by about 50 ms: this was a typical interval preceding EMG onset during which pyramidal tract cells were activated. Was it possible to step further back in time and to analyze neurones in structures which are "upstream" of the motor cortex? Glickstein (1972) found that for visually triggered reactions in trained monkeys, 35–40 ms are used for the signal to reach the visual cortex. Adding this afferent delay of 35 ms to the delay of 50 ms in the efferent pyramidal path, leaves us with an interval of about 30 ms or less for transmission from the sensory cortex to the motor cortex. In such a simple form of "voluntary" movement, this interval is relatively short, but the time increases of course with the more natural situation of a choice reaction. As will be discussed below, the brain processes preceding the release of a self–initiated movement are even much longer.

Grossly speaking, all processes preceding motor output may be subsumed under the term movement initiation, but otherwise this term is not clearly defined. On grounds of psychological reasoning, one may subdivide movement initiation into three categories: (a) sensory evaluation of external or internal signals; (b) decision making; and (c) selection of an appropriate movement strategy. Experimental data reveal that such categories provide reasonable definitions for neural activation processes in a spatial and temporal order. Thus, Rolls (1983), who investigated monkeys grasping for food, was able to establich a sequence of neuronal activation in different brain structures. On the basis of the response pattern and its timing it was usually possible to attribute the neural event and its timing to one of the three categories (e.g., visual evaluation of the food in the inferotemporal cortex, gustatory evaluation in the hypothalamus, etc.).

In influential papers by Kemp and Powell (1971) and Allen and Tsukahara (1974), the principal pathways which lead to the motor cortex have been schematically divided into the corticostriatal, transcerebellar, and corticocortical loops. Has the prediction that the implicated structures are "upstream" of the motor cortex stood up in face of the experimental evidence?

A major source of cortical input to the motor cortex originates in the premotor cortex. Recent findings are indeed compatible with the "upstream hypothesis." I will deal later with the similar position of the supplementary motor area.

With respect to the transcerebellar loop, it appears that neurones of the lateral cerebellum and its output nucleus are activated slightly before the motor cortex (Brooks and Thach 1981). No clear–cut evidence has so far been obtained with respect to basal ganglia output. Reviewing current knowledge, DeLong et al. (1984) recently concluded that the putamen and globus pallidus "are more evidently involved in the execution of limb movement than in the timing of its initiation." However, it is possible that the projection system from association areas to the caudate nucleus and its return path, via the pallidum and thalamus, to the prefrontal association cortex, which has hardly been explored thus far, is concerned with movement initiation.

It should be noted that it is far from clear whether or not the above–mentioned pathways are activated in parallel or whether they subserve specific and exclusive functions in movement initiation.

A completely new dimension was added to the discussion of movement initiation in recent years when it was found that, in man, the medial frontal cortex, possibly the supplementary motor area (SMA), appears to be selectively involved in the process of "movement mentation." The evidence is derived from observations in human patients who underwent measurements of cerebral blood flow (CBF). It was found that a complex finger motor task activated contralateral peri–Rolandic cortex as well as the SMA of both sides. In the following, I will briefly review data, mainly from human studies, which support the view that the SMA has a programming function and that it plays an important role in self–initiated volitional movements. This is in marked contrast with the classical view according to which the SMA is a motor field with a direct access to the spinal cord, acting in parallel with the precentral motor cortex (Woolsey et al. 1952). I will also mention experimental evidence which is more consistent with the older classical view of Woolsey. In conclusion, I will attempt to find an explanation for the seemingly contradictory concepts of the SMA.

THE SMA IS RELATED TO THE SPINAL CORD

The SMA, here defined structurally as the medial portion of cytoarchitectonic area 6, forms, together with the precentral motor cortex of area 4 and the premotor cortex of lateral area 6, the frontal agranular field. The main cytoarchitectonic difference between the SMA and the adjacent area 4 of the motor leg and tail areas is the presence of conspicuous giant Betz cells in the latter and their absence in the former. Like the motor cortex, the SMA contains a continuous row of pyramidal tract cells which are slowly conducting (Macpherson et al. 1982a, b). As in the motor cortex, intracortical microstimulation (ICMS) can evoke motor effects suggesting that the SMA has a relatively direct access to the motoneurones. It should be noted, however, that there are also some characteristic differences in the organization of the SMA projection to the spinal cord (for details, see Wiesendanger 1985): First, the motor effects produced by ICMS are limited to the posterior half of the SMA, and the dominant effects concern shoulder and trunk muscles with relatively rare involvement of distal muscles. Second, the efferent microzones appear to be more intermingled in the SMA as compared with the precentral cortex, and with ICMS the rostrocaudal "simiusculus" representation in the SMA is less evident than in the precentral cortex. Third, the density of the corticospinal projection in the SMA is much less than in the precentral cortex.

THE SMA RECEIVES SENSORY INFORMATION

Although some authors consider the SMA to play a particular role in self–initiated (rather than in sensory–driven) movements, it is now clear that sensory information has access to the SMA. Anatomical studies revealed that somatosensory areas (SI, SII, and parts of the parietal association cortex) feed into the SMA. We have recently found that subsets of SMA neurones indeed react vigorously to passive displacements of the arm in the awake monkey (Wiesendanger et al., 1986). The response patterns were rarely direction–specific and occurred at latencies which, on the average, were delayed with respect to the latencies of precentral neurones studied in the same experiments (the difference in the peak latency was about 10 ms). In earlier experiments on anesthetized monkeys, we found SMA neurones displaying a marked convergence from different limb nerves (Wiesendanger et al. 1973). In human subjects, Libet et al. (1975) recorded long–latency potentials in the SMA which were evoked by light and acoustic stimuli. Finally, it was shown by Tanji et al. (1980) and Tanji and Kurata (1982) that auditory, visual, and somaesthetic signals were quite effective in driving SMA neurones of trained monkeys if these instruction signals called for a movement (often the cell response was more contingent on the sensory signal than on the motor response).

THE SMA GENERATES "COMMAND SIGNALS" FOR MOVEMENTS

Covariance of SMA neurones with a conditioned movement of the arm has now been found by a number of investigators. In two studies, one using self–paced grasp movements (Brinkman and Porter 1979), the other simple movements triggered by sensory cues (Tanji and Kurata 1982), it was found that the lead time of SMA discharges relative to movement onset was long (about 120 ms), suggesting that the SMA is implicated earlier in the process of initiation than is the motor cortex. In a third study, however, Smith (1979) noted a very broad scatter of neuronal onset times, with the majority of SMA cells recruited even after the earliest onset of EMG activity. The task was a self–paced precision grip which had been studied before under the same conditions with recordings from precentral neurones (Smith et al. 1975). The reason for the remarkable difference in the outcome of the above studies is not clear. It does, however, emphasize the caution with which one has to interpret such findings with respect to the problem of serial activation of brain structures during movement initiation. This was well expressed by Fetz (1981) who wrote: "A fundamental problem in attempting to demonstrate serial activation of different motor centers is that any particular region, including the motoneuronal pools, contains cells that are recruited over diverse times, making it difficult to interpret the relative onset times of particular cells in different regions. Moreover, since the duration of most movements greatly exceeds the conduction time between centers, recurrent loops could be "traversed" many times during a single response; thus, the conceptually appealing notion that initiation of movement involved sequential activation seems difficult to prove." The issue of whether or not the SMA is upstream, downstream or parallel to the motor cortex can thus not unequivocally be established. Experimental data are compatible with each of the three possibilities, and this may reflect the anatomical fact that the SMA is reciprocally related to the motor cortex and that it also has its own output to the spinal cord.

THE SMA GENERATES "INSTRUCTION SIGNALS" FOR THE MOTOR CORTEX

Some experimental data do support the notion that the SMA is hierarchically superior to the motor cortex (the upstream theory). Thus, Tanji et al. (1980) discovered a population of SMA cells whose activity was best related to visual instruction (cue) signals indicating to the monkey in which direction a subsequent reaction time movement had to be performed. Usually, these cells did not covary with movement execution. We have found that short-latency, reflex-like discharges of precentral neurones, elicited by muscle stretch, could be suppressed by a preceding electrical stimulus applied to the SMA (Wiesendanger et al. 1975). These results, obtained in anesthetized monkeys have recently been confirmed in an awake monkey. A reflex-like discharge was elicited by passive arm displacements (torque pulse perturbations). Weak intracortical stimulation in the SMA was again found to diminish the subsequently evoked discharge in the motor cortex (Wiesendanger and Hummelsheim, unpublished). We tentatively suggest that the SMA plays a role in "setting" the responsiveness of motor cortical neurones to peripheral feedback signals.

THE SMA RECEIVES INPUTS FROM BASAL GANGLIA, LATERAL CEREBELLUM AND ASSOCIATION CORTEX

The SMA is itself implicated in the three principal transmission pathways thought to be involved in movement initiation (cf. Kemp and Powell 1971; Allen and Tsukahara 1974). The evidence is chiefly anatomical. Injections of the SMA with retrograde and anterograde marker substances revealed labeling in subdivisions of the "motor" thalamus known to receive pallidal outflow (Schell and Strick 1984; Wiesendanger and Wiesendanger 1984, 1985a, b; Jürgens 1984). Schell and Strick (1984) concluded, on indirect grounds, that no significant information from the cerebellum reaches the SMA via the thalamus. The problem, however, is open to discussion since, in our own investigations, we found that the lectin-bound marker substance was transported across cellular boundaries to label also discrete portions of the contralateral dentate nucleus (Wiesendanger and Wiesendanger 1985b). We inferred from these data that lateral cerebellar outflow is also directed via the thalamus to the SMA, an input the SMA is sharing with the motor cortex and the premotor cortex.

The SMA has further complex interconnections with cortical association areas: inputs were found to originate in prefrontal, parietal, and temporal association areas, in the frontal eye field, and in the anterior cingulate cortex. These afferent connections are mainly directed to the rostral portion of the SMA (Jürgens 1984; Wiesendanger and Wiesendanger 1984).

It can, thus, be concluded that the SMA is one of the many "throughstations" relaying activity from association areas to the motor output zones, and that access to the SMA is possible via all three routes indicated in the scheme of Allen and Tsukahara (1974).

THE MOTOR DEFICITS IN SMA-LESIONED MONKEYS ARE DISCRETE

It was clear from the first studies that SMA lesions, whether unilateral or bilateral, do not result in a paresis of limb or trunk muscles. Some authors (e.g., Travis 1955), but not all (e.g., Coxe and Landau 1965) noted an increased muscle tone, and the occurrence of forced grasping was repeatedly mentioned (Penfield and Welch 1951; Travis 1955; Smith et al. 1981). A peculiar bimanual deficit (which was restored after transection of the corpus callosum) was described by Brinkman (1981, 1984). Moreover, this author found, on close inspection of video tapes, other subtle changes in a food retrieval task (lack of hand shaping, bizarre sequence of retrievals, jerky movements). In squirrel monkeys, Kirzinger and Jürgens (1982) made the interesting observation that anterior SMA lesions abolished the "isolation peep," thought to be a learned and not an innate vocalization.

It thus appears that some subtle deficits (which are easily missed on cursory inspection of the lesioned animals and which are often transient) indeed support the contention that the SMA may be implicated in higher-order programming.

IN HUMAN PATIENTS LESIONS OF THE MEDIAL-FRONTAL CORTEX PRODUCE DRAMATIC DEFICITS

Brain tumors and vascular lesions occupying the medial-frontal cortex including the SMA may lead to a pronounced reduction of self-initiated voluntary movements (see Wiesendanger 1985, for detailed references). There are now a number of reports about the poverty of movements and particularly also of spontaneous speech. A relatively good recovery was noted in many cases. The problem of these case reports is that the lesions (often produced by an obstruction of the anterior cerebral artery) were rarely confined to the SMA. In most instances, the exact extent of the destruction could not be established. Nevertheless, it appears that the region of the human SMA, possibly including neighboring structures such as the medial prefrontal and anterior cingulate cortex are crucially involved in the generation of self-initiated voluntary movements (more so than in triggered movements).

THE BRAIN METABOLISM OF MEDIAL-FRONTAL CORTEX IS INCREASED IN HUMAN SUBJECTS PERFORMING COMPLEX FINGER MOVEMENTS AND ALSO DURING MENTATION OF THIS TASK

Measurements of local changes in cerebral blood flow (CBF) are thought to reflect changes in neuronal (and presumably also glial) activity. An implication of the Rolandic cortex (e.g., Olesen 1971) and of the medial-frontal cortex (Roland and Larsen 1976) during motor performance was evident, as one could have expected. More surprisingly, it was found that the medial-frontal focus persisted even if the movement was not executed, but only mentally conceived (Ingvar and Philipson 1977; Roland et al. 1980). Reports of this observation have been received with much enthusiasm because they suggest that the SMA might indeed be involved in programming rather than in execution of movements, and that this area occupies a hierarchically higher position than the motor cortex (as borne out by the term supramotor area, Orgogozo and Larsen 1979). The problem is again that spatial resolution for the metabolic changes was too crude to be sure that they involved only the SMA. However, improved techniques are now available which may more accurately delimit the implicated zones during movement planning (Roland et al. 1982). Furthermore, interpretations of the metabolic changes are hampered by the fact that the method fails to provide the temporal dimension. Lastly, movement "mentation" is hard to control and might, in fact, reflect visual or verbal imagination of the movement, or simply some other sort of attention.

THE PREMOVEMENT "READINESS POTENTIALS" (BEREITSCHAFTSPOTENTIALE) ARE LARGEST OVER MEDIAL – FRONTAL – CORTEX

It is now well known that self-initiated movements are preceded by a positive potential shift which slowly builds up within about 1 s and which can be recorded bilaterally over wide areas of the cerebral hemispheres (Deecke et al. 1969). It was furthermore noted that this premovement slow potential was largest over the vertex, i.e., approximately over the SMA, and that it persisted also in Parkinsonian patients who often fail to display readiness potentials over the Rolandic cortex (Deecke and Kornhuber 1978). From this the authors also concluded that the SMA is crucially involved in the initiation process of voluntary movements. Of course, the same caveat for the exact localization of these potentials, as outlined for the CBF studies, holds also for the slow potential recordings. In this context, it is worth mentioning that similar premovement potentials may be recorded in the monkey brain. Unfortunately, it was found that the potentials in the SMA were relatively small compared, for example, with those recorded from the premotor or prefrontal cortex (Gemba and Sasaki 1984).

A SPECULATION THAT THE SMA EXERTS A LOW – LEVEL AS WELL AS A HIGH – LEVEL CONTROL, AND CONCLUSION

The SMA has a number of properties which it shares with the motor cortex and there can be no doubt that, at least in nonhuman primates, it exerts some direct influence on the motor apparatus. Compared with the motor cortex, however, this influence is considerably less powerful. This is also evident from lesion studies. On the other hand, the available experimental and clinical data which I have outlined above also indicate that the SMA is implicated in "programming" (I believe that the latter assertion cannot be discarded even with a conservative interpretation of the human studies with regard to localization). It thus appears that the traditional concept of Woolsey et al. (1952) and the "modern" concept of a "supramotor area" are not mutually exclusive. What then might be the significance of such a double role?

As proposed earlier (Wiesendanger and Wiesendanger 1984), I suggest that it might be useful were a programming brain structure also capable of directly controlling the motor apparatus. Such a "low-level" control may be seen in the context of postural preparation of a forthcoming movement. The commands for such anticipatory postural changes have to be released at an early stage of movement initiation. This speculation can now be tested in experimental animals and in SMA-lesioned patients. Some investigations in patients with medial-frontal lesions have been reported which suggest that anticipatory postural changes may be lacking in such patients (Elner and Gabibov 1977). However, that study provides no details of experimental data and of the exact localization of the lesion.

In my opinion, it is likely that such a role would not be limited to the SMA and that wide areas of the cerebral cortex are implicated in the process of movement initiation. It is also likely that structures other than the SMA may be involved in the coupling of movements and posture and of movement synergies. There is no doubt, however, that the currently available anatomical and physiological data offer exciting avenues for further research dealing with the general problem of movement initiation.

Acknowledgements. I am grateful to Mrs. S. Rossier for secretarial help.

REFERENCES

Allen GI, Tsukahara N (1974) Cerebrocerebellar communication systems. Physiol Rev 54: 957–1006

Brinkman J (1981) Lesions in supplementary motor area interfere with a monkey's performance of a bimanual coordination task. Neurosci Lett 27: 267–270

Brinkman C (1984) Supplementary motor area of the monkey's cerebral cortex: short– and longterm deficits after unilateral ablation, and the effects of subsequent callosal section. J Neurosci 4: 918–929

Brinkman C, Porter R (1979) Supplementary motor area in the monkey. Activity of neurons during performance of a learned motor task. J Neurophysiol 42: 681–709

Brooks VB, Thach WT (1981) Cerebellar control of posture and movement. In: Brooks VB (ed) Motor control, part 2. American Physiological Society, Bethesda, pp 877–946 (Handbook of Physiology, Sec 1. The Nervous System, vol II.)

Campbell AW (1905) Histological studies on the localization of cerebral function. Cambridge University Press, Cambridge

Coxe WS, Landau WM (1965) Observations upon the effect of supplementary motor cortex ablation in the monkey. Brain 88: 763–772

Deecke L, Kornhuber HH (1978) An electrical sign of participation of the mesial 'supplementary' cortex in human voluntary finger movement. Brain Res 159: 473–476

Deecke L, Scheid P, Kornhuber HH (1969) Distribution of readiness potentials, pre–motion positivity, and motor potential of the human cerebral cortex preceding voluntary finger movements. Exp Brain Res 7: 158–168

DeLong MR, Alexander GE, Georgopoulos AP, Crutcher MD, Mitchell SJ, Richardson RT (1984) Role of basal ganglia in limb movements. Human Neurobiol 2: 235–244

Elner AM, Gabibov GA (1977) Possibility of recovery of the postural synergy after removal of the brain tumor. Agressologie 18: 69–73

Evarts EV (1966) Pyramidal tract activity associated with a conditioned hand movement in the monkey. J Neurophysiol 29: 1011–1027

Evarts EV (1968) Relation of pyramidal tract activity to force exerted during voluntary movement. J Neurophysiol 31: 14–27

Fetz EE (1981) Neuronal activity associated with conditioned limb movements. In: Towe AL, Luschei ED (eds) Motor coordination. Plenum, New York, pp 493–526 (Handbook of Behavioral Neurobiology, vol 5)

Fulton JF (1951) Physiology of the nervous system, 3rd edn., Oxford University Press, New York

Gemba H, Sasaki K (1984) Distribution of potentials preceding visually initiated and self–paced hand movements in various cortical areas of the monkey. Brain Res 306: 207–214

Geschwind N (1974) Selected papers on language and the brain. Reidel, Boston

Glickstein M (1972) Brain mechanisms in reaction time. Brain Res 40: 33–37

Heilman KM, Valenstein E (1979) Clinical neuropsychology. Oxford University Press, New York

Ingvar DH, Philipson L (1977) Distribution of cerebral blood flow in the dominant hemisphere during motor ideation and motor performance. Ann Neurol 2: 230–237

Jasper H, Ricci GF, Doane B (1958) Patterns of cortical neuronal discharge during conditioned responses in monkeys. In: Wolstenholme G, O'Connor C (eds) Neurological basis of behavior. Churchill, London

Jürgens U (1984) The efferent and afferent connections of the supplementary motor area. Brain Res 300: 63–81

Kertesz A (1983) Localization in neuropsychology. Academic Press, New York, p 527

Kleist K (1934) Gehirnpathologie. Barth, Leipzig, pp 343–1408

Kemp J, Powell TPS (1971) The connexions of the striatum and globus pallidus: synthesis and speculation. Phil Trans Roy Soc Lond [Biol] 262: 441–457

Kirzinger A, Jürgens U (1982) Cortical lesion effects and vocalization in the squirrel monkey. Brain Res 233: 299–315

Libet B, Alberts WW, Wright EW, Lewis M, Feinstein B (1975) Cortical representation of evoked potentials relative to conscious sensory responses, and of somatosensory qualities in man. In: Kornhuber HH (ed) The somatosensory system. Thieme, Stuttgart, pp 291–308

Liepmann H (1920) Apraxie. In: Brugsch T, Eulenburg A (eds) Real–Encyclopädie der gesamten Heilkunde. Urban und Schwarzenberg, Berlin, pp 116–143 (Ergebnisse der gesamten Medizin, vol 1)

Luria AR (1980) Higher cortical functions in man, 2nd edn., Basic, New York

Macpherson JM, Marangoz C, Miles TS, Wiesendanger M (1982a) Microstimulation of the supplementary motor area (SMA) in the awake monkey. Exp Brain Res 45: 410–416

Macpherson JM, Wiesendanger M, Marangoz C, Miles TS (1982b) Corticospinal neurones of the supplementary motor area of monkeys. A single unit study. Exp Brain Res 48: 81–88

Olesen J (1971) Contralateral focal increase of cerebral blood flow in man during arm work. Brain 94: 635–646

Orgogozo JM, Larsen B (1979) Activation of the supplementary motor area during voluntary movements in man suggests it works as a supramotor area. Science 206: 847–850

Penfield W, Welch K (1951) The supplementary area of the cerebral cortex. Arch Neurol Psychiat 66: 289–317

Roland PE, Larsen B (1976) Focal increase of cerebral blood flow during stereognostic testing in man. Arch Neurol 33: 551–558

Roland PE, Larsen B, Lassen NA, Skinhoj E (1980) Supplementary motor area and other cortical areas in organization of voluntary movements in man. J Neurophysiol 43: 118–136

Roland PE, Meyer E, Shibasaki T, Yamamoto YL, Thompson CJ (1982) Regional cerebral blood flow changes in cortex and basal ganglia during voluntary movements in normal human volunteers. J Neurophysiol 48: 467–480

Rolls ET (1983) The initiation of movements. In: Massion J, Paillard J, Schultz W, Wiesendanger M (eds) Neural coding of motor performance. Exp Brain Res [Suppl] 7: 77–117

Schell GR, Strick P (1984) The origin of thalamic inputs to arcuate premotor and supplementary motor areas. J Neurosci 4: 539–560

Smith AM (1979) The activity of supplementary motor area neurons during a maintained precision grip. Brain Res 172: 315–327

Smith AM, Hepp–Reymond MC, Wyss UR (1975) Relation of activity in precentral cortical neurons to force and rate of force change during isometric contractions of finger muscles. Exp Brain Res 23: 315–332

Smith AM, Bourbonnais D, Blanchette G (1981) Interaction between forced grasping and a learned precision grip after ablation of the supplementary motor area. Brain Res 222: 395–400

Tanji J, Kurata K (1982) Comparison of movement–related activity in two cortical motor areas of primates. J Neurophysiol 48: 633–653

Tanji J, Taniguchi K, Saga T (1980) Supplementary motor area: neuronal response to motor instructions. J Neurophysiol 43: 60–68

Travis AM (1955) Neurological deficiencies following supplementary motor area lesions in Macaca mulatta. Brain 78: 174–201

Von Monakow C (1914) Die Lokalisation im Grosshirn und der Abbau der Funktion durch kortikale Herde. Bergmann, Wiesbaden

Wachholder K (1928) Willkürliche Haltung und Bewegung. Ergeb Physiol 26: 568–775

Walshe FMR (1935) On the "syndrome of the premotor cortex" (Fulton) and the definition of the terms "premotor" and "motor" with a consideration on Jackson's views on the cortical representation of movements. Brain 58: 49–80

Wiesendanger M (1981) Organization of secondary motor areas of cerebral cortex. In: Brooks VB (ed) Handbook of Physiology, vol II, part I. American Physiological Society, Bethesda, MD, pp 1121–1147

Wiesendanger M (1985) Recent developments in studies of the supplementary motor area of primates. Rev Physiol Pharmacol Biochem 103: 1–59

Wiesendanger M, Wiesendanger R (1984) The supplementary motor area in the light of recent investigations. Exp Brain Res [Suppl] 9: 382–392

Wiesendanger R, Wiesendanger M (1985a) The thalamic connections with medial area 6 (supplementary motor cortex) in the monkey (Macaca Fascicularis). Exp Brain Res 59: 91–104

Wiesendanger R, Wiesendanger M (1985b) Cerebello–cortical linkage in the monkey as revealed by transcellular labeling with the lectin wheat germ agglutinin conjugated to the marker horseradish peroxidase. Exp Brain Res 59: 105–117

Wiesendanger M, Seguin JJ, Künzle H (1973) The supplementary motor area – a control system for posture? In: Stein RB, Pearson KC, Smith RS, Redford JB (eds) Control of posture and locomotion. Plenum, New York, pp 331–346

Wiesendanger M, Rüegg DG, Lucier GE (1975) Why transcortical reflexes? Can J Neurol Sci 2: 295–301

Wiesendanger M, Hummelsheim H, Bianchetti M (1986) Sensory input to the motor fields of the agranular frontal cortex: a comparison of the precentral, supplementary motor, and premotor cortex. Behav Brain Res (in press)

Woolsey CN, Settlage PH, Meyer DR, Spencer W, Hamuy TP, Travis AM (1952) Patterns of localisation in precentral and "supplementary" motor areas and their relation to the concept of a premotor area. Res Publ Assoc Res Nerv Ment Dis 30: 238–264

Comparative Timing of Neuronal Discharge in Cortical and Cerebellar Structures During a Simple Arm Movement in the Monkey*

Y. Lamarre and C. E. Chapman

Centre de Recherche en Sciences Neurologiques, Départment de Physiologie, Faculté de Médicine, Université de Montréal, C. P. 6128, Montréal H3C 3J7, Canada

For movements initiated in response to a sensory cue, where and when in the CNS is the stimulus transformed into a command to move? It is reasonable to assume that task-related neurons in which the activity is mainly related to the triggering stimulus are involved at early stages in the chain of events which transforms a conditioning sensory input into a motor command. On the other hand, neurons in which the activity is time-locked to the performance of the movement, such as motoneurons, are involved in the final execution stage. Moreover, a study of the relative timing of neuronal activity with respect to the stimulus and/or the onset of movement could provide evidence in favor of a hierarchical involvement of different brain areas in a stimulus-response paradigm.

Recordings from primary motor cortex in awake monkeys have firmly established that a majority of neurons modify their discharge frequency 80 to 100 ms prior to the onset of a discrete motor response (Evarts 1966, 1974; Lamarre et al. 1978; Fetz et al. 1980; Georgopoulos et al 1982), suggesting that they are causally related to the initiation of movement. This is supported by the fact that the timing of motor cortical neurons is not modified after deafferentation of the operant limb (Bioulac and Lamarre 1979). Other structures have been found to respond before (Mountcastle et al. 1975; Thach 1975; Strick 1976; Macpherson et al. 1980; Seal et al. 1982; Tanji and Kurata 1982; Weinrich and Wise 1982; Lamarre et al. 1983b; Gibson et al. 1985) or after (Bioulac and Lamarre 1979; Fromm et al. 1981; Fromm and Evarts 1982; Kalaska et al. 1983; Chapman et al. 1984; De Long et al. 1984) the discharge of area 4 neurons might be expected to be modulated. It is, however, difficult to make a precise comparison of data obtained in these various studies since different experimental paradigms as well as methods of analysis were used. Comparisons of timing relations are more meaningful when the studies are performed in the same laboratory and using the same task.

For several years we have been using a paradigm in which monkeys are trained to perform identical movements of the arm in response to three different conditioning stimuli: visual, auditory, and somesthetic. This same paradigm was used to record the activity of neurons in the precentral (area 4) and parietal (SI and areas 5 and 7) cortex as well as in the deep cerebellar nuclei (dentate and interpositus) in five monkeys. In this paper we present the results of a systematic analysis of the responses of the neurons recorded in these different regions of the brain, in order to determine

---

* This research was supported by a Medical Research Council of Canada group grant. Y. Lamarre is a member of the Canadian MRC Group in Neurological Sciences. C.E. Chapman was supported by a Canadian MRC Post-doctoral Fellowship.

Experimental Brain Research Series 15

if the activity is mainly related to the performance of the movement or mainly related to the sensory cue, and to compare the time of occurrence of the responses with respect to the onset of movement.

METHODS

The experiments were performed on five adult macaque monkeys seated in a primate chair. Recordings were obtained in the cerebellar nuclei of one m. mulatta and one m. fascicularis and in the cerebral cortex of three m. mulatta and one m. fascicularis (same monkey also used for cerebellar recordings). Some of the results abtained from these monkeys have already been published elsewhere (Bioulac and Lamarre 1979; Lamarre et al. 1983a, b; Chapman et al. 1984; Lamarre et al. 1985). The upper arm was extended horizontally from the shoulder at an angle of about 25° in front of the coronal plane and supported comfortably in a shallow trough which was hinged about the elbow joint. During training sessions the animal learned to perform flexion and extension movements of the forearm in response to each of the stimuli. Three conditioning stimuli were randomly intermixed: a light (LEDs unless otherwise indicated), a tone (800 Hz, 35 dB above background noise), and a small brief perturbation applied at the elbow via a torque motor. The displacement produced by the torque motor was very small, usually less then 1°. Fruit juice rewards were delivered when movements of a sufficient amplitude (15° – 30°) in the required direction were initiated within 500 ms. There were no reference points for starting or stopping the movement.

After training, the animals were anesthetized with pentobarbital and a cylindrical plastic chamber was attached to the skull under aseptic conditions to allow microelectrode recordings to be made (Lamarre et al. 1970). Multistranded stainless steel wires insulated with teflon were implanted into selected arm muscles and passed under the skin to a multichannel connector attached to the skull. In two animals, electro-oculograms and neck muscle activity were also recorded. Glass-coated tungsten microelectrodes were used to record neuronal activity during the performance of the tasks. The outputs of the angular displacement transducer, electromyogram (EMG) wires, and microelectrode were amplified in a conventional manner and displayed on an oscilloscope. Selected data were recorded on multichannel magnetic tape or directly on photographic paper. On-line data acquisition was performed by a PDP-9 computer which was also programmed to control the task. Neural spike intervals, pulse replicas of the EMG activity (Evarts 1974), and the angular displacement (digitized at 200 Hz) were stored on Dec tapes. Routinely, the onset of changes in neural and EMG activity was calculated on a trial by trial basis both with respect to the onset of the stimulus (RS) and to the onset of elbow displacement (RM), the sum of RS and RM being equal to reaction time (RT), i.e., the time from the GO cue to the onset of movement.

The time of the first change in neural activity was measured initially by hand on individual trials of (a) cells with a high spontaneous discharge frequency in which the response was a decrease in discharge and (b) cells with a low spontaneous rate of discharge in which there was an excitatory response. In order to estimate the latency of the response in a more objective fashion, we have employed a simple method which is described as follows. First, a time "window" corresponding to the approximate duration of the response was determined from peristimulus histograms and/or periresponse histograms constructed from all the trials to be analysed individually. The basic calculation was then to count the number of impulses occurring during that same "window" and to

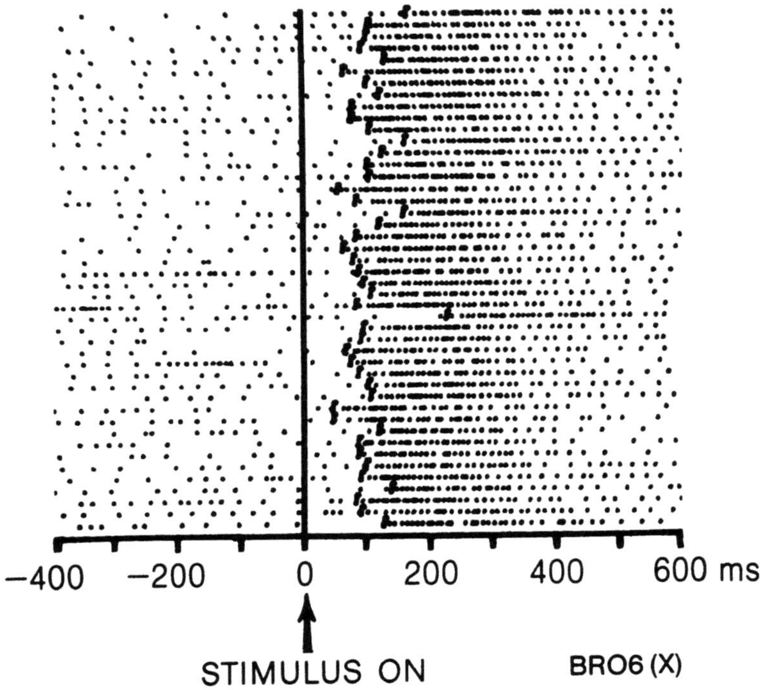

Fig. 1. Activity of a neuron recorded in area 4 during extension (X) of the contralateral elbow. The raster is aligned on the presentation of the stimulus (indicated by the vertical line). The small vertical bars indicate the onset of the excitatory response as determined with the "sliding window" method. Trials are presented in the order of their acquisition; visual and auditory cues are randomly intermixed

subtract this value from the number of impulses occurring during an immediately preceding time "window" of the same duration. A series of such calculations were made, the window being advanced in 2 or 3 ms steps. The analysis routinely started during the control period, before the onset of the stimulus, and was performed for at least 500 ms after the onset of the stimulus. The series of calculated differences, measured successively every 2 or 3 ms, were smoothed with a digital triangular filter to facilitate the detection of maximum and minimum values. The time of occurrence of the maximal peak indicated the onset of an excitatory response while the time of occurrence of the minimum indicates the onset of an inhibitory response. We found this method of computation to be fast and reliable in the majority of cases. The results of this analysis were nevertheless always displayed and controlled by visual inspection of individual raster lines (see Figs 1 and 3).

At the conclusion of the last recording session, the monkeys were killed with an overdose of pentobarbital. For cortical recordings, the dura mater was removed from the depth of the recording chamber and reference points, at known stereotaxic coordinates, were marked on the surface of the cortex by penetrations of a microelectrode dipped in India ink. A photograph was made of the surface of the brain and the coordinates of the point of entry of each microelectrode track were marked upon the photograph. For dentate recordings, small electolytic lesions were left in some of

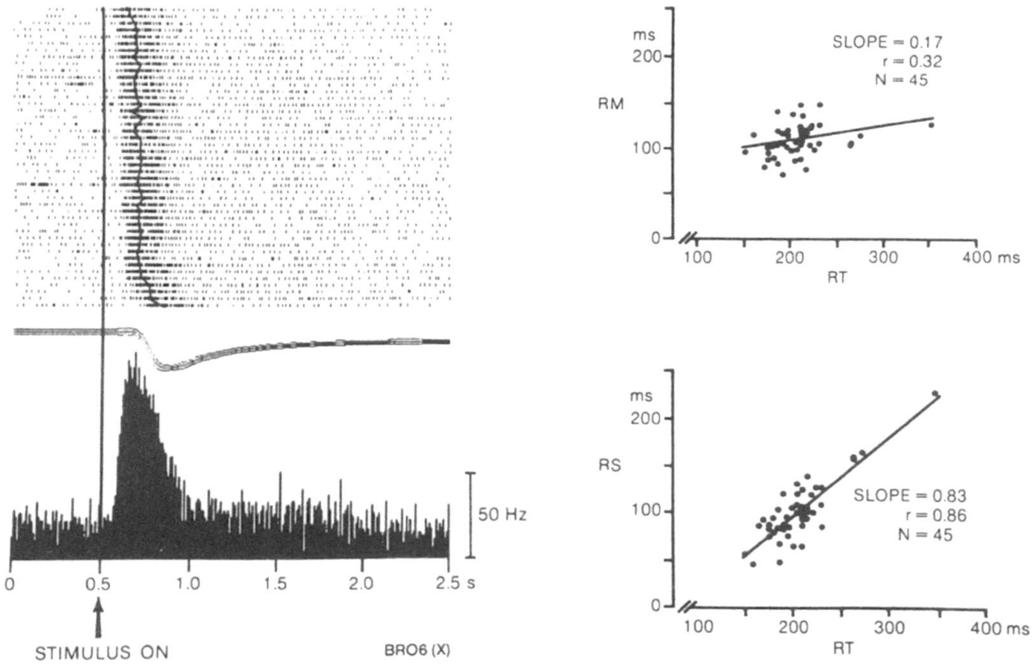

Fig. 2. Temporal relationship of a motor cortical neuron (same cell as Fig. 1) to movement (elbow extension). On the left, the raster and corresponding histogram are aligned on the stimuli. The trials are rearranged here in the order of their response latency. The irregular line indicates the onset of movement. Note that the response latency and RT show almost parallel changes. On the right, the response latency (RS) is plotted as a function of RT (below). The slope was close to 1.0 and r was significant ($P < 0.001$). The converse (RM/RT, above) did not show a significant correlation. One division of the time scale (left) is equivalent to 50 degrees on the movement trace in this and all other figures

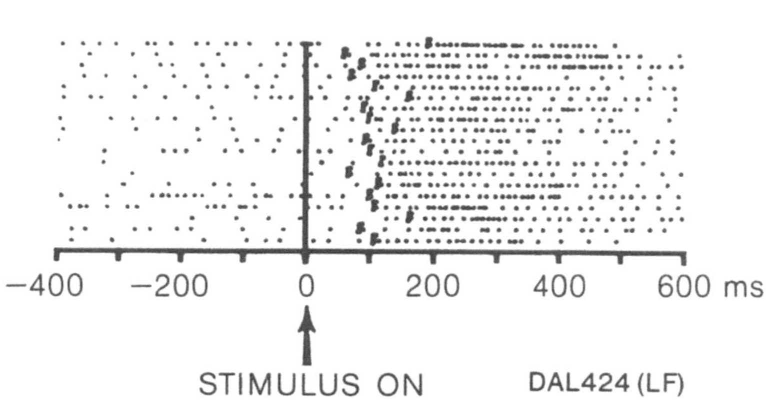

Fig. 3. Activity of a neuron recorded in area 7 during flexion (F) of the contralateral elbow. Only trials triggered by the visual cue are shown. Conventions as in Fig. 1, the small vertical bars indicating the response latency as determined by the "sliding window" method

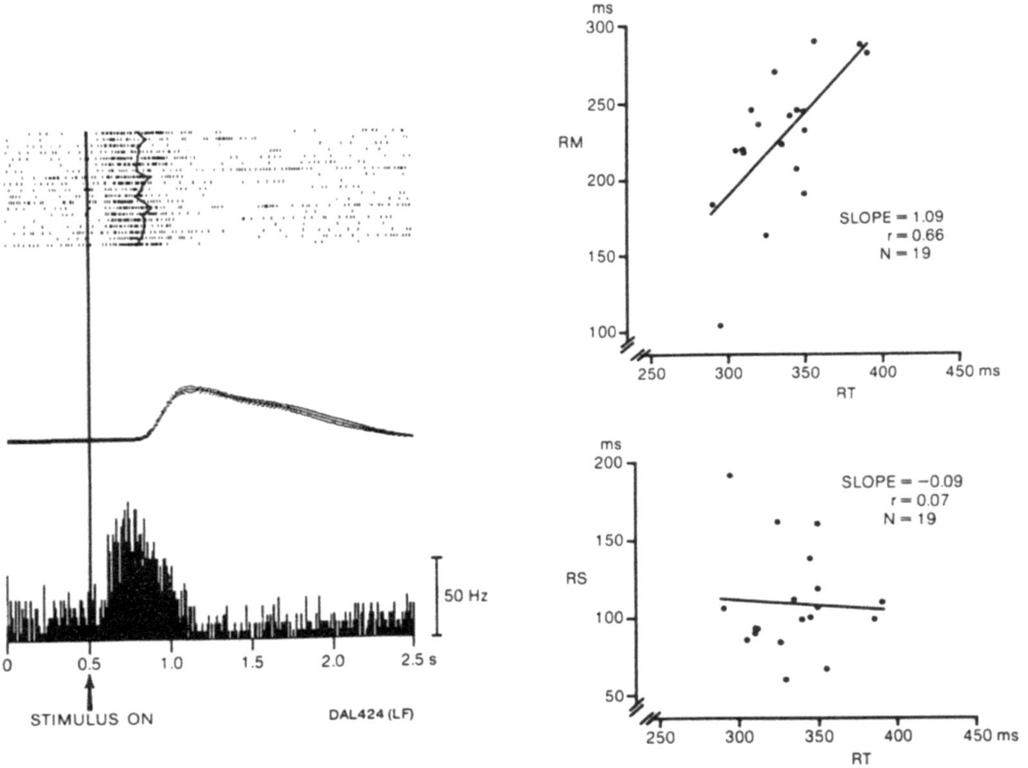

Fig. 4. Temporal relationship of an area 7 neuron to the visual cue (same cell as Fig. 3). Conventions as in Fig. 2, trials being rearranged in order of the response latency, the irregular line indicating the RT. Only the RM/RT plot (above right) was significant, with a slope near 1.0, indicating that the response was stimulus – related

the tracks at known stereotaxic coordinates. The location of electrode tracks was verified in serial 20 ffm sections of the fixed brains. A more detailed description of the methods has been published elsewhere (Lamarre et al. 1983a).

RESULTS

In order to determine precisely whether the neural responses were better related to the stimulus or to the movement, two values were derived from the measurements described in the methods section: RS, the time between the stimulus and the onset of the response, and RM, the time between the response and the onset of movement. RS and RM were then plotted against RT to determine the strength of the correlation coefficient (r) and the slope of the linear regression line. Stimulus – related units would show a significant linear correlation with a theoretical slope value of 1.0 when RM is plotted against RT, and a horizontal line (slope of zero) when RS is plotted against RT. The reverse would be obtained for movement – related units, i.e., the RM/RT plot showing no significant correlation with a slope near zero, and a significant RS/RT correlation with a slope near 1.0. Such a method of analysis has also been applied by Ghez et al. (1983) to characterize the discharge of motor cortical neurons in the cat. Examples of the results obtained in this way are presented in Figs 2 and 4.

Figure 1 is a raster display of the discharge of a cell recorded in area 4. The individual trials are aligned on the stimulus and the trials are presented in the order of their acquisition (only trials cued by the auditory and visual signals are shown). The small vertical bar on each trial indicates the response latency (excitation here) determined with the "sliding window" method. In Fig. 2 the same trials are presented, rank-ordered so that the shortest response latency is on top. The solid, irregular line running across the raster indicates the onset of movement. The discharge of this unit is clearly movement-related since the response latency and RT covary. This relationship is graphically depicted on the right. The RS/RT regression line (lower right) had a slope of 0.83 with a correlation coefficient of 0.86 ($p < 0.001$) while the RM/RT plot (upper right) was not significantly correlated ($r = 0.32$, $p > 0.01$) with a slope of 0.17.

A stimulus-related unit recorded in area 7 (visual trials only) in another monkey is shown in Figs 3 and 4. Figure 3 illustrates the results of the "sliding window" detection of the onset of the response. In Fig. 4 the trials are reordered according to the latency of the response. The small changes in the latency of the response were not associated with any systematic change in RT and thus the RS/RT plot (right) gave a linear regression line with a slope of −0.09. RS and RT were not correlated. The plot of RM vs RT, on the other hand, gave a significant correlation ($p < 0.001$) and a linear regression line with a slope of 1.09.

For the vast majority of neurons analysed in this way, only one of the two correlations was found to be significant ($p < 0.01$). However, for a small number of neurons, all recorded in the dentate nucleus, there was a significant correlation for both the RS/RT and RM/RT plots. It is not possible at the present time to decide whether or not this finding has a true functional meaning and consequently these neurons were excluded from the present study.

Figure 5 and Table 1 summarize and compare the results obtained in different regions of the brain. Timing values obtained from recordings of EMG activity of the prime movers (biceps and/or triceps) are also presented. The column on the left shows the distribution of the mean latency of the responses calculated with respect to the onset of movement (dotted line and time zero on the abscissa). The column on the right shows the distribution of the values of the slopes for the RS/RT regression lines. These data were derived only for those cells which showed a significant correlation between RS or RM and RT. In the distribution of RS/RT slopes shown in Fig. 5, a value around 1.0 (right on the abscissa) signifies that the onset of the neural response is strongly movement-related while a slope value near 0.0 (left on the abcissa) signifies that the response is strongly stimulus-related (the slope of the corresponding RM/RT regression line being near 1.0).

Neurons in area 7 (top histograms) were all recorded in the posterior bank of the intraparietal sulcus. They discharged, as a group, earlier than any other group (mean: 182 ms before the onset of movement). All these neurons were found to have stimulus-related responses with a mean RS/RT slope of 0.10. In contrast, neurons recorded in areas 3b and 1 (bottom histograms, SI) were, on the average, the latest to respond (mean: 6 ms after the onset of movement) and were all movement-related (visual and auditory trials) with a mean RS/RT slope of 0.91. In the dentate nucleus of the cerebellum, we found both types of relationships. A large number of neurons showed stimulus-related responses that occurred on the average 21 ms later than area 7 responses ($p < 0.01$) and 53 ms earlier than movement-related discharge in dentate ($p < 0.001$). Stimulus-related neurons were found more frequently in the caudal two-thirds of the nucleus while movement-related units were found throughout the nucleus.

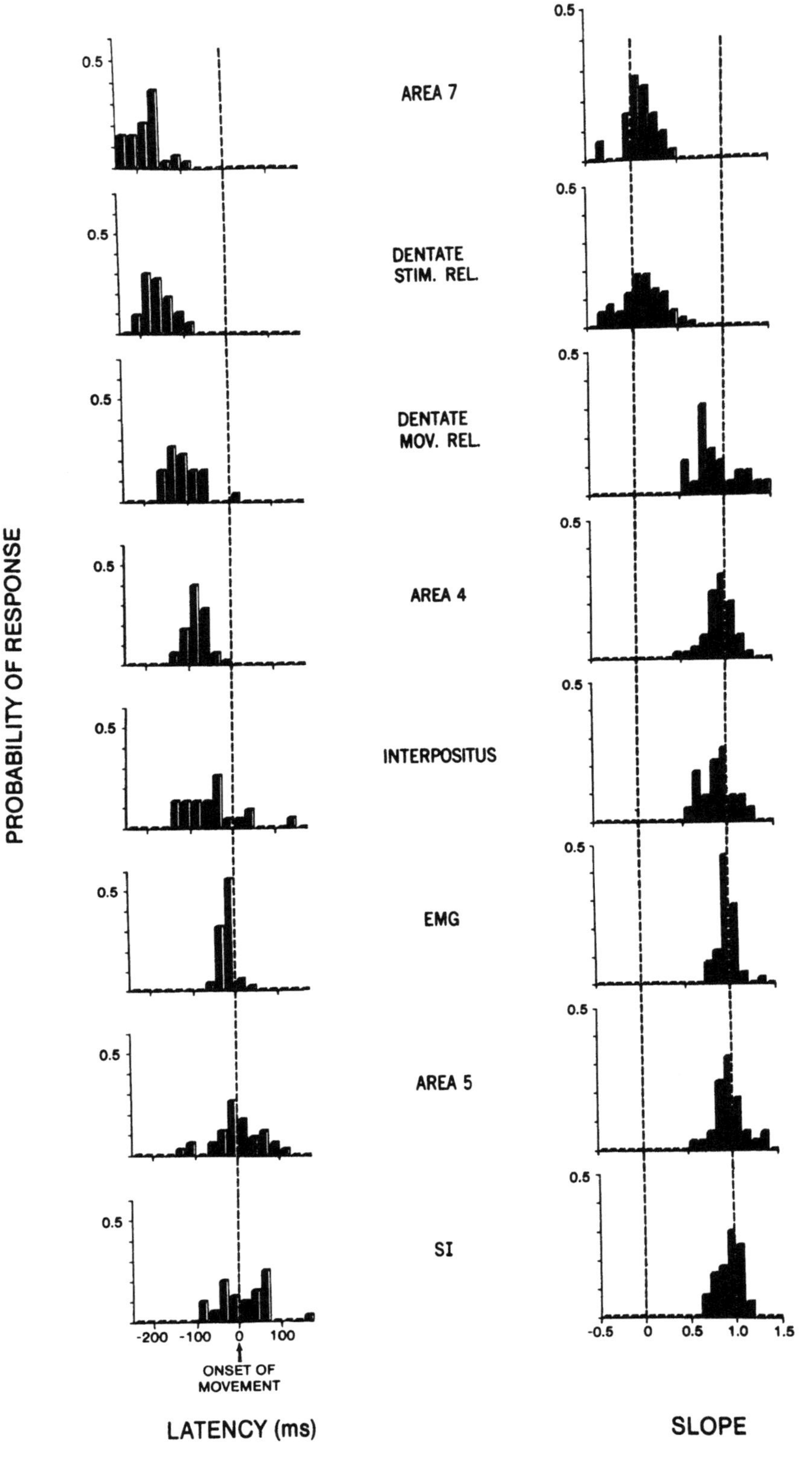

PROBABILITY OF RESPONSE
0.5
AREA 7
DENTATE
STIM. REL.
DENTATE
MOV. REL.
AREA 4
INTERPOSITUS
EMG
AREA 5
SI
-200 -100 0 100
ONSET OF
MOVEMENT
LATENCY (ms)
-0.5 0 0.5 1.0 1.5
SLOPE

The discharge of neurons in areas 3b and 1 (SI), 4 and 5, the interposed nucleus and the prime mover muscles was movement – related, the mean RS/RT slope ranging between 0.87 (interpositus) and 0.96 (EMG). Area 4 neurons responded on the average 85 ms before the onset of movement, i.e., 23 ms later than the movement – related dentate neurons ($p < 0.001$). Interpositus neurons discharged, on the average, 30 ms after area 4 cells (mean: 55 ms before the onset of movement), but the two distributions overlapped extensively and there were interposed neurons that discharged as early as the earliest area 4 neurons. What is clear, however, is that interposed neurons discharge, on the average, significantly later than the movement – related dentate neurons and yet precede EMG activity. The mean onset latency of the responses of area 5 neurons was close to the onset of movement. The distribution of response latencies completely overlapped that observed in SI (areas 3b and 1). In only 10% of the area 5 neurons did the response precede the earliest EMG activity. Most area 5 and most SI neurons responded at the same time as EMG activity appeared in the operant limb. Table 1 lists the mean values of latency, slope, and correlation coefficient calculated for each region.

DISCUSSION

The results presented in this paper support the hypothesis that triggered movements are produced by a hierarchy of mechanisms. The neurons in area 7 and in the dentate nucleus which show stimulus – related activity are involved at the earliest stages of the formulation of the motor command. The functional significance of the movement – related activity can be inferred from the relative timing of this activity with respect to the onset of movement. Movement – related neurons of dentate, along with neurons in area 4 and interpositus, discharge early enough to be causally related to the production of the movement. On the other hand, activity in SI and area 5 occurs later, at a time when feedback from the moving limb could contribute to the observed responses.

For the purposes of this analysis we have pooled the data obtained from five monkeys. Ideally, the comparison of timing relations between different brain areas should be performed in the same animal performing the same task. In one animal, recordings were obtained from all the regions listed in Table 1, except for the cerebellar nuclei. When considering only these data, the timing relations were found to be essentially the same as those obtained when the results of all five animals were pooled. Concerning the cerebellar nuclei, recordings were performed in both dentate and interpositus in two animals, and the results obtained were similar in the two animals. Thus, we feel confident in assuming that our comparisons of timing relations between different brain areas

◀ Fig. 5. Distribution of the response latencies (RM, left) and the slopes of the RS/RT plots (right) in the six different regions investigated, including measures taken from EMG recordings of the elbow flexor and extensor muscles. The histograms have been normalized and the response frequency expressed as a fraction of 1.0. The results are ranked in order of the response latency as measured with respect to the onset of movement, with the earliest responses having occurred in area 7 (above) and the latest in SI (below). The results of the timing analysis are shown on the right where the values of all the RS/RT plots are shown. Area 7 and a large group of dentate neurons were stimulus – related (slope near 0). A smaller group of dentate neurons were active later and their discharge was time – locked to the motor response

Table 1. Comparative timing of neuronal discharge in cortical and cerebellar structures

	Latency (ms)	Slope	r	Number of trials/cell
Area 7 (n = 33)	−182 ± 38	0.10 ± 0.17	0.78 ± 0.15 (RM vs RT)	20 ± 11
Dentate Stim. Rel. (n = 77)[a]	−161 ± 35	0.10 ± 0.23	0.75 ± 0.13 (RM vs RT)	25 ± 17
Denate Mov. Rel. (n = 26)	−108 ± 43	0.88 ± 0.23	0.62 ± 0.11 (RS vs RT)	46 ± 19
Area 4 (n = 50)	− 85 ± 25	0.91 ± 0.15	0.84 ± 0.09 (RS vs RT)	35 ± 10
Interpositus (n = 23)	− 55 ± 64	0.87 ± 0.18	0.66 ± 0.11 (RS vs RT)	32 ± 7
EMG (n = 50)[b]	− 19 ± 18	0.96 ± 0.11	0.88 ± 0.08 (RS vs RT)	34 ± 9
Area 5 (n = 34)	0 ± 57	0.94 ± 0.18	0.75 ± 0.12 (RS vs RT)	24 ± 9
SI (n = 40)	6 ± 55	0.91 ± 0.14	0.80 ± 0.08 (RS vs RT)	23 ± 9

Response latency here was determined in relation to the onset of movement; negative values indicate that response preceded movement. The mean values for the slope of the RS vs RT relationship are shown along with the correlation coefficients (r) of that relationship which was significant (RS or RM vs RT), as is indicated with each measure. In the far right column is given the average number of trials measured for all cells in each region. All values are given ± 1 SD of the mean.

[a] For the stimulus−related cells in dentate, 77 responses in 61 cells were analysed. Some cells responded to both the visual and the auditory cue.

[b] These measures were taken from the EMGs recorded with the motor cortical cells (area 4 above).

based on studies involving different animals are valid and representative of what occurs in different brain areas of the same animal.

Previous studies on timing of changes in neuronal discharge in reaction time tasks usually relied on measurements performed on peri-event time histograms (Evarts 1974). Response latencies can, however, vary from trial to trial so that the values calculated in this way are biased towards the earliest times of change in discharge. Since this variability may not be constant from region to region, it is difficult to compare the timing values between different areas. By measuring the latency of the individual trials, as was done here, a more representative and accurate measure of response latency can be obtained. When applied to the same data, these two methods of analysis can give significantly different results. For instance, area 4 neurons were found to lead movement onset by an average of 100 ms with measurements performed on peri-event time histograms (Spidalieri et al. 1983) as compared with the mean value of 85 ms reported in this paper. This point must be taken into account when trying to compare the present results with previous studies.

The discharge of area 7 neurons in response to the visual cue was found to precede that of the dentate stimulus-related neurons, a high proportion of dentate stimulus-related neurons having responded to the visual cue (76%). The differences in the timing values suggest that area 7, which projects to the lateral cerebellum, might contribute to the visual responses in the dentate nucleus.

Thach (1975, 1978) has previously reported that cerebellar dentate neurons are activated prior to the onset of movement at about the same time, or slightly earlier, than motor cortical neurons. In this study, we found that the discharge of a large proportion of dentate neurons preceded the earliest activity in motor cortical neurons. These dentate cells form a functionally separate group of cells in which the earliest change in discharge is linked to the stimulus, responses to the teleceptive cues being most frequently observed (Chapman and Lamarre 1982; Chapman et al. 1982). A second, and smaller, functional category of movement-related dentate cells was identified, a large proportion of these being coactive with neurons in area 4. Since the bulk of dentate's output is directed towards the precentral motor areas, it is reasonable to suggest that dentate discharge may influence the movement-related discharge in these regions. The timing of movement-related dentate neurons does, however, differ from that of other regions clearly related to the motor response. In all regions but the dentate, the slopes of the RS/RT plots are distributed about 1.0, i.e., they show a close to 1:1 correspondence between response latency (RS) and RT. The distribution of the slopes for the dentate movement-related cells, on the other hand, is somewhat skewed with a mode of 0.7 to 0.8. The corresponding mean correlation coefficient was the lowest when considering all of the movement-related regions. Such observations reinforce the notion that movement-related dentate cells are indeed remote from the final execution stages for movement triggered by sensory cues. Considering both stimulus- and movement-related dentate neurons together, it is evident that this nucleus serves as an important transitional region between those regions involved with the processing of the sensory stimulus (GO cue) and those involved with the elaboration of the motor command.

The discharge of cerebellar interposed neurons, on the other hand, clearly follows that of the dentate neurons both in our experiments and those of Thach (1970, 1978). In our task, however, the discharge of a number of the interposed cells was modulated prior to the onset of movement, at about the same time as motor cortical neurons are active. Although the relative frequency of such precocious responses in the interposed nucleus is not known since the nucleus was not completely investigated in these experiments, such a finding suggests that interposed neurons may play a role in the initiation of movement as well as in the control of ongoing movement (see, for example, Harvey et al. 1979). This suggestion is consistent with the recent results of Gibson et al. (1985) who found that the discharge of neurons in the magnocellular region of the red nucleus, which is one of the major targets for the output of the interposed nucleus, is modulated well in advance of movements of the distal limb.

A comparison of the relative timing of discharge in SI (areas 3b and 1) and area 5, a somatic sensory association area, indicates that these two areas are active at about the same time in this task. The discharge of area 5 could thus represent feedback from the moving limb as relayed by SI. But, movement-related discharge in SI, and not area 5, is abolished by deafferentation (Bioulac and Lamarre 1979; Seal et al. 1982), suggesting that some of the discharge is central in origin, perhaps representing central feedback from precentral motor areas (corollary discharge or efference copy).

In the schemes proposed by Kemp and Powell (1971) and Allen and Tsukahara (1974), the parietal cortex and lateral cerebellum are situated in a position whereby they can influence and control the motor cortex. The data presented here would seem to confirm this prediction, at least in the case of a stimulus-response paradigm. Even though there is some overlap of the distribution of onset times of neural changes in the different brain regions that we have studied, the observed overlap was much less than would have been expected based upon the results of other studies. For instance, previous comparisons of timing betwen the lateral cerebellum and the motor cortex have led to the conclusion that dentate is largely coactive with motor cortex, although a small proportion of dentate neurons respond well in advance of motor cortical neurons (see Brooks and Thach 1981, for a recent review). Our data shows a much more important difference in timing. Indeed, at least 40% of all dentate cells studied changed their firing rate before any change occurred in the motor cortex. When considering only the stimulus-related cells in the dentate, at least two-thirds of these neurons changed their firing rate before the earliest change of activity in area 4. That such a clear separation of timing emerges in our studies is probably due to several factors, one of which could be the systematic comparison of discharge properties in relation to both the stimulus and the motor response. Secondly, the analysis was restricted to the initial change in discharge; later secondary changes were not considered for the purposes of this analysis. Thirdly, the use of three different sensory cues was an important aid in two respects. First, selective responses to one or two of the cues were frequently seen and helped to characterize the response properties of the neurons (e.g., teleceptive responses in the dentate nucleus, visual responses in area 7). Second, RT's systematically varied with the modality of the cue, being longest in response to the visual cue and shortest in response to the auditory cue. Thus, for each cell, we were able to record the neuronal responses associated with a wide range of RTs, facilitating this analysis. If only a single cue had been used, the range of RTs may have been too small to make such a comparison.

The picture that emerges from our data is one of a clear sequence of neuronal activation, with the posterior parietal cortex (area 7) being involved at a very early stage, slightly before the lateral cerebellum, itself being activated before the primary motor cortex. This is not to say that movement initiation always proceeds according to such a strict sequential activation of these different brain regions; for self-initiated movements, the processing of information may be occurring much more "in parallel".

Acknowledgements. The authors wish to acknowledge the participation of Drs. B. Bioulac and G. Spidalieri and Miss L. Busby in some of the experiments. The authors also gratefully acknowledge the assistance provided by the technical staff of the Centre de recherche en sciences neurologiques, and in particular by H. Auzat, M.T. Parent, and R. Bouchoux. The assistance of Dr. J.P. Raynauld for computer facilities was especially appreciated.

REFERENCES

Allen GI, Tsukahara N (1974) Cerebro–cerebellar communication sytems. Physiol Rev 54: 957–1006

Bioulac B, Lamarre Y (1979) Activity of postcentral cortical neurons of the monkey during conditioned movements of a deafferented limb. Brain Res 172: 427–437

Brooks VB, Thach WT (1981) Cerebellar control of posture and movement. In Brooks VB (ed) The nervous system. American Physiological Society, Bethesda, pp 877–946 (Handbook of Physiology, Sect 1, Vol 2, Part 2, Chap 18)

Chapman CE, Lamarre Y (1982) Response of dentate neurones to conditioning stimuli is dependent upon the subsequent performance of a conditioned arm movement in the monkey. Proc Can Fed Biol Soc 25: 26

Chapman CE, Spidalieri G, Lamarre Y (1982) A study of sensorimotor properties of dentate neurons during conditioned arm movements in the monkey. Soc Neurosci 8: 830 [Abstract]

Chapman CE, Spidalieri G, Lamarre Y (1984) Discharge properties of area 5 neurones during arm movements triggered by sensory stimuli in the monkey. Brain Res 309: 63–77

De Long MR, Alexander GE, Georgopoulos AP, Crutcher MD, Mitchell SJ, Richardson RT (1984) Role of basal ganglia in limb movements. Human Neurobiol 2: 235–244

Evarts EV (1966) Pyramidal tract activity associated with a conditioned hand movement in the monkey. J Neurophysiol 29: 1011–1027

Evarts EV (1974) Precentral and postcentral cortical activity in association with visually triggered movement. J Neurophysiol 37: 373–381

Fetz EE, Finocchio DV, Baker MA, Soso MJ (1980) Sensory and motor responses of precentral cortex cells during comparable passive and active joint movements. J Neurophysiol 43: 1070–1089

Fromm C, Evarts EV (1982) Pyramidal tract neurons in somatosensory cortex: central and peripheral inputs during voluntary movement. Brain Res 238: 186–191

Fromm C, Evarts EV, Kroller J, Shinoda Y (1981) Activity of motor cortex and red nucleus neurons during voluntary movement. In: Pompeiano O, Ajmone Marsan C (eds) Brain mechanisms of perceptual awareness and purposeful behavior. Raven, New York, pp 269–294

Georgopoulos AP, Kalaska JF, Caminiti R, Massey JT (1982) On the relation between the direction of two-dimensional arm movements and cell discharge in primate motor cortex. J Neurosci 2: 1527-1537

Ghez C, Vicario D, Martin J, Yumiya H (1983) Sensory motor processing of targeted movements in motor cortex. In: Desmedt JE (ed) Motor control in health and disease. Raven, New York, pp 61-92

Gibson AR, Houk JC, Kohlerman WJ (1985) Magnocellular red nucleus activity during different types of limb movement in the macaque monkey. J Physiol (Lond) 358: 527-549

Harvey RJ, Porter R, Rawson JA (1979) Discharges of intracerebellar nuclear cells in monkeys. J Physiol (Lond) 297: 559-580

Kalaska J, Caminiti R, Georgopoulos AP (1983) Cortical mechanisms related to the direction of two-dimensional arm movements: relations in parietal area 5 and comparison with motor cortex. Exp Brain Res 51: 247-260

Kemp JM, Powell TPS (1971) The connections of the striatum and globus pallidus: synthesis and speculation. Philos Trans R Soc Lond [Biol] 262: 411-457

Lamarre Y, Bioulac B, Jacks B (1978) Activity of precentral neurones in conscious monkeys: effects of deafferentation and cerebellar ablation. J Physiol (Paris) 74: 253-264

Lamarre Y, Joffroy AJ, Filion M, Bouchoux RA (1970) A stereotaxic method for repeated sessions of central unit recording in a paralysed or moving animal. Rev Can Biol 29: 371-376

Lamarre Y, Busby L, Spidalieri G (1983a) Fast ballistic arm movements triggered by visual, auditory and somesthetic stimuli in the monkey. I. Activity of precentral cortical neurons. J Neurophysiol 50: 1343-1357

Lamarre Y, Spidalieri G, Chapman CE (1983b) A comparison of neuronal discharge recorded in the sensori-motor cortex, parietal cortex and dentate nucleus of the monkey during arm movements triggered by light, sound or somesthetic stimuli. Exp Brain Res [Suppl] 7: 140-156

Lamarre Y, Spidalieri G, Chapman CE (1985) Activity of areas 4 and 7 neurons during movements triggered by visual, auditory, and somesthetic stimuli in the monkey: movement-related versus stimulus-related responses. Exp Brain Res [Suppl] 10: 196-210

Macpherson JM, Rasmusson DD, Murphy JT (1980) Activities of neurons in "motor" thalamus during control of limb movement in the primate. J Neurophysiol 44: 11-28

Mountcastle VB, Lynch JC, Georgopoulos AP, Sakata H, Acuna C (1975) Posterior parietal association cortex of the monkey: command functions for operations within extrapersonal space. J Neurophysiol 38: 871-908

Seal J, Gross C, Bioulac B (1982) Activity of neurons in area 5 during a simple arm movement in monkeys before and after deafferentation of the trained limb. Brain Res 250: 229-243

Spidalieri G, Busby L, Lamarre Y (1983) Fast ballistic arm movements triggered by visual, auditory and somesthetic stimuli in the monkey. II. Effects of unilateral dentate lesions on the discharge of precentral cortical neurons and reaction time. J Neurophysiol 50: 1359–1379

Strick PL (1976) Activity of ventrolateral thalamic neurons during arm movement. J Neurophysiol 39: 1032–1044

Tanji J, Kurata K (1982) Comparison of movement-related activity in two cortical motor areas of primates. J Neurophysiol 48: 633–653

Thach WT (1970) Discharge of cerebellar neurons related to maintained postures and two prompt movements. I. Nuclear output. J Neurophysiol 33: 527–536

Thach WT (1975) Timing of activity in cerebellar dentate nucleus and cerebral motor cortex during prompt volitional movement. Brain Res 88: 233–241

Thach WT (1978) Correlation of neural discharge with pattern and force of muscular activity, joint position and direction of intended next movement in motor cortex and cerebellum. J Neurophysiol 41: 654–676

Weinrich M, Wise SP (1982) The premotor cortex of the monkey. J Neurosci 2: 1329–1345

Preparation for Action: Slow Potentials and EMG

C. H. M. Brunia and S. A. V. M. Haagh
Department of Psychology, Tilburg University, P.O. Box 90153, 5000 LE Tilburg, The Netherlands

READINESS POTENTIAL: A MULTISOURCE PHENOMENON

Selfpaced unilateral movements are preceded by a bilateral slow negative wave in the EEG, starting 1500 – 1200 ms prior to agonist EMG onset. Kornhuber and Deecke (1965) were the first to describe this Bereitschafts or readiness potential (BP or RP), which is followed by a small positive wave (premotion positivity: PMP) and a sharp negative deflection (motor potential: MP). Gilden et al. (1966), trying to avoid a functional interpretation of these components, used the more abstract notations N1, P1, and N2, respectively. Nearly identical slow waves have been described in monkeys (Arezzo and Vaughan 1975, 1980; Arezzo et al. 1977; Gemba et al. 1980). On the basis of simultaneously recorded multiple unit activity and slow potentials, Arezzo and Vaughan (1980) concluded that the electrophysiological source of N2 is localized in the precentral gyrus. Using a number of epidural electrodes, Arezzo and Vaughan (1975) had arrived at the same conclusion for N1. Although the evidence was weaker, they suggested that the localization of the sources of N1 and N2 was comparable. A transcortical premovement potential inversion has also been found in the premotor cortex and the somatosensory cortex (Gemba et al. 1979). This suggests the existence of multiple sources contributing to the surface negativity of N1. Support for this notion was provided by Sasaki et al. (1979), who studied the influence of cerebellar hemispherectomy on the RP. This potential was markedly reduced in size in the motor cortex contralateral to the ablation, suggesting a cerebellar role in the emergence of the RP. Although a similar reduction in RP amplitude was found in the contralateral somatosensory cortex, the recovery rate was quite different. In the contralateral somatosensory cortex, recovery occurred within some weeks, while in the motor cortex no recovery was seen after some months. This points to differences in the origin of the RP in these cortical regions.

After a symmetrical onset, the RP becomes asymmetrical from about 500 ms prior to the movement, the larger amplitudes being observed over the contralateral motor cortex. This has been interpreted as an indication of the existence of two components within the RP. To our knowledge, animal research has not yet provided direct support for such a proposition. There is evidence, however, that in monkeys the RP shows a symmetrical distribution over the premotor cortex and an asymmetrical distribution over the motor cortex (Gemba et al. 1980). Surface EEG recordings in man could be a reflection of activity produced by different sources in premotor and motor cortex, resulting in a symmetrical and asymmetrical part of the RP.

Experimental Brain Research Series 15

The two-component hypothesis would be supported if experimental manipulation of kinetic parameters resulted in a differential effect on one, but not on the other component. This is exactly what has been found by Kutas and Donchin (1977) and Hink et al. (1983), who reported a positive relation between the amplitude of the symmetrical component of the RP and the force of an isometric response. the asymmetrical component was unaffected by force manipulation. Additional support for the two-component interpretion stems from magnetencephalographic (MEG) recordings in man (Deecke et al. 1982, 1983). Although a precise localization of the source of movement-related potentials in man is not currently possible, Deecke and Kornhuber (1978) hypothesized that the midline RP has a source different from the one recorded over both hemispheres. They suggested that the former stems from the supplementary motor area (SMA). Supporting evidence for that viewpoint is provided by regional blood flow studies with ^{113}Xe. Roland et al. (1977) found a localized increase in blood flow preceding various movements in two locations simultaneously: (a) in that part of the motor cortex that was involved in the movement execution, and (b) independent of the kind of movement, in SMA. The localized blood flow increase in the SMA was even present when a movement was imagined, but not executed. This suggests, therefore, a different function of motor cortex and SMA. It seems plausible to assume that anticipatory activity in the SMA is related to the concept of abstract response programming which was postulated by for example, Klapp (1977). The actual implementation of such a response program might be represented by the activity in the motor cortex itself.

An important aspect of response programming is the timing of movement onset, a function for which the prefrontal cortex (Brunia et al. 1985) or the SMA (Deecke et al. 1985) may be crucial. In the next section timing aspects of a planned movement are discussed in more detail. Prior to that elaboration, the scope of anticipatory movement-related slow potentials will be extended to the contingent negative variation (CNV).

CONTINGENT NEGATIVE VARIATION: A MOVEMENT-RELATED POTENTIAL?

Simple movements, executed in a reaction time (RT) experiment are also preceded by a slow negative wave, which has been termed contingent negative variation (Walter et al. 1964). If the interstimulus interval between a warning signal (WS) and a response signal (RS), the so-called foreperiod, is sufficiently long ($>$ 3 s), two successive slow waves can be distinguished. The early wave is maximal over the frontal area and reaches its peak value at about 750 ms after WS onset. The late wave has its maximum over the central area and reaches its peak at the moment the RS is presented (Fig. 1).

Arguments in favor of a motor interpretation of the CNV late wave have been reviewed by Rohrbaugh and Gaillard (1983) and can be summarized as follows:

1. The amplitude of the CNV in short foreperiods and of the CNV late wave in long foreperiods is attenuated when no motor response has to be given.
2. During a long foreperiod, the CNV late wave decreases in detection tasks and increases under speed instructions.
3. The CNV amplitude during a short foreperiod is larger when larger muscular effort is required.

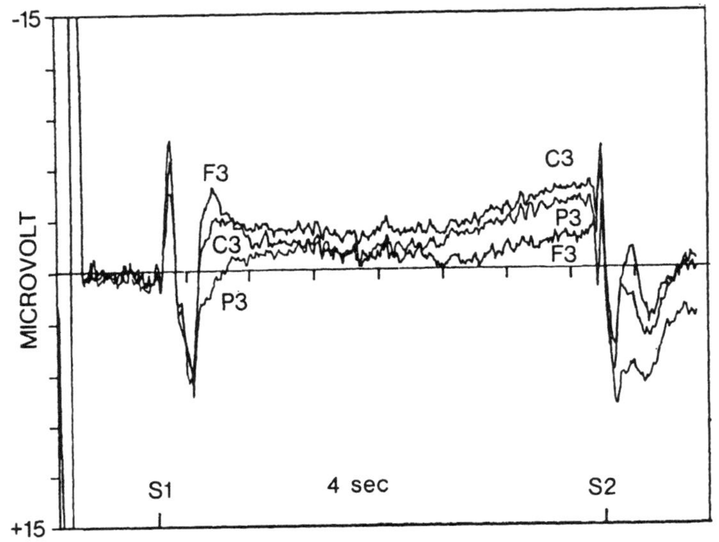

Fig. 1. Average CNV traces from left frontal (F_3), central (C_3) and parietal (P_3) electrode positions. The task was a simple reaction time experiment with a fixed foreperiod duration of 4 s. S1 is the warning stimulus, S2 the imperative stimulus. The response was a plantar flexion of the right foot (n = 25)

On the basis of these and other arguments, Rohrbaugh and Gaillard (1983) concluded that CNV late wave genesis is solely related to kinetic parameters of response execution. CNV late wave and RP show a comparable topographic distribution, as was demonstrated by Brunia (1980), Brunia and Vingerhoets (1980), Boschert et al. (1983b), and Brunia and van den Bosch (1984). Amplitudes of bilaterally recorded RP and CNV late wave are larger over the hemisphere ipsilateral to a foot movement and contralateral to a finger movement (Fig. 2). These results have been tentatively explained by the actual localization of the dipole producing the surface negative shift (Brunia 1980). Prior to foot movements cell columns are activated in the leg area of the motor cortex situated in the medial wall along the longitudinal fissure. The orientation of the cell columns is such that their activity can be more easily picked up by electrodes over the ipsilateral hemisphere. Since the orientation of the functional dipole preceding a finger movement is more or less perpendicular to the lateral convexity of the skull, the hemisphere contralateral to the movement side shows larger amplitudes than the one ipsilateral to the movement side (Fig. 3). Arguments for this interpretation have been given elsewhere (Boschert et al. 1983b; Brunia and van den Bosch 1984; Hari et al. 1983).

The fact that both RP and CNV late wave show the same paradoxical distribution with foot movements (see Fig. 2) is an argument in favor of an important common electrophysiological source. This is not surprising because the same movement is made in both cases. Moreover, it has to be kept in mind that some of the experimental conditions in CNV and RP experiments are identical: subjects are asked to refrain from other movements and to keep their eyes fixated. Furthermore, timing plays a role in both paradigms, although in a different way. In RP experiments uncued movements are made at a certain pace (e.g., every 7 ± 2 s), the subjects being relatively free to choose movement onset. In warned RT tasks, the timing starts by WS presentation. It is

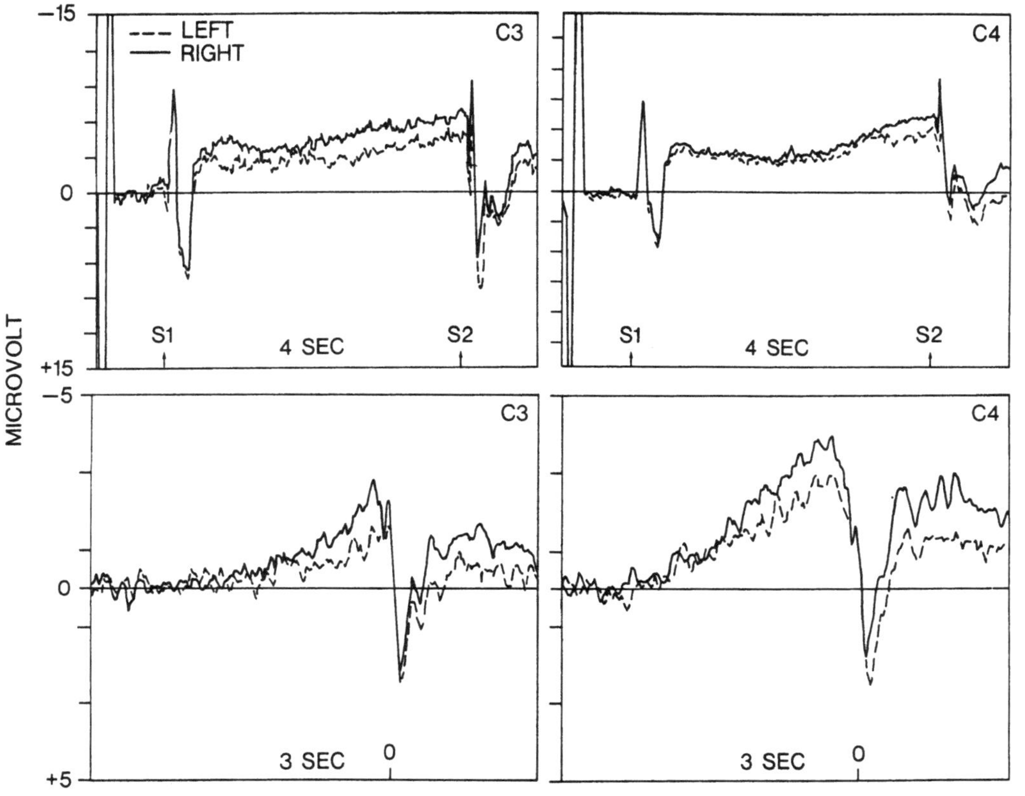

Fig. 2. CNV (n = 10) and RP (n = 20) amplitudes are larger over the left hemisphere (C_3) preceding a right (———) finger movement than preceding a left (······) finger movement. CNV (n = 20) and RP (n = 20) amplitudes are larger over the right hemisphere (C_4) preceding a right (———) foot movement than preceding a left (·······) foot movement. Note the differences in amplitude of CNV and RP

known that time estimation alone is sufficient to induce surface negativity (Peters et al. 1977) and that correct estimation of foreperiod duration enhances RT performance. Loveless (1973) showed that CNV late wave was severely attenuated when correct timing was impossible as a consequence of irregular foreperiods. In the following, we will discuss timing aspects of CNV and RP experiments in more detail.

Stimulus encoding and evaluation are reflected in different components of the evoked potential (EP) after the WS. Initially, the EP reflects activation of the primary sensory areas. From here, or the secondary sensory areas, activity is presumably relayed to frontal and prefrontal areas (see Brunia et al. 1985). This is probably reflected in the early CNV wave, which is known to stem from frontal cortical regions. Research in monkeys has demonstrated that cells in the prefrontal cortex fire intensively during a delay time between a cue signal and the response to be made (Fuster and Alexander 1971). Although a delayed response experiment is different from an RT experiment, time estimation is common to both paradigms. Waiting to respond is accompanied by an increased firing rate of not only the prefrontal cortical cells, but also of cells in the thalamic nucleus dorsalis

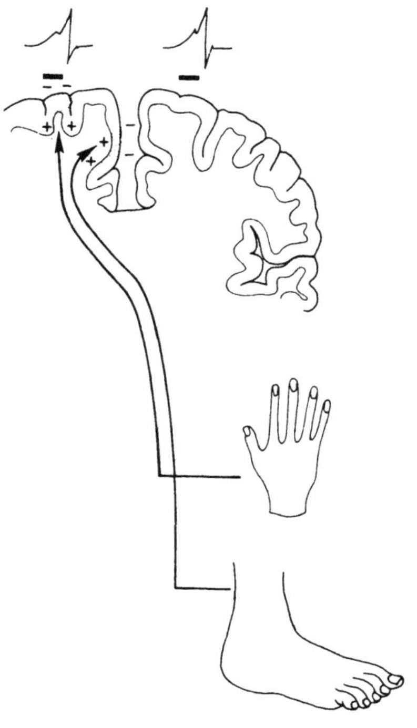

Fig. 3. CNV late wave and RP show a contralateral preponderance in amplitude with finger movements and an ipsilateral preponderance with foot movements. It is suggested that the opposite potential distribution preceding both movements is caused by the different directions of the dipoles in the motor and somatosensory cortex of the hemisphere, contralateral to the movement side

medialis (Fuster and Alexander 1971). The authors interpret their results as reflecting some short-term memory process, but that does not exclude a time estimation process. Therefore, we suggest that this, or some similar activity, is indeed related to time estimation, which could be present over the total length of the foreperiod. The timing of the response is more compelling in a CNV paradigm because of the constraints on response latency. This might account for the finding of generally larger CNV amplitudes as compared with RP amplitudes. This assumption was supported by Rohrbaugh and Gaillard (1983) who reported agonist EMG activity to be larger in a forewarned RT paradigm than in self-paced movements. The implicit suggestion seems to be that slow negative wave amplitudes change in parallel to EMG activity; a point to which we will return later on.

Evidence in favor of a timing process underlying RP genesis has been recently provided by Libet et al. (1982), who distinguished type I and type II RPs. The first is the traditional RP and starts about 1 s prior to the movement. Here the subject really preplans the movement. Type II RPs are much shorter and start only some hundreds of ms prior to the movement; they reflect a sudden impulse to move, without an elaborate timing. Type I seems to coincide with the symmetrical RP component, type II with the asymmetrical component. If our interpretation with regard to time estimation is correct, we would expect type I RPs to show a more frontal distribution than type II RPs. Since Libet et al. (1982) did not study the potential distribution in detail, this suggestion has still to be investigated. Since type II RPs are also very large at Cz, Libet et al. (1982) supposed that apart from areas 4 and 6 other areas contribute to this potential. This could be the SMA, as suggested by Deecke and Kornhuber (1978) and Deecke et al. (1985). Given the evidence presented above, the following statements can be proposed:

1. The CNV late wave comprises an RP as well as additional negativity related to particular task requirements.
2. The CNV late wave reflects activity of various cortical areas such as SMA, premotor, motor, and somatosensory cortex.
3. The CNV late wave reflects various processes over time: those which are absolutely necessary for response execution and those which may enhance performance (e.g., timing, effort, and preprogramming).

Led by the earlier mentioned results of Rohrbaugh and Gaillard (1983) regarding differences in time course and in amplitude of EMG with cued and self-paced movements, a more detailed comparison of anticipatory EMG and slow waves seems promising. However, since slow waves reflect the contribution of various cortical regions to preparation for action, the calculation of a simple correlation coefficient between changes in amplitude of RP or CNV and in EMG activity might yield uninterpretable results. In the next section we will discuss results of our experiments in which EEG and EMG were investigated simultaneously.

CENTRAL AND PERIPHERAL INDICES OF MOTOR PREPARATION

During a fixed 4 s foreperiod, Brunia and Vingerhoets (1980) recorded CNV simultaneously with bilateral EMG from the calf muscles preceding a plantar flexion of the right foot. No changes in EMG on the left side were found. However, mean integrated agonist EMG, expressed in arbitrary units per s, increased systematically over time, the largest values being found when the CNV late wave reached its peak. Moreover, the time course of both CNV and EMG was found to be different in trials with relatively fast and slow responses. Considering separately the 10 trials with the shortest RTs and the 10 with the longest RTs per subject, a larger amplitude of the mean CNV late wave was found preceding fast responses. EMG showed only a trend towards larger amplitudes preceding fast responses. These data seem to indicate that variation in motor preparation was reflected in both CNV late wave and EMG.

It has been also shown that preparing a plantar flexion of the foot is accompanied by activity changes in the activity many other muscles besides the agonist. Haagh and Brunia (1984) recorded EMG activity in a number of muscles in a simple RT task, in addition to EOG and ECG. They found during the foreperiod a systematic increase in EMG activity over time, not only in the right agonist soleus muscle, but also in its antagonist, the tibialis anterior muscle (Fig. 4). Moreover, the ipsilateral peroneus longus and quadriceps muscles showed larger EMG activity during the foreperiod than during the last 4 s of the variable intertrial interval which served as baseline. No differences compared with baseline were found in the contralateral soleus and tibialis anterior or in the deltoideus and trapezius muscles. Finally, mylohyoideus EMG decreased over time, as did EOG activity. Although the number of muscles investigated in this study is still limited, one can conclude that preceding a simple plantar flexion of the foot, more muscles show changes in activity than just the agonist.

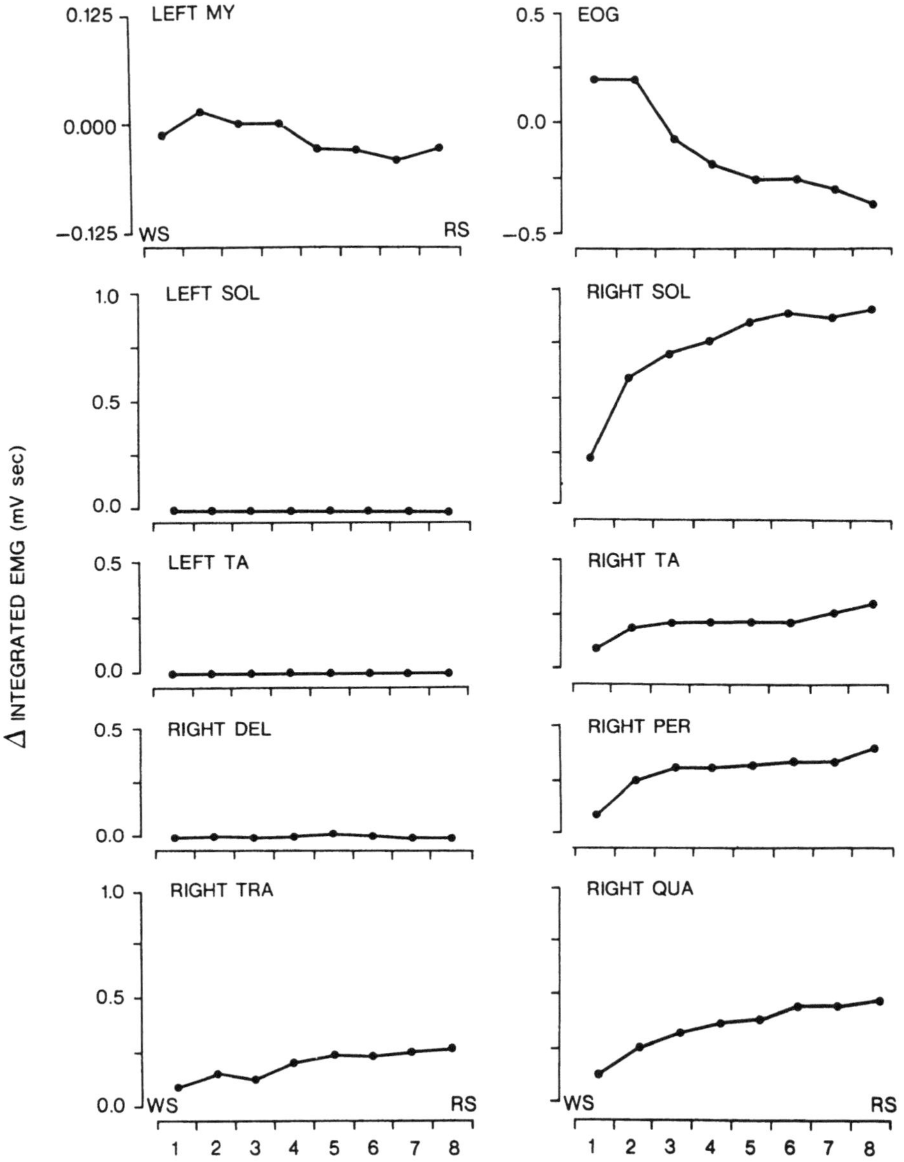

Fig. 4. Mean integrated surface EMG during a 4 s foreperiod of a simple RT task expressed as deviations from the baseline which was the mean value of the last 4 s of the variable intertrial interval. Data from agonist m. soleus (right SOL), antagonist m. tibialis anterior (right TA), m. peroneus longus (right PER), m. quadriceps (right QUA), m. tibialis anterior (left TA), m. deltoideus (right DEL), m. trapezius (left TRA), m. mylohyoideus (left My) and mean integrated EOG are depicted in 8 periods of 500 ms. The response was a plantar flexion of the right foot

To further investigate the possible relation between those EMG changes and behavior, an across–subjects product–moment correlation was calculated between integrated EMG of the last 500 ms epochs preceding the RS and RT. Soleus and tibialis anterior muscles on the right side showed significant negative correlations and so did the right peroneus longus, quadriceps, and trapezius muscles. These results were interpreted as suggesting that different muscles are involved in different aspects of preparation. The latter muscles were assumed to be related to anticipatory postural fixation adjustments, the former muscles to preparation for the movement itself. Moreover, as part of the latter process, the suggestion was made that the increase in antagonist EMG activity served as a brake to prevent premature responses. Thus, a sudden release of the antagonist together with a contraction of the agonist might expedite the response. In conclusion, a number of muscles seem to be tensed at the end of the foreperiod, i.e., at the time that in other experiments the largest CNV late wave amplitudes are found. Subjects tense their muscles as part of a response strategy, aimed at arriving as close as possible to the motor action limit (Näätänen and Merisalo 1977). This we consider a stage in the recruitment order of the relevant motoneuron pool.

Several experiments have been designed to explicitly influence RT by instructing subjects to tense the agonist during the foreperiod of a simple RT task. Thus, Sanders (1980) found a shortening of RT of 40–60 ms and similar results were reported by Haagh et al. (1983). Gaillard et al. (1980) studied the influence of instructed muscle tension on simultaneously recorded CNV and EMG under speed and accuracy instructions. Agonist EMG reached its maximum 1 s after the WS and nearly remained at that level during the remainder of the foreperiod, showing a slight tendency to decline. EMG was significantly smaller in the relaxed condition. Although mean integrated EMG was larger in the speed than in the accuracy condition at all points in time, the authors did not present a detailed description of their RT data in both conditions. The CNV late wave amplitude was larger under speed than under accuracy instructions at Cz, Pz, C_3, and C_4. Muscle tension was not reflected in the late wave. The early wave, however, showed larger amplitudes in the tensed than in the relaxed condition over the left central area. Although, Gaillard et al. (1980) were mainly impressed by the fact that CNV late wave was not attenuated in the relaxed condition, to us the different representation of tension and speed seems to be an equally important result, as will be clarified later on.

Effects of instructed tension in muscles that are irrelevant for the execution of the movement have also been studied. Tecce et al. (1984) recorded CNV and EMG in a relaxed and tensed condition. The response was a key–press with the preferred hand. Muscle tension was, in one condition, invoked by clenching the contralateral fist from several seconds preceding the WS to several seconds after the response. EMG was recorded from the extensor muscles in the arm, although the response was described as a key–press, which suggests that flexor muscles were the agonist. EMG activity was only analyzed after RS. No EMG was recorded from the contralateral response irrelevant muscle. Notwithstanding these methodological shortcomings, the authors concluded that CNV and EMG were dissociated because of the inverse relationship between sustained muscle contraction and CNV late wave. They considered their results an argument in favor of the CNV late wave reflecting attention. Accordingly, the double task asked for divided attention, resulting in a smaller CNV late wave amplitude. It is, however, doubtful whether attentional processes are needed to explain the results of Tecce et al. (1984). Otto et al. (1977) and Otto and Benignus (1978) have found that a sustained muscle contraction results in a prolonged surface–positive potential over the centroparietal areas. Similar findings have been reported by Grünewald and Grünewald–Zuberbier (1983). Since the EEG is a summation of all underlying cortical potentials, a smaller CNV late wave is the only result to be expected when a tonic muscle contraction is executed during a foreperiod. Moreover, instructed contraction of muscles irrelevant for the response execution can be viewed as an obstacle for a good preparation, its disturbing influence being manifest not only as a decrease in CNV late wave amplitude, but also as an increase in RT.

Recently we investigated further the relation between CNV amplitudes and uninstructed voluntary muscle tension in the agonist (Haagh and Brunia 1985). Subjects were divided post hoc into two groups, depending on whether or not they tensed their calf muscle during a 4 s foreperiod of a simple RT task, the response being a plantar flexion of the right foot. EEG was recorded monopolarly from F_z, C'_3, C'_4, C''_3, C''_4, P_3, and P_4, C' positions being 1 cm anterior to C_3 or C_4 and C'' positions 2 cm posterior to C_3 or C_4. EMG activity was measured over a 150 ms epoch in which the individual early wave peak was recorded and over the last 150 ms preceding RS. A between-subjects comparison of EMG and EEG activity in relation to RT was made. Moreover, the 15 fastest and slowest responses per subject were studied separately, in order to compare EMG, EEG, and RT within subjects.

Agonist EMG activity in the tensed subjects was significantly larger than in the relaxed subjects, although this difference did not exceed 3 μV. An early wave was found at F_z, C'_3, C'_4, C''_3, and C''_4, but not in the parietal derivations. A late wave was present over all electrode positions. The early wave was larger in the tensed than in the relaxed group at F_z, C'_3, and C'_4, while the late wave showed a trend to larger amplitudes over central electrode positions ($p < 0.07$). The tensed group did not show a faster RT compared with the relaxed group. Within subjects, the early wave was not related to response speed, but the late wave showed significantly larger amplitudes over all electrode positions preceding fast responses. Agonist EMG did not vary with response speed. This is in contrast to the influence of instructed tension and RT, reported by Sanders (1980) and Haagh et al. (1983). Since it is known that the relation between RT and agonist EMG activity is not always a straightforward one (Sanders 1980), this result might be caused by the fact that EMG activity was too small to expedite response execution.

CONTINGENT NEGATIVE VARIATION: MORE THAN A MOVEMENT-RELATED POTENTIAL

Tecce and Cattanach (1982) have questioned the motor interpretation of the CNV late wave, suggesting that attentional processes are also reflected in it. As we have stated above, we are aware of some difficulties in maintaining an exclusive motor interpretation. For instance, Simons et al. (1979) have found a relatively larger CNV late wave amplitude, when the imperative signal was an interesting slide upon which no motor response was required. (Eye movements were needed to look at the picture, though.) Also data from Elbert et al. (1980) provide evidence for the existence of a late wave under no motor response conditions. Furthermore, CNV amplitude in hyperactive children is attenuated as compared with children with normal concentration abilities (Grünewald-Zuberbier et al. 1978). Even time estimation without an ensuing response can be sufficient to generate a negative wave (Peters et al. 1977). These experiments weaken the extreme motor interpretation advanced by Rohrbaugh and Gaillard (1983). Therefore, we are willing to accept that other than motor functions are reflected in the CNV late wave. In addition, we would like to stress that the early wave, besides reflecting orientation and attention is also influenced by some preparatory processes. We have found that uninstructed voluntary muscle tension is accompanied by an increased early wave amplitude. A similar result was found by Gaillard et al. (1980) for instructed muscle tension. In the study of Haagh and Brunia (1985), significant muscle tension effects were predominantly present over the anteriorly localized electrode positions at the onset of the foreperiod, while a more central localization of this effect was found during the late wave. This suggests that the processes reflected by spontaneous muscle tension are related to rather early preparatory processes. Later on, other processes related to execution of the movement itself may conceal these ongoing early preparatory processes.

In conclusion, we argue that anticipatory slow potentials are generated by response organization per se as well as by various cognitive preparatory processes such as timing, attention, and abstract motor (pre)programming. As we have already discussed, these cognitive processes are related to cortical negativity which starts in the frontal cortex.

REFERENCES

Arezzo J, Vaughan, HG (1975) Cortical potentials associated with voluntary movements in the monkey. Brain Res 88: 99–104

Arezzo J, Vaughan, HG (1980) Cortical sources and surface topography of the motor potential and somatosensory evoked potential in the monkey. In: Kornhuber HH, Deecke L (eds) Motivation, motor and sensory processes of the brain: electrical potentials, behavior and clinical use. Prog Brain Res 54: 77–83

Arezzo J, Vaughan HG, Koss B (1977) Relationship of neuronal activity to gross movement-related potentials in monkey pre- and postcentral cortex. Brain Res 132: 362–369

Boschert J, Brickett P, Weinberg H, Deecke L (1983a) Movement-related potentials preceding toe plantarflexion and dorsiflexion. Hum Neurobiol 2: 87–90

Boschert J, Hink RF, Deecke L (1983b) Finger movement versus toe movement-related potentials: further evidence for supplementary motor area (SMA) participation prior to voluntary action. Exp Brain Res 55: 73–80

Brunia CHM (1980) What is wrong with legs in motorpreparation? In: Kornhuber HH, Deecke L (eds) Motivation, motor and sensory processes of the brain: electrical potentials, behavior and clinical use. Prog Brain Res 54: 232–236

Brunia CHM, van den Bosch WEJ (1984) Movement-related slow potentials. I. A contrast between finger and foot movements in right-handed subjects. Electroencephalogr Clin Neurophysiol 57: 515–527

Brunia CHM, Vingerhoets AJJM (1980) CNV and EMG preceding a plantar flexion of the foot. Biol Psychol 11: 181–191

Brunia CHM, Haagh SAVM, Scheirs JGM (1985) Waiting to respond: electrophysiological measurements in man during preparation for a movement. In Heuer H, Kleinbeck U, Schmidt KH (eds) Motor behavior: programming, control, and acquisition. Springer, Berlin Heidelberg New York Tokyo, pp 35–76

Deecke L, Kornhuber HH (1978) An electrical sign of participation of the mesial "supplementary" motor cortex in human voluntary finger movement. Brain Res 159: 473–476

Deecke L, Weinberg H, Brickett P (1982) Magnetic fields of the human brain accompanying voluntary movement. Bereitschaftsmagnetfeld. Exp Brain Res 48: 144–148

Deecke L, Boschert J, Weinberg H, Brickett P (1983) Magnetic fields of the human brain (Bereitschaftsmagnetfeld) preceding voluntary foot and toe movements. Exp Brain Res 52: 81–86

Deecke L, Kornhuber HH, Lang W, Lang M, Schreiber H (1985) Timing function of the frontal cortex in sequential motor and learning tasks. Hum Neurobiol (to be published)

Elbert T, Rockstroh B, Lutzenberger W, Birbaumer N (1980) Biofeedback of slow cortical potentials I. Electroencephalogr Clin Neurophysiol 48: 293–301

Fuster JM, Alexander GE (1971) Neuron activity related to short–term memory. Science 173: 652–654

Gaillard AWK, Perdok J, Varey CA (1980) Motor preparation at a cortical and at a peripheral level. In Kornhuber HH, Deecke L (eds) Motivation, motor and sensory processes of the brain: electrical potentials, behavior and clinical use. Prog Brain Res 54: 214–218

Gemba H, Hashimoto S, Sasaki K (1979) Slow potentials preceding self–paced hand movements in the parietal cortex of monkeys. Neurosci Lett 15: 87–92

Gemba H, Sasaki K, Hashimoto S (1980) Distribution of premovement slow cortical potentials associated with self–paced hand movements in monkeys. Neurosci Lett 20: 159–163

Gilden L, Vaughan HG, Costa LD (1966) Summated human EEG potentials with voluntary movement. Electroencephalogr Clin Neurophysiol 20: 433–438

Grünewald G, Grünewald–Zuberbier E (1983) Cerebral potentials during voluntary ramp movements in aiming tasks. In Gaillard AWK, Ritter W (eds) Tutorials in ERP Research: endogenous components. North–Holland, Amsterdam, pp 311–327

Grünewald–Zuberbier E, Grünewald G, Rasche A, Netz J (1978) Contingent negative variation and alpha attenuation responses in children with different abilities to concentrate. Electroencephalogr Clin Neurophysiol 44: 37–47

Haagh SAVM, Brunia CHM (1984) Cardiac–somatic coupling during the foreperiod in a simple reaction–time task. Psychol Res 46: 3–13

Haagh SAVM, Brunia CHM (1985) Anticipatory response–relevant muscle activity, CNV amplitude and simple reaction time. Electroencephalogr Clin Neurophysiol 61: 30–39

Haagh SAVM, Spoeltman WTE, Scheirs JGM, Brunia CHM (1983) Surface EMG and Achilles tendon reflex amplitudes during a foot movement in a reaction time task. Biol Psychol 17: 81 – 96

Hari R, Antervo A, Katila T, Poutanen T, Seppänen M, Tuomisto T, Varpula T (1983) Cerebral magnetic fields associated with voluntary limb movements in man. Il nuovo cimento 2D: 484 – 495

Hink RF, Deecke L, Kornhuber HH (1983) Force uncertainty of voluntary movement and human movement – related potentials. Biol Psychol 16: 197 – 210

Klapp ST (1977) Reaction time analysis of programmed control. In: Hutton RS (ed) Excercise and Sport Science Reviews. Journal Publishing Affiliates, Santa Barbara, pp 231 – 253

Kornhuber HH, Deecke L (1965) Hirnpotentialänderungen bei Willkürbewegungen und passiven Bewegungen des Menschen: Bereitschaftspotential und reafferente Potentiale. Pflügers Arch 284: 1 – 17

Kutas M, Donchin E (1977) The effect of handedness, the responding hand and response force on the contralateral dominance of the readiness potential. In Desmedt JE (ed) Attention, Voluntary contraction and event – related cerebral potentials. Karger, Basel, pp 189 – 210

Libet B, Wright EW, Gleason CA (1982) Readiness potentials preceding unrestricted 'spontaneous' vs preplanned voluntary acts. Electroencephalogr Clin Neurophysiol 54: 322 – 335

Loveless NE (1973) The contingent negative variation related to preparatory set in a reaction time situation with variable foreperiod. Electroencephalogr Clin Neurophysiol 35: 369 – 374

Näätänen R, Merisalo A (1977) Expectancy and preparation in simple reaction time. In: Dornic S (ed) Attention and performance VI. Erlbaum, Hillsdale, pp 115 – 138

Otto DA, Benignus VA (1978) Slow positive shifts during sustained motor activity in human. In Otto DA (ed) Multidisciplinary perspectives in event – related barin potential research. US Environmental Protection Agency, Washington DC, pp 131 – 133

Otto Da, Benignus VA, Ryan LJ, Leifer LJ (1977) Slow potential components of stimulus response and preparatory processes in man. In Desmedt JE (ed) Attention, voluntary contraction and event – related cerebral potentials. Karger, Basel, pp 211 – 230

Peters JF, Billinger TW, Knott JR (1977) Event related potentials of brain (CNV and P300) in a paired associate learning paradigm. Psychophysiol 14: 563 – 568

Rohrbaugh JW, Gaillard AWK (1983) Sensory and motor aspects of the contingent negative variation. In: Gaillard AWK, Ritter W (eds) Tutorials in ERP research: endogenous components. North – Holland, Amsterdam, pp 264 – 310

Roland PE, Skinhoj E, Larsen B, Lassen NA (1977) The role of different cortical areas in the organization of voluntary movements in man. A regional cerebral bloodflow study. In: Ingvar DH, Lassen NA (eds) Cerebral function, metabolism and circulation. Acta Neurol Scand 56 [Suppl 64]: 542 – 543

Sanders AF (1980) Some effects of instructed muscle tension on choice reaction time and movement time. In: Nickerson S (ed) Attention and performance VIII. Erlbaum, Hillsdale, pp 59–75

Sasaki K, Gemba H, Hashimoto S, Mizuno N (1979) Influences of cerebellar hemispherectomy on slow potentials in the motor cortex preceding self–paced hand movements in the monkey. Neurosci Lett 15: 23–28

Simons RF, Ohman A, Lang PJ (1979) Anticipation and response set: cortical, cardiac and electrodermal correlates. Psychophysiol 16: 222–233

Tecce JJ, Cattanach L (1983) Contingent negative variation. In: Niedermeyer E, Lopes Da Silva F (eds) Electroencephalography. Urban and Schwarzenberg, Baltimore, pp 543–562

Tecce JJ, Savignano–Bowman J, Dessonville CL (1984) CNV and myogenic functions I. Muscle tension produces a dissociation between CNV and EMG. In: Karrer R, Cohen J, Tueting P (eds) Brain and information. Ann NY Acad Sci USA 425: 283–288

Walter WG, Cooper R, Aldridge VJ, McCallum WS, Winter AL (1964) Contingent negative variation: an electrical sign of sensorimotor association and expectancy in the human brain. Nature 203: 380–384

Characterization of Joint - Interpolated Arm Movements*

J. M. Hollerbach and C. G. Atkeson

Artificial Intelligence Laboratory, Massachusetts Institute of Technology, 545 Technology Square, Cambridge, MA 02139, USA

INTRODUCTION

In considering how the motor control system executes an arm trajectory from a starting point to a goal point, one hypothesis is that a detailed time sequence of arm positions is planned along the trajectory. The two most obvious candidates for the planning variables that specify arm position are (a) joint angles and (b) hand position and orientation. To decide between these possibilities, researchers have sought the variables that give the most parsimonious description of experimental arm trajectories. Morasso (1981) observed that arm movements restricted to a horizontal plane were approximately straight, a finding corroborated by other researchers (Abend et al. 1982; Hollerbach and Flash 1982). Since the joint angle trajectories are apparently complicated functions of the start and goal points, Morasso argued that the manifest simplicity of the hand variable description can be considered as evidence for planning in hand variable space (Fig. 1).

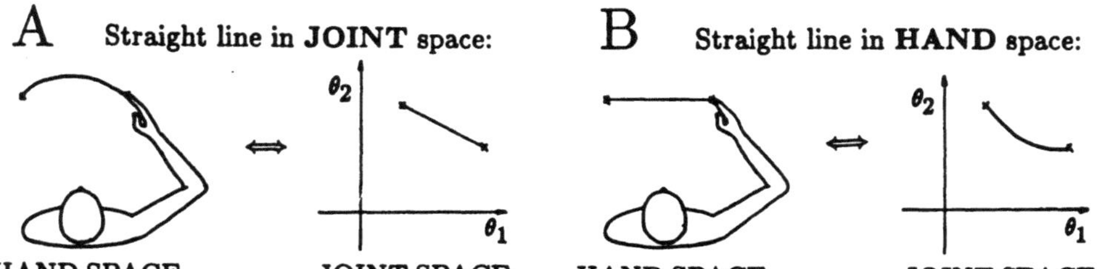

Fig. 1. A. A diagram of a straight line movement in joint space and the corresponding path in hand space. B. A straight line movement in hand space and the corresponding path in joint space. (θ_1 is the shoulder angle and θ_2 is the elbow angle.)

* This paper describes research done at the Department of Psychology and at the Artificial Intelligence Laboratory of the Massachusetts Institute of Technology. Support for this research was provided by NIH Research Grant AM 26710, awarded by the National Institute of Arthritis, Metabolism, and Digestive Diseases, and by an NSF Graduate Fellowship (CGA). The support for the facilities at the Artificial Intelligence Laboratory is provided in part by the Defense Advanced Research Projects Agency under Office of Naval Research contract N00014-80-C-0505.

Experimental Brain Research Series 15

Other researchers have reported evidence for joint space planning. Kots and Syrovegin (1966) found a constant ratio of angular speed for simultaneous movements in elbow and wrist, although these findings have been criticized by Bishop and Harrison (1977). The most intriguing evidence, which triggered the investigations of this paper, is provided in a series of papers on two−joint arm movements in a vertical plane (Soechting and Lacquaniti 1981; Lacquaniti and Soechting 1982; Lacquaniti et al. 1982; Soechting and Lacquaniti 1983). To quote the earliest paper (Soechting and Lacquaniti 1981: 718): "The other consistent findings which characterize the pointing movement examined can be summarized as follows: (1) the ratio of the maximal velocity of the elbow to that at the shoulder is equal to the ratio of the angular excursion at the two joints, (2) the two angular velocities reach a maximum at the same time and (3) their slope is independent of target location as the target is approached. These invariances are all expressed in intrinsic coordinates and, thus, one may pose the question: Is the movement organized in terms of its intrinsic coordinates (θ [shoulder angle] and ϕ [elbow angle]) rather than in terms of extrinsic coordinates (x,y)?"
The reason this evidence is intriguing is that at the same time the joint rate ratios seem constant, inspection of the trajectories shows approximately straight−line hand paths (Fig. 4). It would be expected that these two viewpoints exclude each other, given the complex transformation between joint space and hand space. There are only two degrees of freedom in a two−joint arm and either a hand trajectory or a joint trajectory completely specifies the motion of the arm.

In this paper we reconcile this discrepancy by showing that the joint rate ratio plots are not actually constant despite their appearance. The two viewpoints are generally exclusive and clearly distinguishable, as we show by a theoretical characterization of all possible straight− line joint paths. To explain why the data seem to support both viewpoints, we will focus on the property for the corresponding experimental movements of unidirectional execution towards the workspace boundary. It will be shown that a general kinematic property of two−link mechanisms at the boundary of the workspace is the approach of the joint rate ratios to a constant value. Thus it is unnecessary to invoke motor system planning and control to explain the joint space characteristics.

RESULTS

In this section we derive a succinct mathematical formulation for the set of all possible trajectories obtainable from a strategy of constant joint rate ratio. This formulation allows determination of the extent to which these trajectories are distinguishable from hand−based straight−line trajectories. Next, a surprising kinematic behavior of a two−link mechanism as it approaches the workspace boundary is presented, and is invoked to explain the apparent contradiction of straight−line paths simultaneously in hand and joint space.

Joint Interpolation

In the Introduction we contrasted straight−line motions in hand space to straight−line motions in joint space. We will first show that a strategy of maintaining a constant joint rate ratio is equivalent to executing straight lines in joint angle space. In robotics straight lines in joint space are known as joint interpolation (Brady 1982), defined by the following equation (see Fig. 1a):

$$\boldsymbol{\theta}(t) = \left(\boldsymbol{\theta}(t_f) - \boldsymbol{\theta}(t_i)\right) f(t) + \boldsymbol{\theta}(t_i)$$

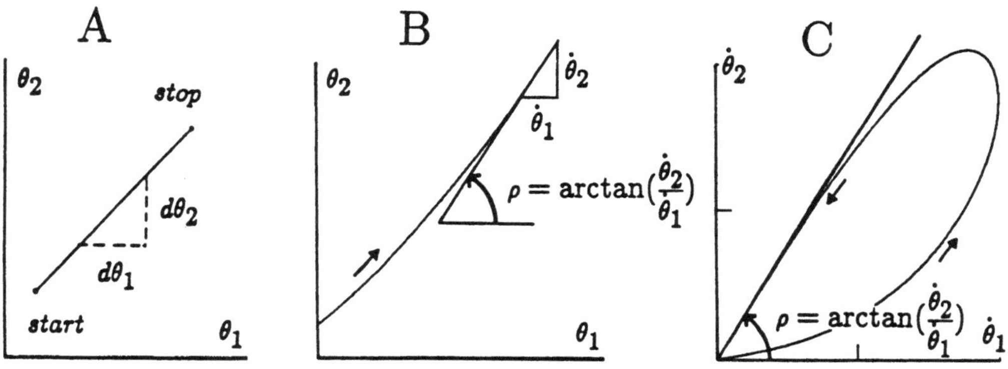

Fig. 2. A. A constant slope in joint space (which leads to a straight line) implies a constant joint rate ratio and vice versa. The slope of the line in joint space is $\dot\theta_2/\dot\theta_1$. B. Segments of the joint space path (in this case a segment of trajectory 1 from Fig. 5) can be fit by straight lines whose slope is given by $\dot\theta_2/\dot\theta_1$ and whose angle from the vertical is given by $\varrho = \arctan(\dot\theta_2/\dot\theta_1)$. C. The same result can be seen in the joint velocity space trajectory where the path approaches a straight line asymptote towards the end of the movement whose slope is $\dot\theta_2/\dot\theta_1$ and whose angle from the origin is $\varrho = \arctan(\dot\theta_2/\dot\theta_1)$ We use the angle ϱ to describe the joint rate ratio since the arctangent of a ratio remains between $-\pi$ and π as the ratio itself can vary between negative and positive infinity

where $\boldsymbol{\theta} = (\theta_1, \theta_2, \ldots, \theta_n)$ is a vector of the n joint angles, $\boldsymbol{\theta}(t_i)$ are the joint angles at the start of the movement, $\boldsymbol{\theta}(t_f)$ are the joint angles at the end of the movement, and f(t) is any monotonic time function ($\dot{f}(t) \geq 0$) with $f(t_i) = 0$, $f(t_f) = 1$. Note that each joint angle will have the same velocity profile except for a scaling factor of distance travelled by that joint. In these trajectories all joints reach maximal velocity at the same time and the ratio of any pair of maximal velocities equals the corresponding ratio of angular excursions of the joints.

To begin the demonstration of equivalence, the ratio of any two joint–angular velocities given a joint interpolation strategy is:

$$\frac{\dot\theta_j(t)}{\dot\theta_k(t)} = \frac{(\theta_j(t_f) - \theta_j(t_i))\,\dot{f}(t)}{(\theta_k(t_f) - \theta_k(t_i))\,\dot{f}(t)} = \frac{\Delta\theta_j}{\Delta\theta_k}$$

which is constant. On the other hand, a constant joint rate ratio strategy implies a constant slope of the trajectory in joint angle space, since by the chain rule:

$$\frac{\dot\theta_2}{\dot\theta_1} = \frac{d\theta_2}{dt}\,\frac{dt}{d\theta_1} = \frac{d\theta_2}{d\theta_1}$$

which is constant (Fig. 2a). Straight lines are not the only possible planning strategy in terms of joint variables, but they are the most common and the most analyzed. Other conceivable strategies include independent joint control, where each joint has a separate time function.

The implication of this analysis is that the features of trajectories that can be used to support joint variable planning are straight-line segments in joint angle space. Soechting and Lacquaniti (1981) actually presented more complicated trajectories for which only a portion of the joint angle plots, in the deceleratory phase, seemed a straight line. Local straight segments of a path can be inferred either by examination of the joint angle plots (Fig. 2b) or by the asymptotic approach in joint velocity space (Fig. 2c). A goal in the subsequent sections is to understand when such straight line segments in joint angle space arise.

The N-Leaved Rose

One possible problem is that the trajectories produced by a joint interpolation strategy may be experimentally indistinguishable from trajectories produced by a hand space strategy. To test for this possibility we have theoretically characterized what paths straight-line movements in joint space will produce in hand space for a planar two-link arm (Fig. 3a). This two-link arm is a reasonable approximation to the forearm and upper arm of a human, since the experimental data typically describes planar movements involving only the shoulder and elbow joints. The features of all straight-line joint space motions of this two-link arm can be described rather simply. For a constant joint rate ratio $K = \dot{\theta}_1/\dot{\theta}_2$ and an initial arm position at full extension ($\theta_1=0$, $\theta_2=0$), we generate a straight line in joint space and a more complex curve in hand space. This can be done for any joint rate ratio, to generate a family of corresponding curves in joint space (Fig. 3b) and in hand space (Fig. 3c). All features of joint interpolated paths have been generated, due to the rotational symmetry of this arrangement with respect to θ_1, in that a shift in the initial condition $\theta_1(0)$ merely shifts Fig. 3b to the left or the right and rotates Fig. 3c about the origin by $\theta_1(0)$ without changing path shape.

It is possible to characterize analytically the family of curves generated by joint-interpolated trajectories in a succint form by using a polar coordinate (r, ϕ) arm description (Fig. 3a), with r the radial distance from the shoulder to the arm tip and ϕ the angle the radial line makes at the shoulder:

$$\phi = \theta_1 + \frac{\theta_2}{2}, \quad r = 2 \cos \frac{\theta_2}{2},$$

$$\dot{\phi} = \dot{\theta}_1 + \frac{\dot{\theta}_2}{2}, \quad \dot{r} = -\dot{\theta}_2 \sin \frac{\theta_2}{2}.$$

Combining equations,

$$\dot{\phi} = \left[\frac{\dot{\theta}_1 + \dot{\theta}_2/2}{\dot{\theta}_2}\right] \frac{-\dot{r}}{\sin(\theta_2/2)}.$$

Suppose $K = \dot{\theta}_1/\dot{\theta}_2$ is constant. Then

$$\dot{\phi} = (K+1/2) \frac{-\dot{r}}{\sin(\theta_2/2)}.$$

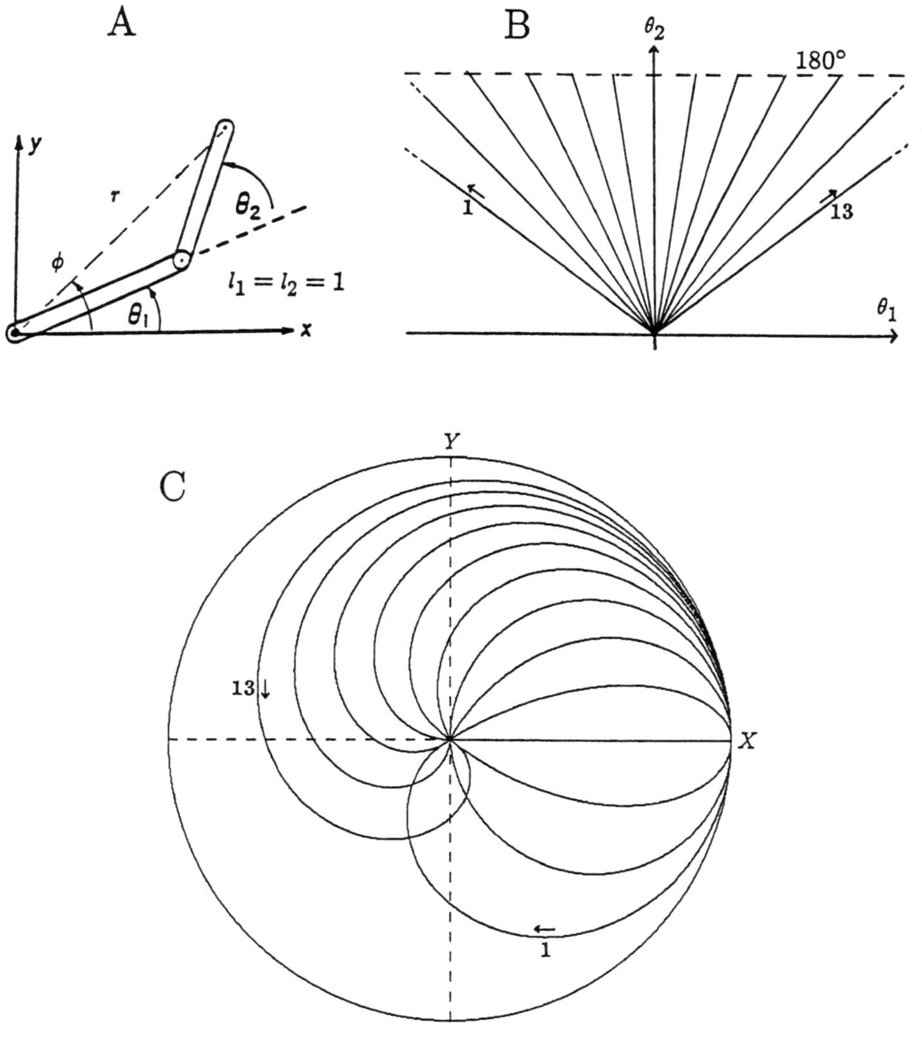

Fig. 3. A. A planar two link arm. (θ_1,θ_2) are the joint angles and (r, ϕ) is the polar coordinate description of the tip position (x,y). For this figure we took the link lengths, l_1 and l_2, to be equal to 1. B. A sampling of the straight lines in joint space from the origin. Note that the origin in joint space corresponds to a fully extended arm whose tip is at x=2, y=0. C. The straight line joint space trajectories of B mapped to hand space

Noting that $\sin(\theta_2/2) = \sqrt{1-(r/2)^2}$ and integrating the equation

$$\int_0^t \dot{\phi}\,dt = \int_0^t \frac{-\dot{r}}{\sqrt{1-(r/2)^2}}\,dt$$

yields

$$\phi-\phi_0 = (2K+1)\left[\cos^{-1}\left(\frac{r}{2}\right) - \cos^{-1}\left(\frac{r_0}{2}\right)\right].$$

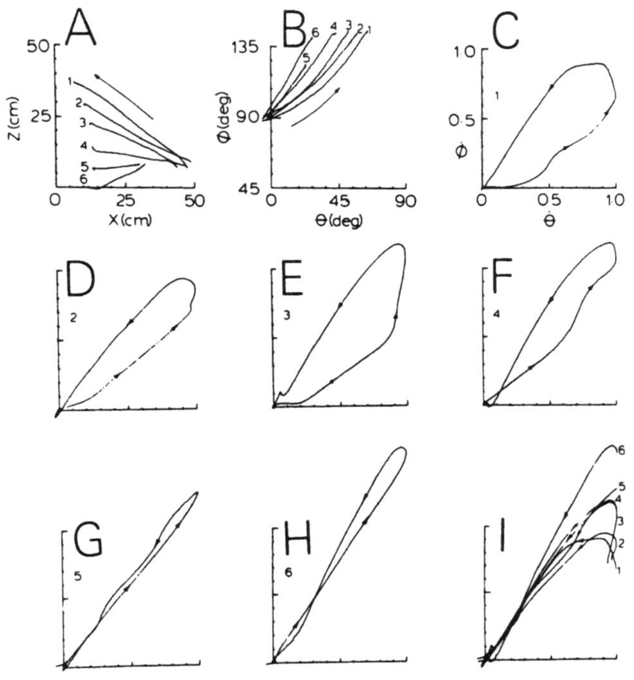

Fig. 4. Dependence of movement trajectory on target location. A. Movement trajectories of individual trials to six target locations described in x, z coordinates. B. The same movements described in θ, ϕ coordinates. Lower target locations require progressively less shoulder flexion. C to H. The trajectories described in the phase plane ($\dot{\theta}$, $\dot{\phi}$) for movements to targets 1 to 6, respectively. Only the portions beginning when shoulder angular velocity is maximal are shown in I. Note that the trajectories form a loop which becomes progressively tighter for lower target locations and that the terminal portions of all trajectories are virtually superimposable. Soechting and Lacquaniti use different joint angles; θ is the shoulder angle: ($\theta = \theta_1 + 90°$), and ϕ is the elbow angle: ($\phi = 180° - \theta_2$). (From Soechting and Lacquaniti 1981)

Taking $\cos^{-1}(r_0/2) = 0$, where the arm is initially fully extended, we obtain the

N-LEAVED ROSE: $$r = 2 \cos \left(\frac{\phi - \phi_0}{2K+1} \right)$$

As can be seen in Fig. 3c, N-leaved roses (Burlington 1942) tend to be strongly curved, especially for movements that are not primarily radial movements. Hence joint-based planning generally yields trajectories readily distinguishable from straight-line hand paths. There is exactly one special case in which straight-line paths exist simultaneously in hand space and in joint space, when $\dot{\theta}_1 = -\dot{\theta}_2/2$ and hence $\dot{\phi} = 0$. The trajectory must then follow a straight radial line through the shoulder. The horizontal straight line in Fig. 3c is an example, and corresponds to a degenerate "rose petal" with no width. In examining the movements of Figs. 4a and 5a, it seems movement 6 does fall into this special category and movement 5 nearly does, but movements 1–4 cannot be explained in this way.

Movements Towards the Workspace Edge

Simulated Movements: Though the previous analysis has shown that, in general, joint– interpolated movements are strongly curved, it nevertheless remains to explain why the data of Soechting and Lacquaniti (1981) apparently show approximately straight–line hand paths and, during the last half of the trajectories, an approach to a constant joint rate ratio (Fig. 4).

One possibility is that the small deviations from hand–space straight lines seen in the data are significant and explain the constant joint rate ratio asymptote. To decide on this possibility, we simulated exactly straight–line movements with approximately the same starting and target points as in a sample of the original data (Fig. 4a). For these simulated movements we used link lengths measured from a human subject and the trajectories were minimum jerk (Hogan 1984). The simulated straight–line movement paths are shown in Fig. 5a and the corresponding joint–space trajectories are shown in Fig. 5b. (We have used standard coordinate systems and joint angle conventions from robotics. This leads to apparent differences between Figs. 4 and 5 in terms of scales and orientation of the plots. We have tried to minimize these apparent differences by rearranging the units of the various axes.) We note that the second half of each of these joint–space trajectories approaches a straight line in both the real and simulated data, which apparently suggests that the last half of the movement is joint interpolated. The joint velocity–space diagrams show the same features in the real and simulated data: an approach to an asymptote and the transition from an oval trajectory to a narrower trajectory as the final target shifted (Figs. 4c–i and 5c).

To pursue further the issue of a constant joint rate ratio, the plot of arctan $(-\dot{\theta}_2/\dot{\theta}_1)$ vs normalized movement distance (the distance traveled so far divided by the total distance) in Fig. 5d evidently shows the approach of the joint rate ratios to a constant. There is clearly something special about these movements that goes against the expectations based on N–leaved roses, and the following section explores why this is the case.

Kinematic Constraints: When the straight lines of Fig. 5a are extended to the workspace boundary, it can be seen that the plots of Fig. 5d actually reach the same constant limit in Fig. 6a. To explain this result, we examine the behavior of the analytic expression for joint rate ratio as the boundary is approached. The general analytic expression is (see Appendix for derivation):

$$\text{Joint rate ratio} = \frac{\text{elbow angular velocity}}{\text{shoulder angular velocity}} = \frac{\dot{\theta}_2}{\dot{\theta}_1}$$

$$= \frac{-\dot{x}l_1 \cos\theta_1 - \dot{y}l_1 \sin\theta_1 - \dot{x}l_2 \cos(\theta_1+\theta_2) - \dot{y}l_2 \sin(\theta_1+\theta_2)}{\dot{x}l_2 \cos(\theta_1+\theta_2) + \dot{y}l_2 \sin(\theta_1+\theta_2)}$$

Near the periphery, the elbow becomes straight and $\theta_2 \rightarrow 0$.

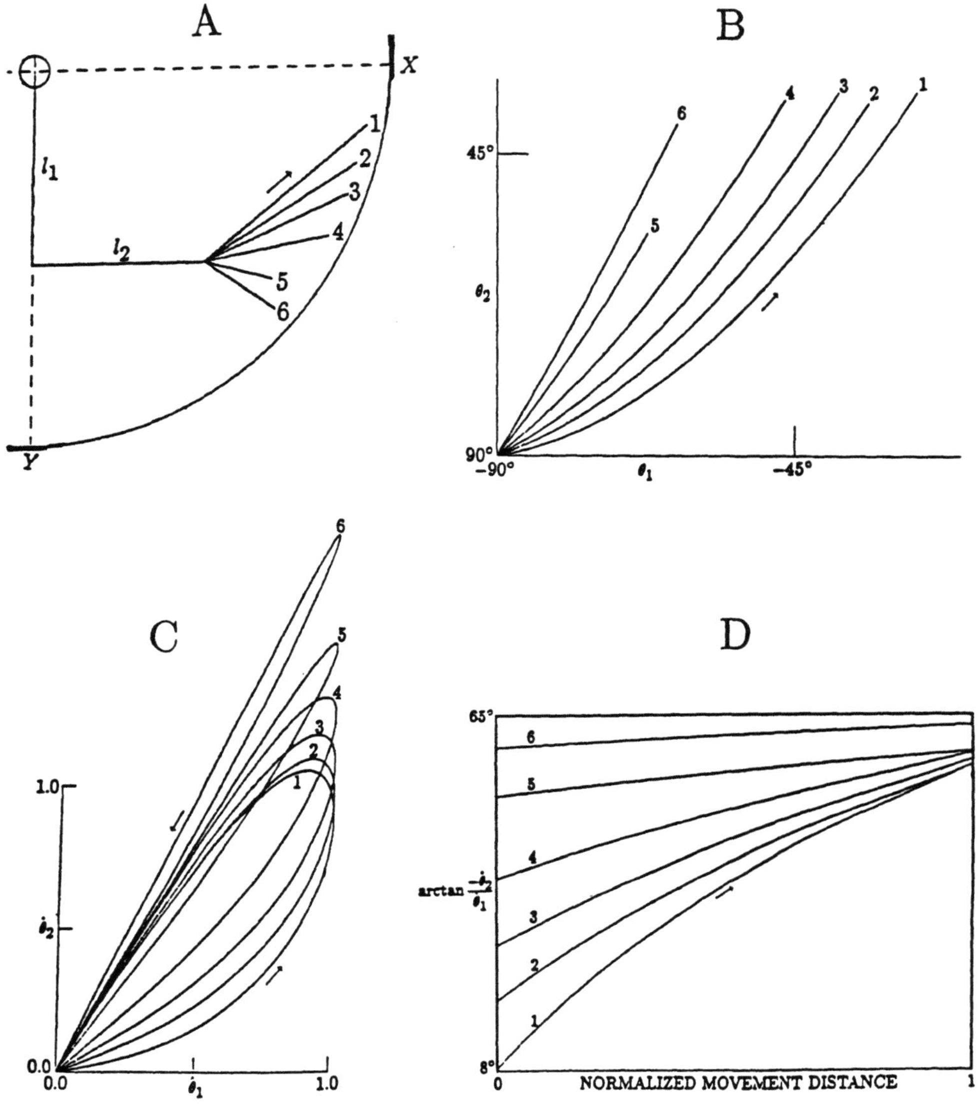

Fig. 5. A. Six simulated straight–line hand space movements. The start and target points for these six movements correspond to those for the six movements of Fig. 4A. The movements are displaced and reversed in direction only due to different coordinate system conventions. B. The corresponding paths in joint space. These paths should match the joint–space paths of Fig. 4B. (We use different joint angle conventions from Fig. 4: $\theta = \theta_1 + 90°$, $\phi = 180° - \theta_2$.) C. The normalized joint velocity–space trajectories. These trajectories should match those in Figs. 4C to 4I. D. The actual joint rate ratios (in angular form) for each movement plotted as the arctan$(-\dot{\theta}_2/\dot{\theta}_1)$ of the points in velocity space vs normalized movement distance. The joint rate ratios all seem to approach a limit, even for these exactly straight–line movements in hand space

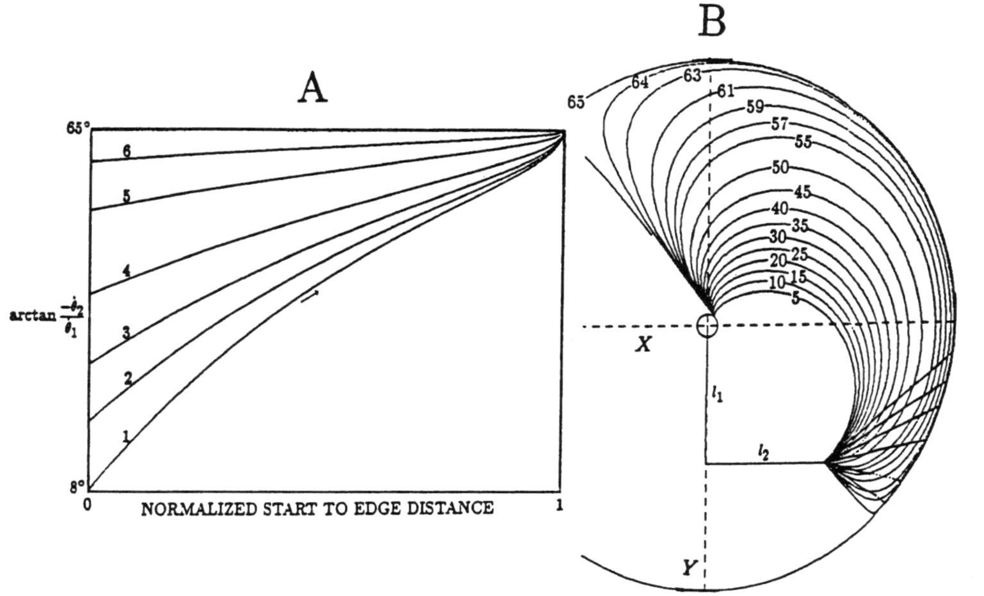

Fig. 6 Effect of joint rate ratio limit at edge of workspace. A. When the joint rate ratio plots for the simulated movements of Fig. 5 are plotted all the way to the edge of the workspace, the joint rate ratios all reach the same limit. B. A contour map can be drawn for the joint rate ratios (in terms of $\arctan(-\dot{\theta}_2/\dot{\theta}_1)$) reached for all straight line hand space trajectories starting at $\theta_1 = -90°$, $\theta_2 = 90°$. Note that the limit of 65° is approached closely in a large portion of the workspace. The simulated movements of Fig. 5 extended to the edge of the workspace are also shown

$$\lim_{\theta_2 \to 0} (\text{joint rate ratio}) = \lim_{\theta_2 \to 0} \frac{\dot{\theta}_2}{\dot{\theta}_1} = -\frac{l_1 + l_2}{l_2}$$

Surprisingly, the joint rate ratio is a constant, dependent only on the link lengths. The constancy holds for any point on the workspace boundary and for any trajectory that approaches the boundary. Before the workspace boundary, the joint rate ratios are not constant despite appearances. It is critical to distinguish the statements "is a constant" vs "tending towards a constant." The trajectories of Soechting and Lacquaniti (1981) therefore do not actually demonstrate a constant joint rate ratio in the deceleratory phase, and there is no contradiction with the trajectory features of hand–space straight–line paths.

To show that this is a general kinematic phenomenon for two–link planar mechanisms, we plot a contour map of constant $\arctan(-\dot{\theta}_2/\dot{\theta}_1)$ lines connecting the shoulder point to the movement starting point (Fig. 6b). Overlayed onto this contour map are the extended straight–line hand–space movements of Fig. 5a. As these movements traverse towards the boundary, each crosses the contour lines in a particular order before finally reaching the boundary value of 65°. The experimental movements of Fig. 4a reached at least the 57° contour.

DISCUSSION

The misleading constancy of joint rate ratio in the data of Soechting and Lacquaniti (1981) is due to an unfortunate choice of experimental movements. If they had chosen movements in the reverse direction, from the boundary to their original interior starting point, then they would have observed an apparent constancy of joint rate ratio in the acceleratory rather than deceleratory phase of movement. If they had chosen start and goal points completely interior to the boundary, then they might have observed straight–line hand paths corresponding to complicated functions of joint rate ratio.

While our conclusion is that this experimental data is consistent with straight–line hand paths rather than with straight–line joint paths, we do not wish to imply that humans select only hand–space trajectory plans. What our analysis shows is that if one is interested in the question of joint–space vs hand–space planning, then one should stay away from the workspace boundary. Movements near the boundary will always appear to have a constant joint rate ratio, and there is not a clear enough distinction from hand–path trajectories.

Explicit vs Implicit Trajectories

The point–by–point evolution of an arm trajectory need not be the result of an explicit trajectory plan, as would be hypothesized for planning in joint or hand space. An alternative viewpoint is that trajectory shape is an implicit function of control. One example is end–point control, where the muscle actuation is proportional to error between a current arm state and the end state; the trajectory evolves implicitly through the interaction of movement dynamics and the end–point dependent actuation. Hollerbach (1981) proposed a coupled oscillator theory of handwriting in which an individual writing stroke is generally shaped by parameters of the oscillation, but is not directly specified. The various proposals for final position control are a form of end–point control where muscle ensemble equilibrium point is changed in a step–like fashion from start to goal; the muscle properties with or without proprioceptive feedback loops set up force fields that mechanically direct a movement toward an equilibrium goal (Crossman and Goodeve 1963; Fel'dman 1974; Bizzi et al. 1976; Kelso 1977; Kelso and Holt 1980).

Evidence has been accumulating against end–point control and for intermediate trajectory points. Delatizky (1982) concluded from simulation that it is unlikely such a simple mechanism could generate observed straight–line hand paths with bell–shaped velocity profiles. Bizzi et al. (1982) inferred the existence of intermediate trajectory points through perturbation experiments. An alternative hypothesis is that trajectory generation is the outcome of some continuous process or internal (mental) dynamic system, as described in Saltzman and Kelso (1983).

Explicit trajectory planning has still not been conclusively demonstrated, but there are reasons for hypothesizing its existence. Planning in hand space allows handling external task constraints such as motion along geometric constraint surfaces (Hollerbach 1982), obstacle avoidance (Lozano–Perez 1982), and minimization of load perturbation (Flash 1983). The cost of hand space planning is that it requires transformations between hand position variables and the directly sensed and controlled variables such as joint angles and muscle lengths. Simple paths in hand space such as straight lines can lead to complex trajectories in joint space (Fig. 1b).

The advantage of planning in joint space is that it simplifies the planning process. Each joint has an initial angle and a final angle, and a smooth transition is simply generated from the initial to the final angle for each joint individually. Coordinating the joints need only be done by coordinating the timing for joint motion start and stop. A simple joint plan such as this will lead to straight line paths in joint space if the movements of each joint are amplitude scaled versions of each other. The problem with this form of arm motion planning is that it often generates complex trajectories of the hand (Fig. 1a), which may cause collision with obstacles or unwanted forces on a hand–held load.

The question of what spaces or sets of coordinates are used in the planning and control of human movements cannot be settled on the basis of kinematic data alone, given that the kinematic data is in general complex and the shapes of movements can reflect a wide variety of task demands. However, discovering simple models that explain a wide variety of kinematic behavior give strong indications of what types of signals and processes exist in human motor control. Since the simple description of point–to–point trajectories as handspace straight line motions seems to apply throughout the workspace, we take this as evidence of trajectory planning at a level higher than muscle or joint variables.

SUMMARY

1. A strategy of constant joint rate ratio of shoulder and elbow joints is formally equivalent to joint interpolation, i.e., straight line paths in joint space.

2. For a given joint rate ratio, all hand trajectory points lie on a polar coordinate curve called an N–leaved rose whose locus is independent of the exact time profile. These N–leaved roses tend to be strongly curved, so that joint based planning generally yields trajectories readily distinguishable from straight line hand paths.

3. However, in movements towards the edge of the workspace the ratio of shoulder angular velocity and elbow angular velocity tends towards a constant value no matter how the edge of the workspace is approached. This is a general property of two–link mechanisms, arising solely from kinematics.

4. We note that the experimental movements provided as evidence for joint based planning were unidirectional towards the boundary of the workspace. In this situation even a straight line path in hand space approaches a constant joint rate ratio. Thus it is inappropriate to suggest a planning strategy in terms of joint angles on the basis of this evidence, which is an artifact of arm kinematics.

5. All the data presented so far is consistent with and can be most concisely explained by planning straight line movements in hand space.

APPENDIX

In this appendix we derive an analytic expression for the joint rate ratio given the tip velocity for a simple 2 joint arm (Fig. 3A). The direct kinematics are

$$x = l_1\cos(\theta_1) + l_2\cos(\theta_1 + \theta_2)$$
$$y = l_1\sin(\theta_1) + l_2\sin(\theta_1 + \theta_2)$$

where (x,y) is the tip position, θ_1 and θ_2 are the joint angles, and l_1 and l_2 are the link lengths. In order to find the joint velocities we differentiate the above equations:

$$\begin{bmatrix} \dot{x} \\ \dot{y} \end{bmatrix} = \begin{bmatrix} -l_1 \sin(\theta_1) - l_2 \sin(\theta_1+\theta_2) & -l_2 \sin(\theta_1+\theta_2) \\ l_1 \cos(\theta_1) + l_2 \cos(\theta_1+\theta_2) & l_2 \cos(\theta_1+\theta_2) \end{bmatrix} \begin{bmatrix} \dot{\theta}_1 \\ \dot{\theta}_2 \end{bmatrix}$$

where the 2 × 2 matrix relating these two vectors is the *Jacobian* **J**. The joint rate ratio can be found by inverting **J**.

$$\begin{bmatrix} \dot{\theta}_1 \\ \dot{\theta}_2 \end{bmatrix} = \frac{\begin{bmatrix} l_2 \cos(\theta_1+\theta_2) & l_2 \sin(\theta_1+\theta_2) \\ -l_1 \cos(\theta_1) - l_2 \cos(\theta_1+\theta_2) & -l_1 \sin(\theta_1) - l_2 \sin(\theta_1+\theta_2) \end{bmatrix}}{\text{determinant } (\mathbf{J})} \begin{bmatrix} \dot{x} \\ \dot{y} \end{bmatrix}$$

and therefore

$$\text{Joint rate ratio} = \frac{\text{elbow angular velocity}}{\text{shoulder angular velocity}} = \frac{\dot{\theta}_2}{\dot{\theta}_1}$$

$$= \frac{-\dot{x}l_1\cos\theta_1 - \dot{y}l_1\sin\theta_1 - \dot{x}l_2\cos(\theta_1+\theta_2) - \dot{y}l_2\sin(\theta_1+\theta_2)}{\dot{x}l_2\cos(\theta_1+\theta_2) + \dot{y}l_2\sin(\theta_1+\theta_2)}$$

REFERENCES

Abend W, Bizzi E, Morasso P (1982) Human arm trajectory formation. Brain 105: 331–348

Bishop A, Harrison A (1977) Kots and Syrovegin (1966) – a demonstration of modular units in motor programming?. J Hum Movem Stud 3: 99–109

Bizzi E, Polit A, Morasso P (1976) Mechanisms underlying achievement of final head position. J Neurophysiol 39: 435–444

Bizzi E, Accornero N, Chapple W, Hogan N (1982) Arm trajectory formation in monkeys. Exp Brain Res 46: 139–143

Bizzi E, Accornero N, Chapple W, Hogan N (1984) Posture control and trajectory formation during arm movement. J Neurosci 4: 2738–2745

Brady JM (1982) Trajectory planning. In: Brady JM, Hollerbach JM, Johnson TL, Lozano–Perez T, Mason MT (eds) Robot motion: planning and control. MIT Press, Cambridge, pp 221–244

Burlington RS (1942) Handbook of mathematical tables and formulas. Handbook Publ Inc, Sandusky, Ohio, p 33

Crossman ERFW, Goodeve PJ (1963) Feedback control of hand–movement and Fitts' Law. Paper presented at the meeting of the experimental psychological society, Oxford, July 1963, Published: Q J Exp Psychol (1983) 35A: 251–278

Delatizky J (1982) Final position control in simulated planar arm movements. Soc Neurosci (Abstr) 8: 283

Fel'dman AG (1974) Change of muscle length due to shift of the equilibrium point of the muscle–load system. Biofizika 19: 534–538

Flash T (1983) Organizing principles underlying the formation of hand trajectories. PhD Thesis, Harvard/MIT Division of Health Sciences and Technology

Hogan N (1984) An organizing principle for a class of voluntary movements. J Neurosci 4: 2745–2754

Hollerbach JM (1981) An oscillation theory of handwriting. Biol Cybern 39: 139–156

Hollerbach JM (1982) Computers, brains, and the control of movement. Trends Neurosci 5(6): 189–192

Hollerbach JM, Flash T (1982) Dynamic interactions between limb segments during planar arm movement. Biol Cybern 44: 67–77

Kelso JAS (1977) Motor control mechanisms underlying human movement reproduction. J Exp Psychol: Hum Percept Perform 3: 529–543

Kelso JAS, Holt KG (1980) Exploring a vibratory systems analysis of human movement production. J Neurophysiol 43: 1183–1196

Kots YM, Syrovegin AV (1966) Fixed set of variants of interaction of the muscles of two joints used in the execution of simple voluntary movements. Biofizika 11: 1061–1066

Lacquaniti F, Soechting JF (1982) Coordination of arm and wrist motion during a reaching task. J Neurosci 2: 399–408

Lacquaniti F, Soechting JF, Terzuolo CA (1982) Some factors pertinent to the organization and control of arm movements. Brain Res 252: 394–397

Lozano – Perez T (1982) Task planning. In: Brady JM, Hollerbach JM, Johnson TL, Lozano – Perez T, Mason T (eds) Robot motion: planning and control. MIT Press, Cambridge, pp 473 – 498

Morasso P (1981) Spatial control of arm movements. Exp Brain Res 42: 223 – 227

Saltzman E, Kelso JAS (1983) Skilled actions: a task dynamic approach, SR – 76. Haskins Laboratory Status Report on Speech Research, pp 3 – 50

Soechting JF, Lacquaniti F (1981) Invariant characteristics of a pointing movement in man. J Neurosci 1: 710 – 720

Soechting JF, Lacquaniti F (1983) Modification of trajectory of a pointing movement in response to a change in target location. J Neurophysiol 49: 548 – 564

Precuing Methods in the Study of Motor Programming

H. N. Zelaznik[1] and D. D. Larish[2]

[1] Motor Behavior Laboratory, Department of Physical Education, Purdue University, West Lafayette, IN 47907, USA
[2] Motor Behavior Laboratory, Department of Physical Education, Arizona State University, Tempe, AZ, USA

In situations in which time is of the essence, it is important for a performer to have a small reaction time (RT). To be successful, athletes and those taking part in games must select and execute appropriate actions quickly and efficiently. Coaches and teachers of skills invariably attempt to increase the RT of their opponents for their own advantage. In the laboratory, however, researchers attempt to manipulate RT of their subjects, not to win competitions, but rather to understand the underlying processes involves in speeded performance.

Specifically, motor skills researchers are interested in understanding the temporal organization of processes involved in the selection and execution of discrete motor acts. Some scholars have termed this process motor programming (Keele 1968, 1981; Klapp 1977; Larish and Frekany 1985; Rosenbaum 1980, 1983; Schmidt 1976; Zelaznik 1981; Zelaznik et al. 1982), with the implication that programming itself involves the selection and execution of response parameters prior to movement execution. We would like to defer discussion of this critical question of action (see Goodman and Kelso 1980), and take a similar position to that of Rosenbaum (1983), in that regardless of the construct of control, certain cognitive decisions about the upcoming act must be made prior to action. And, that reaction time methods can be utilized to study these processes. In the present paper we shall focus upon one of these methods – precuing.

BASIC ASSUMPTIONS

In order to examine the processes involved in the selection and execution of motor actions, it is necessary to be able to manipulate the durations of these processes. The precuing technique manipulates these durations by varying the amount of stimulus and/or response information prior to the presentation of a signal to initiate a movement. For example, suppose that a subject was required to produce one of eight possible responses which were determined by crossing three dimensions of movement, each with two values. In an aiming task they could be arm (right/left), direction (towards/away), and extent (near/far).

To study the organization of the selection processes for arm, direction, and extent, subjects are placed in a reaction time situation in which prior to the arrival of the imperative stimulus, information is provided in the form of a precue. This precue reduces the uncertainty of the upcoming response on none, one, two or all three dimensions of movement. The reductions in uncertainty of these motor dimensions should result in savings in reaction time. By analyzing the

Experimental Brain Research Series 15

pattern of reaction time data as a function of the number and kind of precue information it is possible, theoretically, to ascertain the temporal organization of the processes reponsible for response initiation.

As Rosenbaum (1980, 1983) has discussed in detail, there are a few critical assumptions of the precuing method that must be adhered to if the method is to be useful. First, and foremost, the model assumes that response selection involves a feature selection process. In other words, providing advance information concerning a dimension of movement alters the duration of a process responsible for selecting that dimension. And, furthermore the model assumes that there is at least more than one process involved in movement initiation. If this second condition was not met it would be misleading to ask questions about independence, serial and parallel nature of these response selection processes.

Goodman and Kelso (1980) have criticized these two assumptions of the precuing method, and instead have suggested that programming is a whole process, not capable of being broken down into parts. We think that part of this controversy stems from problems in the definitions of programming and response selection. The precuing method asks questions about how subjects select what to do, not how to do the specific act. In other words it is truly in the domain of cognitive behavior, and as such it appears reasonable to ask question about serial or parallel nature of these cognitive decisions. On the other hand, the actual organization of the already selected act, what many term programming (Schmidt 1975), might in fact be a whole process so that it is unreasonable to ask about the parts of programming. The present paper will focus upon an examination of precuing methodology, rather than question the validity of the paradigm as a research tool. The latter part of the paper shall suggest how precuing methodology can be utilized to ask some additional questions about motor control.

Assume that in a precuing experiment three dimensions of movement (A, B, and C), each having two levels, are manipulated (we are trying to maintain the same convention as Rosenbaum 1980). The basic model of precuing assumes that for any condition:

$$T(A) = a(A) + b(A) + g(A) \tag{1}$$

in which T(A) represents the reaction time when dimension A has to be specified. The lower-case a represents the time to process the imperative stimulus, b represents the time to select the appropriate response dimensions, and g the time necessary for response execution.

A necessary condition for some of the tests of the precuing methodology assumes that the T's for the various dimensions are not equal, but instead are ordered. In other words, when A is selected, RT is lengthened, as is the case for B and C. And, for example, the ordering of times could be: $T(A) < T(B) < T(C)$. Rosenbaum assumes that the reason for the difference in RT is due to a difference in b, the time for motor specifications, and that the other times are of equal value since there are equivalent numbers of stimulus-response alternatives, and the response sets are equivalent.

Assume that the inequality of specification times is true. Rosenbaum then argues that if two processes each responsible for specifying a different dimension, for example, B and C, are selected serially, then the following inequality should hold:

$$T(BC) - T(B) \geq T(C) - T(A) \tag{2}$$

This inequality is derived from the following assumptions:
(a) The difference in RT between any two conditions having an equal number of stimulus-response alternatives is due to differences in the b specification times. For example, $T(C) - T(A) = b(C) - b(A)$. (b) Since $b(A) > 0$, then $T(C) - T(A) < b(C)$. Rosenbaum's test for a serial specification of two dimensions, for example, B and C, is:

$$b(BC) \geq b(B) + b(C) \tag{2.1}$$

which implies that the additional requirement to specify C should increase reaction time by at least the effect of specifying C.

Since one assumes that the stimulus and response execution times are equivalent in the "specify only one dimension" case, the substitution of the constituent times into the original inequality (2) generates the following expression:

$$a(BC) + b(BC) + g(BC) - a(B) - b(B) - g(B) \geq b(C) - b(A) \tag{3}$$

How might this inequality hold? Clearly there are at least two conditions that one can specify so that this inequality holds. First, if $a(BC)+g(BC)$ is almost equal to $a(B)+g(B)$ and, if the inequality holds, it is true that $b(BC) - b(B)$ is greater than $b(C) - b(A)$. On this basis we can test the validity of Eq. (2).

However, what if where is a large increase in stimulus processing time so that $a(BC) - a(B)$ dominates the left side of Eg. (2)? If this difference is larger than $T(C) - T(A)$, you could conclude that B and C are specified serially, when in fact you are examining the increase in stimulus processing time. This is a very strong possibility since in the "specify two dimensions" case there are four stimulus–response alternatives, while in the "specify one dimension" case, only two stimulus–response alternatives are present. Zelaznik (1978; Zelaznik et al. 1982) has argued that this confounding of stimulus–response information with the precued dimensions causes great difficulties in the interpretation of precuing data.

Rosenbaum (1980) was aware of this problem and attempted to examine this issue in his second experiment. Instead of producing the movement that the imperative stimulus required, the subject was required to state as quickly as possible whether the imperative stimulus was a valid member of the precued set or not. It was clear that the pattern of results of this experiment did not match that of the first one, and as such, Rosenbaum concluded that stimulus processing could not be the cause of the difference observed in the first experiment.

That interpretation is correct within the same level of stimulus–response uncertainty, e.g., all the "specify one or two dimensions" cases. However, since RT increased as the number of alternatives increased, it remains likely that the increase in T(BC) is the result of the increased time for stimulus processing, rather than the extra time to specify B when C already is uncertain. Therefore, in the two experiments that we would like to discuss (Larish and Frekany 1985; Zelaznik and

Hahn 1985) the effects of having the number of S – R alternatives covary with the number of precued motoric dimensions were examined.

FIXED VS VARIED S – R ALTERNATIVE EXPERIMENTS

In the first experiment (Larish and Frekany 1985) subjects performed aimed hand movements of either the right or left arm, towards or away from their body, and to near or far targets. Thus, the three variables were arm, direction, and extent. In both the varied and fixed S – R alternative methods of precuing, the precue consisted of lists of possible upcoming responses. In the fixed method there always were two possible responses. By changing the pairing between two possible responses, it was simple to effectively "precue" the upcoming movement dimensions. For example, suppose that the arm was to be precued, with direction and extent uncertain. The pairing of the two S – R pairs could be left – away – near, and left – towards – far. Thus, in this case, the hand is common between the two responses, while the direction and extent are not common. Thus, arm was precued, while direction and extent were not.

In the varied precue condition, the responses were precued by illuminating all possible responses that could be required on the upcoming trial. For example, if the left arm was to be precued, all of the stimuli corresponding to the left arm movements (four of them) would be displayed during the precue interval. Thus, in the "precue two dimensions" situation, the procedure was identical to that in the fixed method, however in the precue none, and one dimension cases, the number of precue stimuli increased; Table 1 presents the relevant results.

First, examining the one dimension conditions, Larish found that the time to specify direction was longer than arm and extent, the latter two being of equivalent duration. This was true in both the fixed and varied conditions. Since the three specification times are not unequal, we cannot check for the serial nature of all three processes. Rather, we can check for the serial nature of direction and arm, and direction and extent. According to the precepts of the Rosenbaum model, if direction and arm are specified serially, the following inequality should be observed:

Table 1. Mean reaction time (ms) as a function of precue method and dimensions to be specified during reaction time interval. For example, A means that arm was unknown at the time of the imperative stimulus, and as such must be specified during reaction time

	Dimension(s) to be specified							
Precue method	N	A	D	E	AD	AE	DE	ADE
Varied	270	314	333	305	356	334	353	368
Fixed	270	304	322	301	330	316	332	330

N, none; A, arm; D, direction; E, extent

$$T(AD) - T(A) \geq T(D) - T(E) \tag{4}$$

In the varied method the comparison is between 42 and 28 ms for the left and right sides, respectively, while in the fixed method it is between 26 and 21 ms. Thus, both methods of precuing suggest that arm and direction are specified serially. For testing direction and extent for serial processing, the following inequality should hold:

$$T(DE) - T(E) \geq T(D) - T(E) \tag{5}$$

In the varied method the difference is 48 and 28 ms on the left and right sides, respectively, and in the fixed method it is 31 on the left and 21 on the right side. Thus, both methods of precuing, varied and fixed, infer that direction is specified prior to arm and extent. More importantly it seems clear that using the inequalities of Rosenbaum (1980) both methods of precuing provide the same conclusion.
However, in the "specify three dimensions" conditions there is a clear difference in the pattern of results. In the varied method the ADE condition has a grater RT than all of the other "specify two dimensions" conditions, but in the constant condition ADE is equal to the two dimensions conditions, in which direction is to be specified. The interpretation of these results we think is clear. Direction is specified prior to arm and extent, which are specified in parallel. The extra increase in RT in the varied case for the ADE condition is the result of the increase in stimulus processing time for the extra S–R alternatives. Thus, analysis of the Larish and Frekany experiment suggests that if two dimensions are specified serially, then both the varied and fixed S–R alternative case provides the same insights into the underlying organization of these dimensions. Furthermore, it suggests that in S–R compatible situations, the stimulus identification process times do not change that much when the number of stimulus–response alternatives increases.

While it is true that the above analysis suggests that direction and arm, and direction and extent are programmed in a serial fashion, it is also true that the Larish and Frekany data support a model in which direction is programmed in parallel with arm and extent, and direction is the longest process. The purpose of the previous paragraphs was not to state that direction is programmed serially with respect to arm and extent, but rather to highlight the fact that both the fixed and varied methods of precuing produce the same inference. This conclusion suggests that the stimulus uncertainty argument raised by Zelaznik (1978) might not be empirically important.

The next experiment (Zelaznik and Hahn 1985) utilized a less compatible S–R mapping to examine the varied and fixed methods of precuing of hand, digit, and duration of a key press.

Zelaznik and Hahn Experiment

Three movement dimensions were manipulated in this experiment. The hand (left or right), digit (index or thumb), and duration (100 or 300 ms) of a key press were the movement dimensions. Klapp (1977) suggested that the duration and digit of a key press response could be programmed in parallel. However, in his experiment 1, which provides a major line of support for the parallel model, only the varied precuing technique was employed.

Table 2. Mean reaction time (ms) as a function of precue method and dimensions to be specified during reaction time interval. For example, H means that the hand was unknown at the time of the imperative stimulus, and as such must be specified during reaction time.

Precue method	Dimension(s) to be specified N	H	D	T	HD	HT	DT	HDT
Varied	334	608	460	613	731	894	718	924
Fixed	321	543	464	585	557	579	584	583

N, none; H, hand; D, digit; T, duration (time of key press)

In the present experiment the stimuli were alphanumeric characters. The upper case letters, L,R,I,T represented the left and right hands, and index and thumb responses, respectively. The 0 and 2 represented the 100– and 300–ms key press duration. Thus, the "LT0" stimulus represented the left–thumb, 100–ms key press response.

In the varied method the precue consisted of three characters. If the dimension was to be specified, then its value was presented; if not, an "X" appeared in its place. In the fixed method, two sets of stimuli corresponding to the two sets of upcoming responses were presented as precues (for further details, see Zelaznik and Hahn 1985); Table 2 presents the results of the experiment.

As can be seen from the table, it appears that the patterns of reaction time data are different for the fixed and varied method of precuing. First, in the "specify one dimension" case for the varied method, the ordering of specification times is digit < hand = duration, while in the fixed method the ordering is digit < hand < duration. For the varied method, when two dimensions were to be specified, the conditions in which digit was uncertain had equivalent RT values which were less than the specify hand–duration condition. This effect was not observed in the fixed method in which the two choice conditions with duration to be specified had a greater RT than the specify hand–and–digit condition.

Since the Rosenbaum method depends upon having an ordering of processing times, the varied method only allows for an examination of the serial nature of hand and digit, and duration and digit. We shall ask the same question for the fixed method. The inequality that must be satisfied if hand and digit are specified serially is:

$$T(DH) - T(H) \geq T(H) - T(D) \tag{6}$$

For the varied method, the inequality is 123 vs 148 for the left and right sides, respectively. Scheffe post–hoc analysis shows this difference not to be significant. In the fixed method the inequality is 14 vs 81 ms for the left and right sides, respectively, a difference which is statistically significant.

For digit and duration specifications, the test for serial processing is:

$$T(TD) - T(D) \geq T(T) - T(D) \quad (7)$$

In the varied method the inequality is 258 vs 153, while for the fixed method it is 119 vs 121 ms. Thus, it appears that the two different methods of precuing produce a different pattern of results.

DISCUSSION

Why is it that in the Larish and Frekany experiment it appears that both precuing methods produce support for serial processing, while support is equivocal in the Zelaznik and Hahn experiment? Perhaps it is due to a procedural difference in that Zelaznik and Hahn precued dimensions in the varied method, and responses in the fixed method, whereas Larish and Frekany always specified responses. Rosenbaum (1980, 1983) has shown that when multiple responses are specified, you do not produce the same pattern of results as when dimensions are specified.

There is another plausible explanation for the present results. The Rosenbaum model assumes that the three major processes during reaction time are independent and serial. The requirement is necessary in order to conduct the analysis of difference equations for serial organization. Furthermore, as we showed in a previous section, it is necessary that changes in stimulus processing be small across different numbers of precued dimensions. It is possible that the requirements are met in situations in which the S−R mapping is compatible so that overall RT is small, and changes due to stimulus uncertainty also are small. One of us (Larish 1985) has provided evidence that S−R compatibility does have large effects upon the pattern of RT data in precuing experiments, and has argued that a necessary condition for the implementation of precuing experiments is a compatible S−R mapping (see also, Goodman and Kelso 1980).

In the Zelaznik and Hahn experiment, there is no natural and compatible mapping between stimuli and responses and as such the basic assumptions of precuing methodology are not adhered to. This leaves us with the thorny issue of how to conduct these types of precuing experiments. The solution we think is simple. Include both a fixed and varied method in the experimental design. This will allow for a check as to the consistency of the RT pattern in both methods. Consistent patterns can then be attributed to motor programming effects, and inconsistent patterns to stimulus and translation effects. In our final few paragraphs we would like to adress the role that precuing experiments can play in the advancement of theory in motor control work.

One of the long−range goals of motor programming research, and reaction time research in general, is to discover variables that are essential for the central motor control systems (Gelfand et al. 1971; Kelso 1982). Some writers define essential as being involved in the structural prescription for movement, such that all of the relative motions in space and time are contained in it, while the scaling of these motions, called the metrical prescription, is carried out by other processes.

The precuing paradigm can be used to examine and/or discover variables that are in the structural prescription. The rationale is simple. It seems reasonable to infer that essential variables must be selected prior to scaling variables, and as such one should observe a hierarchy in the selection process. For example, consider a case in which a subject is given a precue concerning the overall speed of a movement, but that upcoming movement could be either a series of button presses, or an abduction of the arm to reach a spatial target. Although there might be some overall arousal effect of knowing the speed of the movement, there should not be any benefits of motor preparation concerning the speed of movement since these two movements can be argued to possess different structural prescriptions. In terms of motor programming theory it would be argued that these two movements are governed by different generalized motor programs, and that the program must be selected prior to selection of the overall rate parameter.

The precuing paradigm is well suited to answer this kind of question. One would show no benefit of precuing the nonessential dimensions of movement until the essential one(s) were precued. Unfortunately it is still unclear to what extent the precuing methodology is able to obtain stable, reliable effects.

The lack of consistent ordering results in precuing experiments might be the result of the fact that the variables manipulated in precuing experiments are not of an essential nature, and that these precuing experiments are studying more transient effects. Furthermore, the sensitivity of the precuing paradigm to effects of number of S–R alternatives, S–R compatibility, and instruction set (prepare dimensions or movements) further bolsters the notion that the manipulated dimensions of movement are far removed from the actual underlying structure and function of the motor system. Future work needs to incorporate better theoretical accounts of the motor system with the precuing paradigm.

REFERENCES

Gelfand IM, Tsetlin ML, Shik ML, (1971) Mathematical modeling of mechanisms of the central nervous system. In: Gelfand V, Gurfinkel S, Fomin S, Tsetlin M (eds) Models of the structural–functional organization of certain biological systems. MIT Press, Cambridge, pp 329–345

Goodman D, Kelso JAS (1980) Are movements prepared in parts? Not under compatible (naturalized) conditions. J Exp Psychol 109: 475–495

Keele SW (1968) Movement control in skilled motor performance. Psychol Bull 70: 387–403

Keele SW (1981) Behavioral analysis of movement. In: Brooks V (ed) Handbook of physiology: motor control. American Physiological Society, Washington DC

Kelso JAS (1982) Human motor behavior: an introduction. Erlbaum, Hillsdale

Klapp ST (1977) Response programming as assessed by reaction time does not establish commands for particular muscles. J Mot Behav 9: 301–312

Larish DD (1985) Influence of stimulus response translations on response programming: examining the relationship of arm, direction, and extent of movement. Acta Psychol (to be published)

Larish DD, Frekany GA (1985) Planning and preparing expected and unexpected movements: re-examining the relationship of arm, direction, and extent. J Mot Behav 17: 168-189

Rosenbaum DA (1980) Human movement initiation: specification of arm, direction and extent. J Exp Psychol 109: 444-474

Rosenbaum DA (1983) The movement precuing technique: assumptions, applications, and extensions. In: Magill RA (ed) Memory and control in motor behavior. North-Holland, Amsterdam, pp 231-274

Schmidt RA (1975) A schema theory of discrete motor skill learning. Psychol Rew 82: 225-260

Schmidt RA (1976) Control processes in motor skills. Exerc Sport Sci Rev 4: 229-262

Zelaznik HN (1978) Precuing response factors in choice reaction time: a word of caution. J Mot Behav 10: 77-79

Zelaznik HN (1981) The effects of force and direction uncertainty on choice reaction time in an isometric force production task. J Mot Behav 13: 18-32

Zelaznik HN, Hahn R (1985) Reaction time methods in the study of motor programming: the precuing of hand, digit, and duration. J Mot Behav 17:190-218

Zelaznik HN, Shapiro DC, Carter MC (1982) The specification of digit and duration during motor programming: a new method of precuing. J Mot Behav 14: 57-68

Programming of Rapid Finger Sequences*

D. A. Rosenbaum, V. Hindorff, and E. M. Munro

School of Communication and Cognitive Science, Hampshire College, Amherst, MA 01002, USA

A fundamental issue regarding action patterns is how their central plans, or motor programs, are constructed in real time. We have investigated this issue by studying the timing of keyboard responses in choice reaction time tasks. Keystrokes are easy to record in the laboratory and are also likely to be important functional units for the motor programming of keyboard performance. Thus, the timing of keystrokes may reveal how keyboard patterns are programmed, which in turn may bear on the programming of other kinds of action patterns.

Recently, a number of studies concerning the programming of keyboard performance have focused on the timing of brief bursts of keyboard responses in reaction time experiments. A major result from these studies is that the time to initiate such bursts after the appearance of a start signal increases with the number of keystrokes to be produced (Rosenbaum et al. 1984b; Sternberg et al. 1978). This outcome has been taken to suggest that the control of keyboard sequences is based on detailed programs constructed prior to the first response in the sequence. Sternberg et al. suggested, further, that motor programs for highly prepared keyboard sequences occupy a special program "buffer" (see Monsell, this volume).

A major question that these studies leave unanswered is how programming is achieved for extended sequences of keyboard responses. For example, if programs are assumed to occupy a buffer, what happens when all the subprograms in the buffer have been executed? Common experience suggests that it is unnecessary to introduce prolonged pauses during extended trains of keyboard responses (e.g., during virtuosic piano playing), as might be expected if the buffer had to be replenished after all its stored instructions were executed. We hypothesize that the problem of programming extended sequences of responses is solved by selecting forthcoming responses while earlier responses are being performed. With such a parallel system, the subprogram for a given response can be ready for execution when that response is required. The evidence we present for this hypothesis takes the opposite form from the previously obtained "length effects" on initiation time. In the special circumstances of our experiments, the time to initiate a sequence of keyboard responses *decreases* with the number of responses to be produced. We show that this result, as well as the

* This study was supported by Research Career Development Award 1KO4 NS00942–01 from the National Institute of Neurological and Communicative Disorders and stroke, and by grants BNS–8120104 and BNS–8408634 from the National Science Foundation. The report was prepared while the first author was a Visiting Scientist at the MIT Center for Cognitive Science, under a grant from the AP Sloan Foundation's Particular Programm in Cognitive Science.

previously reported increases of initiation time with sequence length, are consistent with a parallel model.

EXPERIMENT 1

Subjects in our first experiment made responses on a simple button panel consisting of six buttons arranged like the s, d, f, j, k, and l keys of a Qwerty typewriter. With their fingers (but not their eyes) resting on the buttons, subjects performed memorized patterns of button presses. At the start of each block of trials, the subject learned two finger sequences and the reaction signals – the letter X or O – with which each sequence was paired. After the subject practiced the sequences, the experimental block began. On each trial, one of the two letters appeared in the center of a video screen after a brief warning interval. As soon as the signal appeared, the subject was supposed to perform the corresponding response sequence as quickly and as accurately as possible without any simultaneous responses (i.e., with all latencies exceeding 20 ms). We recorded the latency and identity of each button press and provided feedback to motivate subjects to maximize accuracy and minimize the time between the reaction signal and completion of the required sequence. Each subject participated in one 1–h session on each of two consecutive days. Details of the procedure used in similar experiments can be found in Inhoff et al. (1984) and Rosenbaum et al. (1984 a,b).

Table 1 lists the sequences required in each of the choice conditions. Each subject performed in each choice condition in two successive blocks of 22 trials each, the first two trials being for

Table 1. Choices used in experiment 1

Choice[a]	Number of responses	Number of response after Uncertain response
Rri Rrm	3	0
Rri Rmi	3	1
RRri RRrm	4	0
RRri RRmi	4	1
RRRRri RRRRrm	6	0
RRRRri RRRRmi	6	1

[a] R denotes the button press made with the right ring finger, and i, r, and m denote button presses made with the left index, ring, and middle fingers, respectively.

practice only. The sequences had 3, 4, or 6 responses, and the uncertain response was located either in the last or next–to–last serial position.

We varied sequence length and the distance of the uncertain response from the end of the sequence in order to test a model of sequence choice performance recently proposed by Rosenbaum et al. (1984a). According to this Hierarchical Editor, or HED model, choices between two response sequences are achieved by identifying the reaction signal and then making two passes through a hierarchically organized motor program. The first, Edit, pass allows for the selection of all initially uncertain responses and proceeds serially from the first uncertain point in the program to the end. The second, Execution, pass begins after the Edit pass is over and proceeds serially from the beginning of the program to the end, allowing for the physical production of responses whenever terminal elements are encountered in the program hierarchy. The HED model has been shown to account for sequence–choice data from a number of experiments (see Rosenbaum et al. 1984a).

The HED model makes two clear predictions about performance in the present experiment. Because these predictions concern the mean latency of the first response, T_1, we refrain from discussing latencies of subsequent responses in this paper. The first prediction is that T_1 should always be longer when the uncertain response is in the next–to–last serial position than when it is in the last serial position. This prediction derives from the assumption that extra editing is needed when the uncertain response is one step removed from the end of the program. The second prediction is that T_1 should increase, or at least remain constant, with the length of the sequence to be performed. This prediction derives from the assumption that increasing the length of the sequence increases the complexity of the Execution pass, including the complexity of those operations in the Execution pass that precede the first response.

Since the HED model assumes that execution of a motor program does not begin until the entire program has been edited, the model runs into the problem that a choice between two lengthy response sequences could take an inordinately long time to complete. The model may, therefore, only apply to rather short sequences, in which case auxiliary assumptions may be needed to explain how programming is achieved for sequences containing more responses. However, since the sequences used here have as many keystrokes as the sequences for which the model was originally developed (Rosenbaum et al. 1984a), we would expect the predictions of the HED model, outlined above, to hold in this experiment.

An alternative is to question the HED model's assumption that execution of a program cannot begin until the entire program has been edited. Suppose, instead, that editing of the components of a motor program *continuously* precedes execution of those components, that is, editing and execution go on in parallel. Such a system solves the problem of programming extended motor sequences, but it also runs into the problem that since the time to select responses is much longer than the time to execute responses, as is well established by previous reaction time studies, it would be necessary to pace production of already–selected responses so that long delays do not occur after those responses have been performed and later responses are being selected. A consequence of such pacing is shown in Fig. 1. First consider panel (a), which illustrates a choice between two sequences of four responses, with only the fourth response uncertain. Suppose the subject has learned how long it takes to choose between the subprograms for the two possible fourth responses, designated P_4 and P'_4, and suppose the subject chooses to minimize the delays between responses, thereby minimizing the duration of the entire performed sequence. Given this state of affairs, the

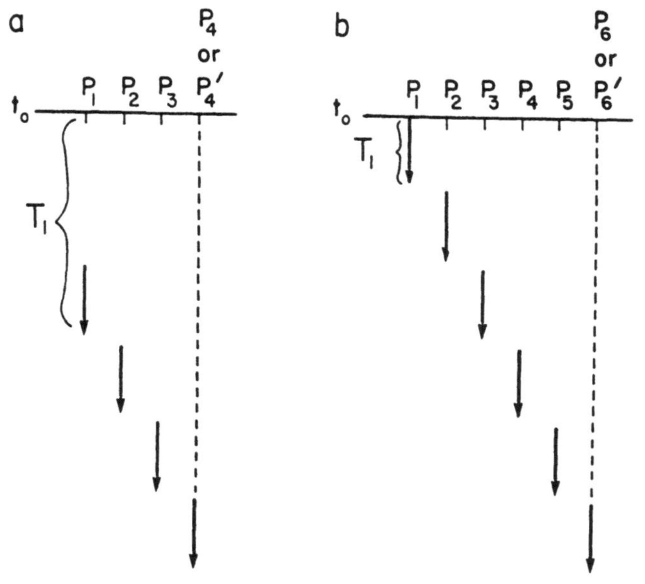

Fig. 1. The parallel selection model. Solid lines denote efferent delays following initiation of responses corresponding to motor subprograms $P_1 - P_6$. Dashed lines denote the act of making a choice between two possible subprograms, P_4 or P_4 in (a) or P_6 or P_6 in (b), corresponding to two possible responses

subject can schedule the initiation of the first, second, and third responses so that the fourth response is ready for execution just after the third response has been performed. Panel (b) shows a choice between two sequences of six responses with the uncertain response in the last serial position. Here there are more responses to perform while the choice is being made, and so T_1 can be much shorter than it was before. These examples show that when responses can be selected and executed in parallel, T_1 can decrease as the uncertain response recedes from the beginning of the sequence, in which case the time to start the sequence can actually *decrease* with sequence length if sequence length and the distance of the first uncertain response from the beginning of the sequence are related.

Figure 2 shows the obtained results. As predicted by the parallel model, T_1 decreased as the length of the required sequence increased, $F(2,8) = 13.65$, $p < 0.003$. This result violates the prediction of the HED model and also runs counter to the well-known result that the time to initiate a sequence of responses increases with length of the sequence. To our knowledge, this is the first report of an inverse length effect in reaction time performance.

The second important result is that the latency of the first response was uniformly longer when the uncertain response was in the penultimate serial position, that is, when two Edit steps were theoretically required, than when the uncertain response was in the final serial position, that is, when one Edit step was theoretically required, $F(1,4) = 11.26$, $p < 0.03$ (The interaction between sequence length and position of uncertain response was not statistically significant, $p > .25$) The effect of the position of the uncertain response supports the prediction of the HED model. How-

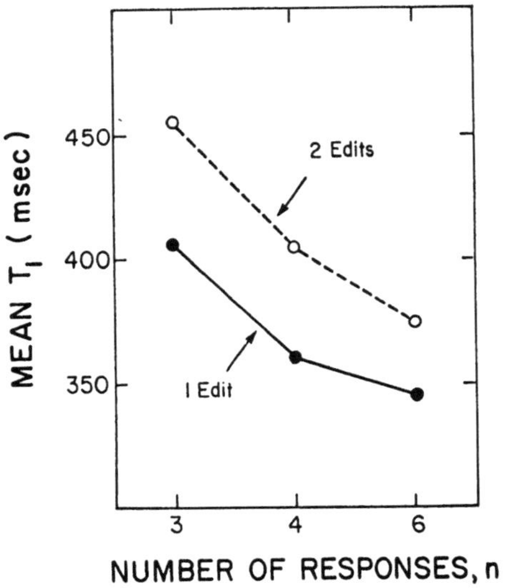

Fig. 2. Mean latency of the first response when 3, 4, or 6 responses were to be performed and the uncertain response was in the last serial position (1 Edit step) or next–to–last serial position (2 Edit steps)

ever, before too much stock can be placed in this result, it must be noted that the distance of the uncertain response from the end of the sequence was confounded with its distance from the beginning of the sequence. Thus, when two Edit steps were theoretically required and there were four responses to perform (RRri vs RRmi), the number of responses preceding the uncertain response (two) was the same as when one Edit step was theoretically required and there were three responses to perform (Rri vs Rrm). In fact, as seen in Fig. 2, T_1 was very similar in these two conditions. The correlation between T_1 and number of responses preceding the first uncertain response, -0.93, was statistically significant, $p < 0.001$, as predicted by the parallel model.

EXPERIMENT 2

How can the parallel model be tested further? One prediction that arises from it is that T_1 should be reduced if response selection and response execution are minimally interfering. Table 2 shows an experiment that tested this prediction. Here subjects chose between sequences with the uncertain response in the final serial position and with two or four responses preceding that position. We varied the length of the sequence to see whether we could replicate the inverse length effect of the last experiment. To make sure the inverse length effect was not specific to repetitions of single finger responses, we compared sequences that started with single finger responses with sequences that started with response couplets. The most important manipulation was whether the certain and uncertain responses were performed with the same hand or with different hands. The prediction of the parallel model was that T_1 would be elevated if the uncertain responses used the same hand as the certain responses. This prediction was based on the assumption that decisions about which fingers to use at a later point in the sequence would take longer if those fingers were engaged in ongoing performance than if they were not so engaged. The HED model made no definite pre-

Table 2. Choices used in experiment 2

Hands used by starting (certain) and ending (uncertain) responses	Starting responses	Number of responses before uncertain response[a] 2	4
Same	Singleton	rrm rri	rrrrm rrrri
	Couplet	mrm mri	mrmrm mrmri
Different	Singleton	rrM rrI	rrrrM rrrrI
	Couplet	mrM mrI	mrmrM mrmrI

[a] i,r, and m denote button presses made with the left index, ring, and middle fingers, respectively. I and M denote button presses made with the right index and middle fingers, respectively. Sequences between which choices were made always had the same number of responses before the uncertain response.

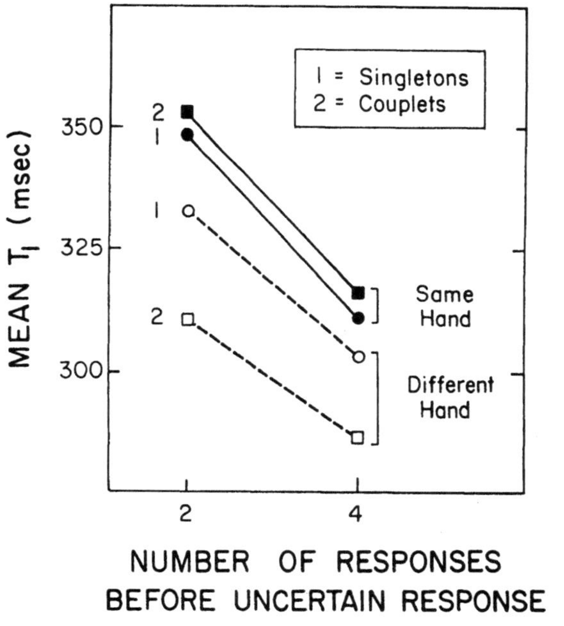

Fig. 3. Mean first response latencies in Experiment 2

dictions about the effects on T_1 of having the certain and uncertain responses use the same hand or different hands.

As seen in Fig. 3, T_1 decreased with the number of responses to be performed, $F(1,7) = 3.87$, $p < 0.09$, replicating what we found in the first experiment. Since the inverse length effect was obtained with singleton as well as couplet fillers, the effect generalizes beyond the case of single finger repetitions. When the uncertain and certain responses used different hands, the latencies were significantly shorter than when they used the same hand, $F(1,7) = 7.55$, $p < 0.03$, as predicted by the parallel model. In addition, latencies were shorter when couplet fillers were used than when singleton fillers were used, but only in the different-hands condition, $F(1,7) = 9.78$, $p < 0.02$. This interaction can be taken to mean that when certain and uncertain responses use different hands subjects have more flexibility in the way they select and execute responses, a proposal that is consistent with the assumptions of the parallel model.

CONCLUSION

This paper has addressed the question of how the selection and execution of rapid finger movements are coordinated, especially in extended trains of finger movements. Two choice reaction time experiments were reported which showed that when subjects chose between two finger sequences with an uncertain response at a given serial position, the time to initiate the required sequence decreased as the serial position of the uncertain response receded from the beginning of the sequence. Because of the composition of the sequences the time to initiate the sequences also decreased with sequence length, making this the first report of an inverse length effect in reaction time performance.

The model we adopted to account for these results assumes that selection of forthcoming responses occurs in parallel with execution of earlier responses. This constitutes a modification of the model that led to the present experiments, the Hierarchical Editor model of Rosenbaum et al. (1984a), which assumes that execution of a sequence does not begin until the entire program for the sequence has been edited, that is, until all of its responses have been selected and all of its constituents checked. The data here require only a single change in the Hierarchical Editor model, namely, that execution begins before editing is completed.

How can our inverse length effect be reconciled with the positive length effects that have been reported in previous studies? We attribute these opposing length effects to two sources that can normally operate together. The positive length effect is attributable to the complexity of the "unpacking" process that precedes each response, including the first one, during the execution pass (see Rosenbaum, 1985, for a general disussion of this hypothesis). The inverse length effect results from pacing the start of the execution pass so that selection of the initially uncertain response is completed by the time the preceding response has been performed. Pacing of response production to compensate for delays in selection, and parallelism of response selection and execution, are likely to be important components of skilled performance, both for rapid finger sequences and for other kinds of action patterns.

REFERENCES

Inhoff AW, Rosenbaum DA, Gordon AM, Campbell JA (1984) Stimulus – response compatibility and motor programming of manual response sequences. J Exp Psychol: Human Percep Perform 10: 724 – 733

Rosenbaum DA (1985) Motor programming: a scheduling theory. In: Heuer H, Kleinbeck U, Schmidt KH (eds) Motor behavior: programming, control, and acquisition. Springer, Berlin, pp 1 – 33

Rosenbaum DA, Inhoff AW, Gordon AM (1984a) Choosing between movement sequences: a hierachical editor model. J Exp Psychol: Gen 113: 372 – 393

Rosenbaum DA, Saltzman E, Kingman A (1984b) Choosing between movement sequences. In: Kornblum S, Requin J (eds) Preparatory states and processes. Erlbaum, Hillsdale, NJ, pp 119 – 134

Sternberg S, Monsell S, Knoll RL, Wright CE (1978) The latency and duration of rapid movement sequences: comparisons of speech and typewriting. In Stelmach GE (ed) Information processing in motor control and learning. Academic, New York, pp 117 – 152

Programming of Complex Sequences: Evidence from the Timing of Rapid Speech and Other Productions*

S. Monsell

Department of Experimental Psychology, University of Cambridge, Downing Street, Cambridge CB2 3EB, England

If one is interested in levels of control at which action sequences spanning many "simple" movements are specified and patterned, it is hard to know what to measure. In effect, peripheral processes interpose unknown layers of buffering and filtering between the observer and the output of central processes responsible for planning and/or controlling whole sequences of actions. There is a grievous shortage of known, measureable, and manipulable phenomena capable of granting us a window through these peripheral layers of control.

In this paper I describe research on a pair of behavioral phenomena that appear to be of this desirable kind. They involve striking regularities in the behaviour of the onset latencies and the durations of well-prepared brief action sequences, as we manipulate their length and content. Saul Sternberg and I began to study these phenomena in speech some years ago at Bell Labs. Since then they have been extensively explored by Sternberg, Knoll and Wright (who have extended them to typewriting) at Bell Labs, and in my laboratory at Chicago. We interpret these phenomena in terms of an idea which, though of long standing in the physiological and psychological literature (e.g. Lashley 1951), has continued to lack a firm foundation of empirical support: the idea that an extended sequence of movements may be executed under the control of a plan or "program" constructed in advance of initiation. Our use of the "program" construct does not (as did Lashley's and many others') imply exclusion of control by feedback, but simply the idea of a structured specification of the identity and order of elements. Our claim is to be able to link the temporal phenomena we observe under certain conditions to the sequential retrieval and execution of discrete sub-units of such a program as the sequence is produced. If our model, which I describe more fully below, is correct, we should be able to use the phenomena, interpreted via the model, to ask questions about the units, capacity and maintenance of the program. First we need to develop a sufficient basis for confidence in the model. We have devoted a substantial part of our work to two kinds of test of the model, which I will illustrate: (a) developing and testing alternative accounts of (at least part of) the effects observed, and (b) trying to derive and test unique predictions from the model. Because the model has so far run this gauntlet more or less intact, we have increasingly used it to pursue questions about units, capacities, etc.: I will review some current conclusions. However, we must remain alert to the possibility that someone will come up with a different and superior account of these phenomena. I will therefore make a point of emphasising aspects of the conclusions that depend rather little upon details of the model.

* This paper is based on work at the University of Chicago, supported by the National Science Foundation under grant no. BNS-8121372, and at Bell Laboratories.

Experimental Brain Research Series 15

Fuller accounts of this work and its background are in preparation or available in Sternberg et al. (1978, 1980, 1983), Monsell and Sternberg (1981), and Monsell and Nelson (1983). Because of space limitations and because my own research is on speech, I will make only passing reference to the work on typewriting, and will focus mainly on what can be learned from relatively crude measures of the latency and duration of spoken utterances.

THE EXPERIMENTAL PARADIGM

Most experiments involve only a few subjects, each receiving many trials of the following form. The action sequence to be prepared on that trial is displayed on a TV monitor. In a speech experiment, it is specified as a word, or series of words, to be spoken as a single fluent *utterance*. In a typing experiment, it is specified as a string of letters to be realised as a fluent *burst* of key-strokes. The display is removed after 2 or 3 s and the subject is allowed 2 or 3 s more to prepare for a "go" signal. To maximise preparation, we usually precede the "go" signal with a fixed foreperiod terminating in two rhythmic warning signals, and to discourage anticipation we omit the "go" signal on some "catch" trials. However, these details are not critical: in the first experiment to be described, the foreperiod varied randomly between 2.6 and 5.4 s and there were no warning signals or catch trials. Instructions and feedback encourage the subject to produce the sequence (a) *correctly* (i.e. clearly articulating the specified words in the specified order), (b) *fluently* (i.e. without stumbles or pauses), and (c) *aiming to minimise completion time* (i.e. the interval between the "go" signal and the end of the utterance). With one exception, the data to be described are from subjects who had received substantial practice.

The principal variable manipulated is the number of sequential *elements* in the utterance: what is meant by "element" will become clear below. For speech the *latency* (go signal to speech onset) and *utterance duration* are measured automatically and online by a sensitive algorithm operating upon the digitised acoustic waveform. We sometimes also measure within-utterance *element durations* from recordings, but these measures depend somewhat upon subjective judgement and are much more labour-intensive. For typing, of course, inter-response times can be measured as easily as the latency.

THE LATENCY AND RATE EFFECTS

Figure 1 shows data from one of our earliest experiments, in which the utterances were lists of from one to six digit-names or letter-names. A new random list was presented on each trial, and a randomly variable foreperiod preceded the "go" signal. The data are averaged over four subjects, who served for two sessions, each of about 200 trials, following practice. There are two basic effects.

The Latency Effect

Mean latency (Fig. 1, panel a) increases approximately linearly with the number of words in the utterance. The size of the effect is expressed by the slope of the fitted function, $L_n = \eta + \theta n$.

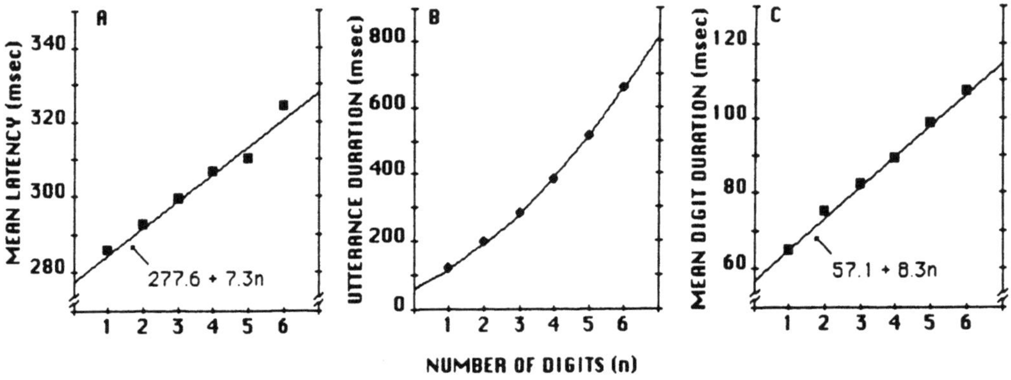

Fig. 1. Mean latency (*A*), utterance duration (*B*), and estimated mean word duration (*C*), as a function of number of words, for random sequences of digit or letter names

The value of θ observed in this experiment is typical, but exhibits considerable variability over experiments and subjects.

The Rate Effect

Mean utterance duration (Fig. 1, panel b) does not increase linearly with utterance length, as one might expect, but with an upward curvature well fitted by a function of the form, $D_n = \alpha + \beta n + \gamma n^2$. Upward curvature implies a slower average rate for longer sequences. The good fit of this function implies that the average element duration, d, increases approximately linearly with utterance length. Because small degrees of curvature are hard to see and compare, we generally transform a set of observed utterance durations (D_1, D_2, ..., D_n, ...) into estimates of the "mean element durations" (d_1, d_2, ..., d_n, ...) as follows: fit a quadratic function, as above (using a least squares estimation procedure that adjusts for the increase in variance as elements are added); subtract the constant α from each D_n and divide by n to yield the estimated mean element duration d_n. To the extent that the utterance duration observations are well fit by the quadratic function, these estimates of d will be well fit by the linear function: $d_n = (D_n - \alpha) / n = \beta + \gamma n$. Figure 1 shows the duration data thus transformed in panel c; the linear function clearly fits well, and the slope (γ) is again typical.

Is this behaviour of the *mean* element duration representative of the durations of individual elements? Direct measures of element durations in other experiments have shown that the effect of increasing the number of words is distributed across the durations of all the words (e.g. Sternberg et al. 1980, p 520). (Intrinsic differences in duration among words are averaged out by using lists as utterances and rotating or randomising the occupants of each serial position.) The same is more easily demonstrated for inter-keystroke intervals in burst typing (Sternberg et al. 1978, p 144). The effects of serial position, per se, on element duration are complex, and somewhat different in form for speech and typing. In contrast, the effects of number of elements on *mean* element duration are both simple and identical in form for speech and typing. This has encouraged us to focus on the latter – which will be termed the "rate effect".

The basic latency and rate effects have been observed with a variety of sequence types: novel and arbitrary lists of words and nonwords, familiar sequences like "one–two–three–four" or "Monday–Tuesday–Wednesday", and even reiterant sequences like "Monday–Monday–Monday" (Sternberg et al. 1978). As already noted, similar effects occur in typewriting. The effects are robust with practice, in two senses. They are observed both in naive subjects (see below) and in subjects who have had many hours of practice. They survive repeated production of the same sequence; if the same utterance is produced on three trials in a row, the duration is significantly reduced, but the form and slope of the latency and rate effects are unaffected. Sternberg and colleagues have observed non–zero effects even after twenty consecutive repetitions of the same typewritten sequence.

Here, then, is an impressive degree of order and regularity. How is it to be accounted for? Effects of response "complexity" on latency for initiating a prepared action sequence had been observed before, and attributed to a process of "programming" the whole sequence during the latent interval (Henry and Rogers 1960); the rate effect, however, was new, as was our finding concerning the form of the latency effect. One rationale for our rather different model was this: since the form and size of the latency and rate effects are so similar, it seemed sensible to try to account for both phenomena in terms of a common mechanism. A second rationale was the need to account for the following pattern of additivities that we have repeatedly observed.

ADDITIVITY OF UTTERANCE LENGTH AND WORD LENGTH EFFECTS

Figure 2 shows data from an experiment in which the utterances were random sequences of 1, 2, 3 or 4 nouns. Half the lists were constructed by sampling from a vocabulary of mono– syllabic nouns (e.g. *Bay*, *Mark*), half from a vocabulary of stress–initial disyllabic words (e.g. *Baby*, *Market*) related to the words of the first vocabulary by the addition of an unstressed syllable. The data shown are from four practiced subjects. Linear latency and mean element duration functions

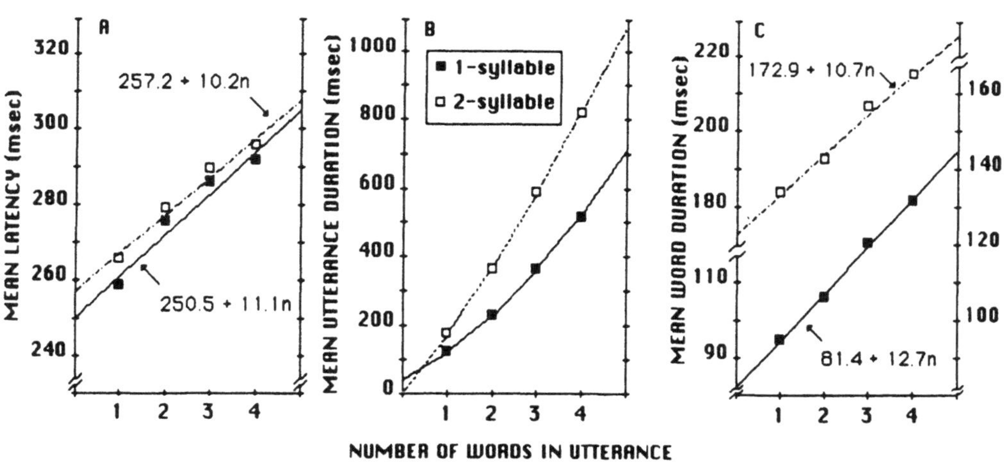

Fig. 2. Mean latency (*A*), utterance duration (*B*), and estimated mean word duration (*C*), as a function of of the number of monosyllabic or of disyllabic words

describe the data well. Two additivities are apparent. First, although adding an unstressed syllable to each word certainly increased the mean word duration, d, this effect was independent of the effect of number of words on d – i.e. the rate effects are well described by fitting a common slope (γ) to the two sets of mean element durations. Secondly, the effects on latency of the number of words (θ) is similarly uninfluenced by word length, though there is a small but significant main effect of word size on mean latency (4.5 ± 1.3 ms).*

Since this last effect, of word size (syllabic length) on mean latency, is small and controversial, it should be noted that we have observed it in other studies. One example is supplied by a recent experiment that I conducted to study the behaviour of very long sequences. Six subjects each learned a cycle of 12 monosyllabic words and a cycle of 12 stress–initial trisyllabic words, until they could rapidly recite the whole cycle, and arbitrary parts of it, with ease. The mono– and tri–syllable cycles were matched across subjects in a way designed to match the transitions between words: the monosyllabic words were "expanded" into trisyllabic words by inserting two syllables in the middle of each (e.g. "...pine sea bus" expands to "...porcupine subsidy blunderbuss"). Within subjects, the mono– and tri–syllabic sequences were similar in phonetic content. The subjects were then required to prepare and produce, under our standard rapid speech conditions, sub–sequences varying in length between one and twelve words; each starting point in the cycle was used equally often for each length, thus balancing the occupants of each serial position. The latency data are shown in Fig. 5 in the later section of the paper where the long sequences are discussed. The present point is that for short sequences of up to four or five words, the increase in latency per word added to the sequence is, as before, largely independent of word size, but there is again a small but significant main effect of word size. (The intercept diffence between the initial limbs of the functions, fitted as described below, is 8.6 ± 3.2 ms.)

To summarise, there are five robust properties of the phenomenon to be accounted for, with the important restriction that, so far, an utterance "element" has been defined as a word with one stressed syllable: (a) the linear increase in mean element duration with number of elements (θ), (b) the linear increase in latency with number of elements (γ), (c) the similar size of these effects, (d) the independent effects on mean element duration of element size and number of elements, and (e) the independent effects on latency of element size and number of elements (at least up to some limiting utterance length).

THE SUB–PROGRAM RETRIEVAL (SPR) MODEL

The model (Fig. 3) makes the following assumptions.

1. During the preparation interval the speaker constructs, at some yet–to–be–defined level of abstraction, a *program* specifying the elements of the utterance, their sequence and, presumably, global patterning such as the intonation contour. When the "go" signal is detected, the initial unit or *sub–program* is *retrieved*, its specifications are *unpacked* and then a *command* process initiates downstream activity resulting ultimately in a pattern of vocal tract activity appropriate to

* Here and elsewhere, the mean is followed by the standard error estimated from the variability in the parameter over subjects.

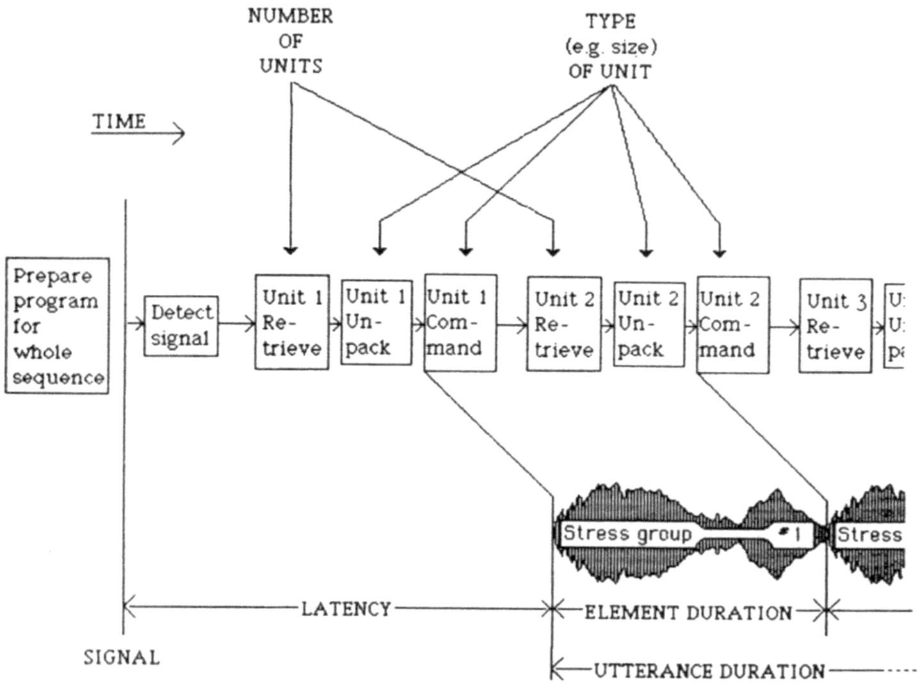

Fig. 3. Sequence of processing stages in the sub–program retrieval model

the speech element specified by that sub–program. This cycle of three processing steps repeats until every unit has been executed.

2. The duration of the retrieval stage is a function of the number of units in the program. The more units are simultaneously in a state of preparation, the more time taken to "locate", "retrieve", "select" or "activate" each one. (Any of the these terms might supply the most appropriate metaphor for *retrieval*.) Why the function is approximately linear, we do not yet know.

3. The duration of the retrieval stage is *not* influenced by the composition of the sub–programs, only by the *number* of sub–programs.

4. The duration of the unpacking and command stages depend only upon the composition of the sub–program being processed, and *not* on the number or composition of the other sub-programs, if any.

5. These are "processing stages" in the strict sense defined by the assumptions of Sternberg's (1969) "additive factors" method. Under the constraints of the requirement of maximally rapid completion (and perhaps *only* under such constraints) the initiation time for each element is determined by the sum of the time taken to detect the signal, go through the complete cycle for all the preceding elements, and to retrieve and unpack the sub–program for the present element.

Hence the latency includes the summed durations of the one retrieval stage (influenced by the number of units) and one unpacking stage (influenced by unit size) preceding speech onset. Likewise, the average interval between each word's onset and the next is the sum of average duration of the remaining retrieval stages (influenced by number of units) and of the unpacking and command stages (influenced by unit size).

A Soothing Comment on Terminology

Use of terms like "program", "retrieval", etc. tends to provoke aversive reactions among some action theorists, especially those advocating a "dynamic" perspective (e.g. Kelso, this volume). Though these terms are clearly "computational" in origin, and refer to "representational" concepts, they should be interpreted in a suitably abstract, but physiologically realisable, manner. The program "represents" the utterance, but only in the sense, and to the extent, that it specifies to lower levels of control the parameters they need to realise the utterance. We are willing to accept that the detailed patterning of vocal tract activity is taken care of by the evocation and parameterisation of learned "routines" or "coordinative structures" at lower levels of control, and that these are designed to take advantage of the dynamic biomechanical properties of the respiratory system and vocal tract. We are even willing to contemplate the possibility, as for example suggested by Fowler (1983), that speech is essentially produced by evocation and then cyclic "reparameterisation" of a system coordinating the activities of jaw, tongue, etc. to produce a stressed vowel, and that the rapid closures and occlusions required to realise consonants and unstressed vowels are somehow superimposed upon, or modulations of, this basic cycle. Some mechanism must nevertheless specify to such a system the *sequence* of parameterisations required, and that (no more, no less) is the function we attribute to our "program" construct. It is also worth noting that the discrete processing stage structure of program execution in the SPR model is not an *a priori* assumption derived from a naive mis–application of the concept of step–by–step program execution as manifest in digital computers, but a rather surprising (to us) postulate to which we are led by patterns in the data. Nor would we wish necessarily to attribute such a temporal structure to processing carried out at other levels of control in the system.

REASONS FOR REJECTING OTHER CLASSES OF EXPLANATION

One approach to testing a model like ours is to develop promising alternative classes of explanation to a point where critical empirical comparisons can be made. I cannot exhaustively review the efforts we have made in this direction, but will summarise briefly our principal reasons for rejecting four classes of alternative explanation.

An Artefact of Practice or Subject Selection?

We have previously reported data from small sets of highly practiced subjects. These were, moreover, selected from a somewhat larger random sample of subjects, rejecting those whose data in a screening session exhibited high variability, poorly defined speech onsets and offsets, dysfluency, and lack of responsiveness to the rapid completion instruction. Our motives were impeccable: by the normal standards of mental chronometry, these effects are numerically small and easily swamped by trials spoiled by inattentiveness, lack of preparation, non-speech noises, or articulatory difficulties. But could we have either (unconsciously) selected only subjects who exhibited the effects, or somehow trained the phenomenon into them? Reassurance is available from a recent study in which I ran (as a prelude to a more selective experiment) 56 unselected, naive volunteers, in a single session experiment. The utterances were subsequences of one to five words from the cycle "two-four-six-eight-ten-two-, etc.." and the data analysed were from 150 trials following just 50 practice trials (about the minimum needed to learn the task adequately). The average data exhibit highly significant latency and rate effects, of characteristically linear form. A linear trend accounted for 99.3% of the variance in latency due to utterance length, with a slope (θ) of 4.8 $\pm$ 0.6 ms. The equivalent values for the mean element duration function were: 98.6% and $\gamma = 4.0 \pm 0.5$ ms.

Programming Time Theories

As noted above, increases in the latency of a prepared action sequence, as response elements are added to it, have previously been attributed to time taken to "load" (Henry and Rogers 1960) or "assemble" (Glencross 1972) a program for the entire sequence during the latent interval: the more elements to be programmed, the longer the loading or assembly time. As an account of the latency effect alone, such a model is not rejected by our data, though it gives no obvious account of the additivity of the effects of word size and number of words. The main counter-argument must be one of parsimony: unlike our model, these programming time theories can give no account at all of the rate effect; it cannot be necessary to *re*program the *whole* sequence prior to the initiation of every element within it!

Appeals to Special Properties of Speech-Motor Control

It is possible to come up with ingenious accounts of at least components of the phenomena in terms of the biomechanics of speech. For example, suppose that our subjects tend to take bigger breaths when preparing for longer utterances; the higher sub-glottal pressure and laryngeal tension might then require a more controlled, and hence slower, release to initiate the utterance. We have explicitly tested some of the more plausible accounts of this kind (e.g. Sternberg et al. 1980, p 514). However, by far the most powerful counterargument is this: Much the same set of phenomena, including the "localisation" of the rate effect to be described in the next section, occur in a response modality with very different biomechanics, namely typewriting (Sternberg et al. 1978, 1983).

This should not lead us to expect that *any* response modality must exhibit the complete phenomenon. For example, Hulstijn and Van Galen (1983) have reported that prepared sequences of handwritten characters produced rapidly under similar conditions to ours exhibit a latency effect, but not a rate effect. This dissociation of latency and rate effects is not necessarily a problem for the model. Even if a cycle of processes like that proposed for speech and typing also underlies production of a prepared handwritten sequence, a rate effect should be observed only when it is the durations of these central processes that limit the production rate. An analogy: The rate at which a highway is constructed can be limited either by the efficiency of the bureaucrats responsible for obtaining planning permission, purchasing land, etc., or by the work rate of the construction crews.

Only in the former case would measurements of the *rate* of construction tell us anything about the bureaucracy (as we claim for speech and burst typing). Possibly the biomechanical constraints on handwriting make it a case of the latter kind.

General Capacity Explanations

We ask our subjects both to remember a sequence and to execute it rapidly. It could be argued that these are separate tasks competing for the same limited general-purpose processing capacity (GPPC). The longer the sequence, the greater the memorial demand upon capacity, and the less available for detection of the signal, production of each element etc.; hence these processes are accomplished more slowly.

Appeals to GPPC, though popular in psychology, tend to be over-powerful unless more precisely specified (cf. Allport 1980): almost any pattern of data can be interpreted post-hoc in terms of capacity demand here, no capacity demand there. Two kinds of evidence supply general arguments against a GPPC account in the present case. First, as mentioned above, we generally obtain parametrically similar data for random, arbitrary, novel sequences – which are in some sense "hard to remember" – and for sequences that, because they are already well learned, or are re-iterant, seem easier to maintain while waiting to speak. Second, if GPPC is required to maintain the relevant representation of the sequence, we should increase the demand on GPPC by requiring the subject also to maintain an *extra* sequence – a memory load – while preparing and producing the utterance.

Sternberg and I tested this prediction in an experiment in which, while preparing and then rapidly producing an utterance consisting of a random sequence of between 1 and 5 digit names, the subject had also to retain for later recall another list of between 0 and 4 digit or letter names. The total number of items never exceeded five. The effect of utterance length upon latency was a highly significant, and linear, 11.1 ± 2.9 ms item, while the effect of adding items to the concurrent load was a negligible 0.4 ± 1.3 ms/item. Results from an experiment I recently completed using more draconian loads are summarised in Fig. 4. In this case four subjects (whose memory span was

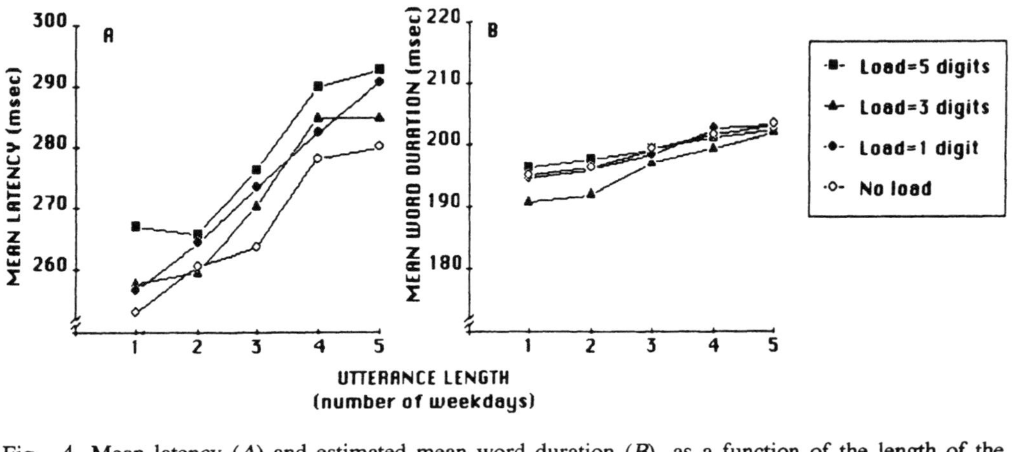

Fig. 4. Mean latency (*A*) and estimated mean word duration (*B*), as a function of the length of the utterance (a sequence of weekday names) and the size of a concurrent digit load

between 7.2 and 8.3 digits), prepared and produced familiar sequences of one to five weekday names (e.g. "Tuesday Wednesday Thursday") while retaining a memory load of 0, 1, 3 or 5 random digits. As before there was a substantial effect upon latency of utterance length (7.8 ± 2.2 ms/word), but a small and insignificant effect of load (1.2 ± 1.3 ms per digit added in the dual task conditions). Also the rate effect observed, though smaller than usual (γ = 2.2 ± 0.7 ms/word), was significant, while adding digits to the load did not increase the mean element duration at all. Even in the worst case – an utterance of five weekdays followed by recall of a five-digit load – recall of the load was correct on 78% of the trials. It is not easy to maintain a concurrent load, but it does not seem to interfere with the preparation and execution of the rapid utterance as a GPPC account would predict.

CONFIRMATION OF A UNIQUE PREDICTION FROM THE MODEL

A second approach to validating the model is to seek to confirm predictions from it which seem otherwise counter-intuitive or unexpected. The timing of elements in the speech stream is, *ex hypothese*, controlled by a cycle of three discrete stages: retrieval, unpacking, and the command stage. Only the duration of the first of these is supposed to be influenced by the number of units in the sequence. Thus, unless the temporal "smoothing" contributed by downstream processes is extreme, one might expect to see the effect of number of units localised in the duration of only part of the speech element corresponding to each sub- program unit. Sternberg et al. (1980, p 523) explored this prediction as follows. They had subjects produce sequences of one to five disyllabic words chosen so that the syllable onsets were relatively prominent in the amplitude envelope. Using a sophisticated recognition algorithm (validated against measures of spectrograms), they measured the durations of each of the consonant and vowel segements in each utterance. Mean word duration was, as usual, a linear function of number of words. However, almost all of this effect was localised in the duration of the second (unstressed) vowel. In a later experiment on typewriting, the same researchers were able to monitor the trajectory of one finger movement embedded in bursts of between 3 and 5 keystrokes. Again, the effect of sequence length was localised in the duration of only one part of the inter-keystroke interval: the later part of the lifting movement prior to the "strike". These are exactly the sort of peripheral manifestations of high level "intermittency" or "discreteness" predicted by the model, but otherwise not expected.

EXPLOITING THE PHENOMENA (INTERPRETED VIA THE SPR MODEL)

So far, I have focused on the adequacy of the SPR model. Now I illustrate the process of trying to use the phenomena, interpreted (for the most part) via the model, to discover properties of the utterance program and its execution.

Units

The question is often posed: "What are the units of speech production?" The answer must, however, be complex, since multiple levels of processing, each characterised by its own units, must be involved. No one technique will reveal (all) the units at every level. Here I summarise what our latency and rate effects have so far told us about units.

I described above data from two experiments in which the number of syllables per word was varied independently of the number of words in the utterance. If, and only if, utterance length was defined as the number of words, then we observed the pattern of additivities characteristic of the model, and similar values for the sizes of the latency and rate effects (i.e. $\theta \cong \gamma$). Our model-dependent conclusion was that (for these utterances) the sub-program unit of execution corresponded to the word. A conclusion that is somewhat weaker, but does not depend upon the model, can be argued as follows. Many different measures of utterance length – number of syllables, number of phonemes, duration, etc. – are *a priori* equally reasonable as a description of the phenomena. However, *only* when we plot latency or element duration in terms of number of words do the data approximate parallel functions, describable with a *single* slope parameter across conditions. Thus, parsimony alone suggests that (for the utterances so far) the word corresponds to *some* kind of underlying processing unit.

However, in the sequences used so far, each word always contained one stressed syllable, thus confounding number of "stress groups" (or "feet") and number of words. Further experiments have led Sternberg and me to conclude that it is the number of stress groups that is critical. In one study, we had subjects produce counting sequences of from one to five equally stressed digits, such as "ONE-TWO-THREE": these were the control sequences. In other conditions, subjects produced these same sequences with *unstressed* connective words inserted between each digit and the next, to yield sequences such as "ONEandTWOandTHREE". If the latency and rate effects are expressed in terms of the number of stressed syllables, the data for sequences with the words "and" or "minus" inserted were well described using linear functions approximately parallel to the (linear) functions for the control sequences – much more parallel than if plotted as a function of the number of words or the number of syllables. On grounds of parsimony alone, therefore, the "stress group" seems a better unit for predicting the data than the word or syllable. If the assumptions of the SPR model are made, more precise predictions can be derived for the various alternative unit descriptions. Consider, for example, the effect of inserting "minus" between each digit and the next. If each "minus" added one *word* sub-program, the slope of the function relating mean stress group duration to number of stressed syllables should be four times the slope for the control sequences. (There would be two sub-programs per stress group, with the retrieval time for each being increased by 2 γ for each stress-group added to the utterance.) If each "minus" added two *syllable* sub-programs to the program, the multiplier should be nine. Both of these can be decisively rejected.

Another condition of this experiment yielded an outcome that, though anomalous, seems to validate the logic of this test. When subjects were asked to insert the word "by", unstressed, between each digit and the next, the increase in stress group duration per stress group added, though still nicely linear, had approximately three times the slope obtained for the other sequences – closer to the value predicted by a word unit than by a stress group unit. One interpretation is that the phonetic properties of "by" make it difficult to de-stress it enough for it to fall within the scope of an adjacent stress group. Though necessarily post-hoc, this interpretation points to a possible empirical criterion for objective definition of degrees of syllabic stress.

Converging evidence against the word (*qua* lexeme or morpheme) as the appropriate unit comes from an experiment by Sternberg, Knoll and Wright, comparing sequences of monosyllabic, stressed, words to carefully matched sequences of non-words (Sternberg et al. 1980, p 528). The latency data, and the slope and forms of the mean element duration functions were identical for the two kinds of sequence. The program appears to be constructed at a level of representation subsequent to any retrieval of phonology from the lexicon.

So far, we have rejected first the syllable (and, *pari passu*, any sub-syllabic segment), and then the word or morpheme, as corresponding to the sub-program retrieved during program execution, in favour of the stress group. Now entertain the following idea, inspired by the hierarchical descriptions of action sequences assumed by (for example) MacKay and Rosenbaum in their contributions to this symposium. The experiments described so far have all used utterances structured as lists; this permits randomisation or balancing of the occupants of serial positions in the utterance. But perhaps the number of stress groups has so far determined the latency and rate effects only because that is the *highest* level of structure available in a list-structured utterance. In sentential utterances, higher-order units of structure can be discerned, and they might prove to be the critical units instead of the stress group. To test this hypothesis, and to generalise the phenomenon beyond list-type utterances, I recently compared sentence and list utterances as follows. Utterance length was varied by deleting or leaving intact the parenthetic portion of sentences such as "Barbara tricks a (rather rueful) dean", to yield short (three stress groups) and long (five stress groups) versions with the same beginning. All were of (NP (V, NP)) form, and the deletion changed only the second NP, leaving the superordinate structure (according to almost any linguistic theory) unchanged. (The locus of sentence stress was explicitly controlled with a context question.) The length of a parallel set of list utterances, with the same beginnings as, and the remainder phonologically modelled on, the sentences – e.g. "Barabara Trixie (Arthur Reuben) Dean" – was manipulated in the same way in separate blocks of trials. Each utterance was produced on four successive trials, not counting catch trials. The latency effects observed for sentences and lists were almost identical. Omitting the first production of each utterance, on which most of the errors occurred, the latency for short and long sentences differed by 13.3 ± 1.8 ms, and for short and long lists by 11.3 ± 2.0 ms.

Hence, present evidence suggests that each cycle of the process executing the utterance deals with one stress group, and this is so whether the utterance is a sentence or a list. That is not to deny the psychological reality of hierarchical structural descriptions, merely to assert that they do not seem to determine the units of execution of a prepared utterance program.

The Capacity of the Program

Presumably, if we continue to increase the length of the utterance we ask the subject to prepare, a limit will be reached, and the effect of utterance length upon latency will break down. The utterance length at which this limit is reached offers a measure of the maximum capacity of the utterance program. However, there is a well-known and logically distinct limit, measured by memory span, in our ability to retain, for any purpose, an unfamiliar and arbitrary sequence of words. In the hope of measuring the first limit unconstrained by the second, we have looked at the behaviour of long utterances derived from highly over-learned sequences.

In our first experiment of this kind, Sternberg and I had four practiced subjects prepare and produce sequences of from two to sixteen digit or letter names. These were cyclic counting sequences beginning with "one" or "five" (e.g. "five six seven eight nine ten one two") or sections

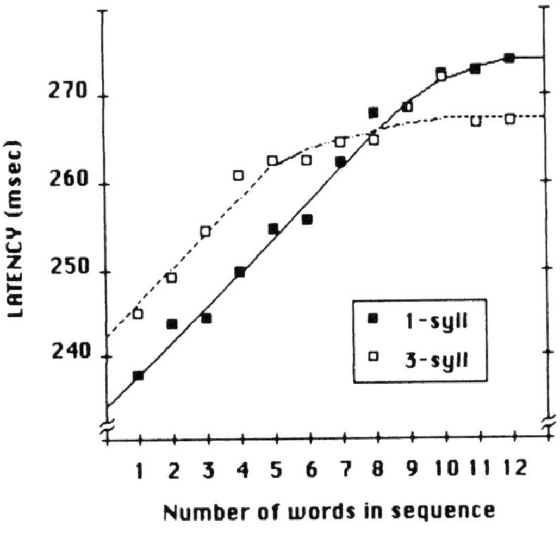

Fig. 5. Mean latency for overlearned sequences of monosyllablic or of trisyllablic words as a function of utterance length, and averages of the fitted bilinear functions

of the alphabet beginning with "A" or "H". For each subject, the latency data were reasonably well described by fitting a linear function up to a critical number of words, and then a constant (i.e. flat) function thereafter. The mean of this critical utterance length was 7.8 ± 0.5 words.

It was to explore this issue further that I conducted the experiment described earlier, in which six subjects prepared and produced sequences of from one to twelve mono– or tri–syllabic words derived from well–learned word cycles (Monsell and Nelson 1983). The question was: would the capacity limit be a constant number of "execution units" – i.e. stress groups – regardless of their size, or would capacity depend upon the size of the stress groups. The latency data shown in Fig. 5 clearly support the latter conclusion. There is a significant ($p < 0.01$) interaction between utterance length and stress group size. For trisyllabic words, latency approaches asymptote around five or six words, whereas for monosyllabic words, the function remains linear up to about ten words. The smooth fitted functions shown are obtained by averaging over subjects the simple bilinear functions (increasing then flat, with a common slope fitted for the initial segments) fitted to individual subjects' data. (Note that averaging several bilinear functions with a range of different values for the location of the "elbow" yields a smoothed curve over that range.) The fitted functions provide a reasonable description of the data. Hence the number of execution units that can be held in the programmed state appears to depend upon the complexity of the utterance fragment each represents.

Maintenance of the Program

How is a program, once constructed, maintained? For the psychologist, a natural approach to this question is to ask how the utterance program relates to temporary storage mechanisms identified on the basis of other evidence. Tasks such as memory span, requiring retention of the order of verbal items over short intervals, have yielded evidence for a "phonological buffer", capable of representing a sequence of several words in a phonological code (Baddeley et al. 1975; Vallar and Baddeley 1984). It has also been suggested that the natural function of this buffer is to encode a speech plan (e.g. Ellis 1979). I have no quarrel with either claim, but will sketch two arguments, based on evidence already mentioned in this paper, that the utterance program presently hypothesised is maintained at a level of processing separate, and presumably "downstream", from the phonological buffer.

The first argument comes from the two experiments that required the subject to maintain a concurrent memory load while maintaining preparation for, and then executing, a rapid utterance. According to standard theory, the load demands capacity in the phonological buffer (or the "articulatory rehearsal loop" of which it is part). Yet in neither experiment was there a significant effect of number of load items on the retrievability of sub-program units – as estimated by the size of the latency and rate effects. The subjects in the second experiment had memory spans measured at between 7.2 and 8.3 random digits – equivalent to perhaps 5 or 6 disyllabic words. Yet on 78% of the trials they were able to maintain unimpaired preparation for and produce a familiar sequence of five weekday names while maintaining for perfect recall a five-digit load – hardly the mutual attrition one would predict from competition for a common storage capacity.

The second argument comes from the experiment on long learned sequences discussed in the previous section. Mean latency for monosyllable sequences continued to increase linearly up to about ten words. Yet these same subjects' mean memory span for random sequences of the same words (i.e. the list length at which only 50% of such lists were recalled correctly) was significantly smaller: 7.4 ± 0.6 words. Even if one attributes all rather than (like Baddeley et al. 1975) only part of memory span to the phonological buffer, the maximum length of the utterance program seems to be substantially in excess of the estimated capacity of the phonological buffer.

CONCLUSIONS

I have described a pattern of effects of the length and content of a prepared burst of fluent speech upon the temporal parameters of its production, and interpreted this pattern as evidence for control of a multi-word utterance by an abstract "program" prepared before the speaker begins to utter. I described how our proposed account of the phenomena has so far survived two kinds of empirical test. (Regrettably, I lack space to discuss the minor anomalies encountered to date.) Further exploration of the phenomena suggests that: the program is maintained at a level of articulatory representation more peripheral than the "phonological", post-lexical, level revealed by span-type tasks, that the program is executed in chunks based on stressed syllables, but that the number of stress groups for which preparation can be maintained diminishes as their syllabic complexity increases. My broader intention has been to demonstrate, with a concrete example, the possibility of finding behavioural phenomena with which we can penetrate peripheral levels of control to identify and explore central processes responsible for the patterning of extended movement sequences, and to suggest one way in which the classical idea of a motor program may be given empirical substance.

Acknowledgements: I thank Erich Nelson and Priscilla Purnik for assistance, my colleagues at Bell Labs for their permission to describe our joint work, and Herbert Heuer, David Rosenbaum and Saul Sternberg for comments on a draft.

REFERENCES

Allport DA (1980) Attention and performance. In: Claxton G (ed) Cognitive psychology: new directions. Routledge and Kegan Paul, London, pp 112–153

Baddeley AD, Thomson N, Buchanan M (1975) Word length and the structure of short-term memory. J Verb Learn Verb Beh 14: 575–589

Ellis AW (1979) Speech production and short-term memory. In. Morton J, Marshall JC (eds) Psycholinguistics 2: structures and processes. MIT Press, Cambridge, Mass, pp 157–187

Fowler CA (1983) Converging sources of evidence on spoken and perceived rhythms of speech: cyclic production of vowels in monosyllabic stress feet. J Exp Psychol: Gen 112: 386–412

Glencross DJ (1972) Latency and response complexity. J Mot Beh 4: 251–256

Henry FM, Rogers DE (1960) Increased response latency for complicated movements and a 'memory drum' theory of neuromotor reaction. Res Q Am Assoc Health Phys Ed Recreation 31: 448–458

Hulstijn W, Van Galen GP (1983) Programming in handwriting: reaction time and movement time as a function of sequence length. Acta Psychol 54: 23–49

Lashley KS (1951) The problem of serial order in behavior. In: Jeffress LA (ed) Cerebral mechanisms in behavior. Wiley, New York, pp 112–146

Monsell S, Sternberg S (1981) Speech programming: a critical review, a new experimental approach, and a model of the timing of rapid utterances: part 1. Unpublished technical memorandum, Bell Laboratories [available from the author]

Monsell S, Nelson EM (1983) Control of rapid speech: limits on utterance programming capacity. Paper presented to the Psychonomic Society, San Diego, November [available from the author]

Sternberg S (1969) The discovery of processing stages: extensions of Donders' method. In: Kosters WG (ed) Attention and performance II. Acta Psychol 30: 276–315

Sternberg S, Monsell S, Knoll RL, Wright CE (1978) The latency and duration of rapid movement sequences: comparisons of speech and typewriting. In: Stelmach GE (ed) Information processing in motor control and learning. Academic, New York, pp 117–152

Sternberg S, Wright CE, Knoll RL, Monsell S (1980) Motor programs in rapid speech: Additional evidence. In: Cole RA (ed) The perception and production of fluent speech. Erlbaum, Hillsdale NJ, pp 507–534

Sternberg S, Knoll RL, Monsell S, Wright CE (1983) Control of rapid action sequences in speech and typing. Invited address to the American Psychological Association, August 1983 [available from the author]

Vallar G, Baddeley AD (1984) Fractionation of working memory: neuropsychological evidence for a phonological short-term store. J Verb Learn Verb Beh 23: 151–161

Intermanual Interactions During Programming of Finger Movements: Transient Effects of 'Homologous Coupling'*

H. Heuer

Fakultät für Psychologie und Sportwissenschaft, Universität Bielefeld, Postfach 8640, 4800 Bielefeld 1, FRG

In this paper two different but related issues are addressed. The first concerns "homologous coupling" of the hands with respect to the fingers. The term "homologous coupling" denotes the factors that create the tendency to move homologous fingers of both hands, when roughly simultaneous movements are required. Thus, "homologous coupling" is used as an explanatory term here, but not as a descriptive one. It refers to some processes of facilitation/inhibition which cause simultaneous movements with the left and right hand to be biased towards simultaneous activation of homologous muscle groups.

The second issue concerns the validity of a rather widely used explanatory principle that I shall call the "principle of advance specification". This principle is used as some kind of theoretical guideline in interpreting the results obtained with a particular methodology in the study of motor programming. In the context of the problem of homologous coupling it will be shown that at least one of the assumptions that is implied by this principle is not necessarily valid.

INTERMANUAL INTERACTIONS DURING MOTOR PROGRAMMING AND THE PRINCIPLE OF ADVANCE SPECIFICATION

Intermanual interactions are a well-known phenomenon (for review see Heuer and Wing 1984). A second well-established phenomenon is preparation that precedes execution of a movement (for review see Requin 1980; Rosenbaum 1985; Brunia et al. 1985). Indicators of preparatory processes seem to be rather generalized and independent of the properties of the forthcoming movement partly, but partly they are specific to the response performed. This suggests that there are preparatory processes which specifically determine the characteristics of the forthcoming movement, and these are usually called "programming". The concept of a motor program, that is established before execution of a movement and controls it during execution, implies some kind of continuity of processes (Heuer 1985): characteristics of a response that show up when it is executed, should at least in part be represented before execution takes place. The two facts of intermanual interactions and response-specific preparatory processes taken together suggest, that it might be worthwhile to look for intermanual interactions that occur during simultaneous programming of left-hand and right-hand responses (Heuer 1984b).

* This research was in part supported by Deutsche Forschungsgemeinschaft grant no. He 1187/3-1.

Intermanual interactions during programming can be studied using a reaction time task which requires a choice between left- and right-hand responses. In different experimental conditions the movements assigned to the left and right hand can be the same or different. In such a task only one response is performed at a time, but before presentation of the imperative signal and for some time thereafter subjects do not know whether this will be a left-hand or a right-hand response. In order to minimize reaction time they should thus program both responses simultaneously to the extent that this strategy is possible. This hypothesis is supported by data outlined below, which relate to the observable effects of intermanual interactions during the postulated simultaneous programming.

In response to the question about the observable effects of intermanual interactions during the postulated simultaneous programming, reference can be made to the explanatory principle of advance specification. Application of this principle presupposes a particular concept of motor programming. It is postulated that several parameters of control have to be specified, that are somehow related to the characteristics of the forthcoming response. There is no necessary one-to-one correspondence between the parameters and kinematic or dynamic variables of the response; for the sake of simplicity, however, parameters are often discussed as if this were the case.

Note that the term parameter is used here in a rather broad sense. There is no distinction, for example, between "program" and "parameter" (e.g., Schmidt 1980). Rather, different programs are treated in the same way as different parameter values. Basically parameters are thought of as "control" variables to distinguish them from physical variables that can be used to describe response characteristics. It is obvious that, although a one-to-one correspondence between parameters and observable response characteristics is not required by this approach, some specificity of the mapping is.

The explanatory principle of advance specification is based on an assumption about how control parameters are specified (Rosenbaum 1980; Rosenbaum and Saltzman 1984). Basically it is postulated that some parameters are specified in advance of the imperative signal. These are parameters that are related to characteristics that are the same for the choice responses. Specification of other parameters, related to kinematic or dynamic variables on which the responses differ, must wait until after presentation of the imperative signal. This rationale obviously neglects the possibility of specifying different parameter values at the same time – a situation, however, that can easily be remedied (Heuer 1984a).

Consider how the explanatory principle of advance specification applies to the choice between left-hand and right-hand responses. If responses are the same, they can be specified in advance of the imperative signal. If they differ on a particular variable, two different results are to be expected depending on whether the hands are coupled with respect to this variable or not:

1. If coupling exists, it should be impossible to specify different parameter values for the left and right hand at the same time, or at least the simultaneous specification of parameter values should be impaired. Reaction time should thus be increased due to the necessity of specifying parameters after presentation of the imperative signal (Heuer 1982a,b,c).

2. If no coupling exists, it should be possible to specify different parameter values for the left and right hand at the same time. In this case of "multiple preparation" (Rosenbaum 1980; Goodman and Kelso 1980; Heuer 1982a,b; Rosenbaum and Kornblum 1982) of different responses, the time needed to choose between them should not be longer than in the case of same responses.

By taking the deductions from the explanatory principle of advance specification as a basis for inductions one obtains a foundation for the study of intermanual interactions during motor programming. Specifically, an increased reaction time for different responses can be taken as an indicator of a coupling of the hands with respect to that variable on which the responses differ, while no increase in reaction time in spite of a difference between the responses on a particular variable suggests multiple preparation, i.e., an independence of the hands with respect to that kinematic or dynamic variable.

In line with the explanatory principle of advance specification, choice reaction time data indicate temporal coupling of the hands as has been found previously in studies of simultaneously performed movements (e.g., Kelso et al. 1979; Klapp 1979; Schmidt et al. 1979). If responses that are assigned to the left and right hand differ with respect to duration and relative timing, a triad of effects occurs, namely reaction time increases and becomes more variable, while frequency of choice errors (confusion between hands) becomes smaller (Heuer 1982a,b; 1984a,b). These results have been found for simple up-and-down movements of a finger that differ in total duration, and for pairings of up-and-down movements with mainly horizontal fro-and-back movements, that are either of the same duration or are performed as quickly as possible, so that up-and-down movements are faster. It should be noted that the joint effect of a difference between the responses on mean reaction time and frequency of choice errors cannot be explained by a simple change in the speed-accuracy criteria of the subjects (Heuer 1983).

THE PROBLEM OF HOMOLOGOUS COUPLING: DISCREPANT CONCLUSIONS AND A RECONCILING HYPOTHESIS

There is one difference between responses with the left and right hand that does not lead to any of the three effects described above. This is a difference in the responding finger (Heuer 1982a,b,c; 1984b; Rosenbaum and Kornblum 1982; an exception is reported by Zelaznik et al. 1982). Although the lack of effect has been taken as evidence for multiple preparation, an obvious problem exists. Provided that processes during preparation and execution are at least in part identical (Heuer 1985) one should expect convergent conclusions from different kinds of data, just as they exist for the temporal coupling of the hands. The conclusion based on choice reaction time experiments, however, that fingers can be specified independently, is clearly at variance with conclusions that are based on other sets of data.

The first kind of data that indicate homologous coupling of the hands, are from studies of simultaneous finger movements. Rabbitt et al. (1975) had their subjects produce finger chords as quickly as possible in response to a signal. In one of their experiments, index fingers and middle fingers of both hands were used. Within the six two-finger chords, reaction time was longest when nonhomologous fingers had to be used; reaction time was fastest for both index fingers, and all other two–finger chord latencies were between these extremes.

Homologous coupling in simultaneous movements is also indicated by accuracy data. MacKay and Soderberg (1971) had their subjects tap simultaneously with the fingers of both hands in a sequence from left to right or right to left, e.g., left little finger–right index finger, left ring finger–right middle finger, left middle finger–right ring finger, etc. The most frequent errors were "homologous intrusions", i.e., simultaneous responses with homologous fingers. One can easily demonstrate to oneself that the task is simple as long as it is done slowly, but that homologous intrusions become more likely when it is speeded up.

The second kind of data that indicate homologous coupling again by reaction time as well as by accuracy criteria, are from studies of successive finger movements. In an experiment on the psychological refractory period Wakelin (1976) had his subjects first choose between index and middle fingers of the left hand and a short time thereafter between the same fingers of the right hand. Reaction times of both responses were faster when they were performed with homologous fingers than in the case of nonhomologous fingers.

Evidence for homologous coupling based on an error criterion is reported by Rabbitt and Vyas (1970). Erroneous responses in serial reaction time tasks are more often performed with a finger that is homologous to the one used for the preceding correct response than with nonhomologous fingers.

How can the discrepant conclusions based on choice between the left and right hand and on other kinds of data be reconciled? The conclusion that fingers can be specified independently for the two hands was derived from a particular experimental result together with the explanatory principle of advance specification. In view of the evidence on the existence of homologous coupling, it appears that something must be wrong with the principle of advance specification.

The explanatory principle of advance specification can be decomposed into different assumptions that need not necessarily be correct. The assumption that is crucial for the present problem is that interference between simultaneous processes of programming different responses shows up in the results of programming, i.e., in the partial programs that are set up at the time of the imperative signal.

It is obvious that there are at least two ways in which interference during simultaneous processes of motor programming can manifest itself. First, as is implied by the principle of advance specification, the final results of motor programming can be impaired. Second, interference could leave the results that are finally achieved unaffected, but could show up as a transient impairment of the program during the process of programming only. The present suggestion is that homologous coupling of the hands with respect to fingers acts in the latter way.

The hypothesis of transient effects of homologous coupling during finger specification is in line with available experimental data. First, consider the choice task in which no effect of assigning homologous or nonhomologous fingers to the hands has been observed. In these experiments the choice conditions were blocked. There is thus ample time to achieve the final state of finger specification that is not affected by interference due to homologous coupling. For example, finger specification could take place at the start of a block of trials and be maintained throughout that block.

Second, consider the tasks in which evidence for homologous coupling has been found. In these tasks finger selection and response execution seem to occur in close temporal contiguity. For example, if finger chords have to be produced (Rabbitt et al. 1975), fingers have to be selected after presentation of the imperative signal and to be activated as fast as possible. The same situation exists whenever the imperative signals inform subjects about which of at least two fingers of a hand has to be used for the response. Similarly, if sequences of finger movements have to be produced (MacKay and Soderberg 1971), selection of the appropriate finger is immediately followed by execution except when the sequence is performed slowly.

Given that there is a transient effect of homologous coupling and not much time until execution, the time needed for finger specification and thus total reaction time should increase. Further, if it is assumed that response execution does not have to wait until the final state of finger specification is achieved, temporary biases toward the current state of the finger selection process for the other hand will show up in overt performance as "homologous errors". Therefore, homologous coupling of the hands with respect to fingers should be indicated by a time and/or an error criterion.

Of course, such post hoc explanations are liable to spurious harmony with experimental data. That the hypothesis of transient effects of homologous coupling during finger specification is in line with previous results is, therefore, no guarantee that it is correct. The following experiments are designed to examine the hypothesis directly. The prediction is that the time needed to choose between the left and right hand does depend on whether the fingers assigned to the hands are the same or different, provided that finger specification is not yet completed at presentation of the imperative signal. An attempt was made to induce incomplete finger specification by lack of time in the first experiment and by lack of motivation in the second.

EXPERIMENT 1

In the first experiment simple key press responses with the index fingers and middle fingers of both hands were used. In two choice conditions the responses involved homologous fingers and in two other conditions nonhomologous fingers. There were two further conditions in the experiment in which choice was between fingers of the same hand. These conditions were included since the available reaction time data on choice between vs within hands are strikingly discrepant (Kornblum 1965; Rosenbaum and Kornblum 1982; Miller 1982; Reeve and Proctor 1984). They are not of particular interest for the present problem, however, and the results will not be reported in detail.

The crucial feature of the experiment is that choice conditions are varied in a random sequence across trials rather than blocked. On each trial the choice condition is indicated by a precue at a particular time before the imperative signal. When the precuing interval is long, the situation is basically the same as when choice conditions are blocked, since there is sufficient time to achieve the final state of finger specification. Therefore, choice reaction time should not depend on whether the choice condition involves homologous or nonhomologous fingers, and "homologous errors" should not occur.

The same result could be expected when the precuing interval is zero, since the task is basically a four–choice task, no matter which responses are precued. This situation, however, is somewhat complex. For example, some processing of the precue may occur during the reaction time interval. Therefore, some effects of the relationship between responses might occur.

Of main interest are the intermediate precuing intervals. Here, specification of fingers has been started, but is not yet completed at the time of the imperative signal. Transient effects of homologous coupling could show up in a twofold way. First, reaction time could be increased when nonhomologous fingers have to be selected, since finger selection will be less advanced at presentation of the imperative signal. Second, homologous biases during the process of finger specification could cause a higher frequency of erroneous responses with nonprecued fingers if nonhomologous fingers have been precued.

Method

Signals were two horizontal rows of light–emitting diodes on a slightly inclined panel, located directly behind the response board. Four yellow LEDs of the upper row were used as precues, four green LEDs of the second row 1 cm below the first were used as imperative signals. Lateral center–to–center separation of the LEDs was 2 cm, their diameter was 4 mm. On the response board, four keys were placed in positions that allowed a comfortable placement of the fingers (width of keys 1 cm, lateral edge–to–edge separation 2 cm). On top of the keys there were contact plates that made it possible to monitor correct finger placement.

A single trial began with an auditory warning signal (1330 Hz, 60 dBA) when the fingers had been placed on the response keys for 1 s. The warning signal was presented by a loudspeaker for 200 ms; 500 ms later the precue came on, followed by the imperative signal after the appropriate delay. Both precues and imperative signals were response–terminated. Errors were indicated to subjects by five 100 ms flashes of the upper or lower row of LEDs. The upper row was used in the case of anticipatory responses, the lower row in the case of incorrect responses.

Precuing intervals of 0, 250, 800, and 2000 ms were presented in separate blocks of trials. In each test session four blocks of trials were presented, one for each precuing interval. The order of precuing intervals was varied across subjects according to a digram–balanced latin square. A block consisted of 12 warm–up trials followed by 120 test trials. On test trials the six choice conditions were presented in a random sequence with equal frequencies. Within each choice condition, the two precued imperative signals were again presented in a random sequence with equal frequencies.

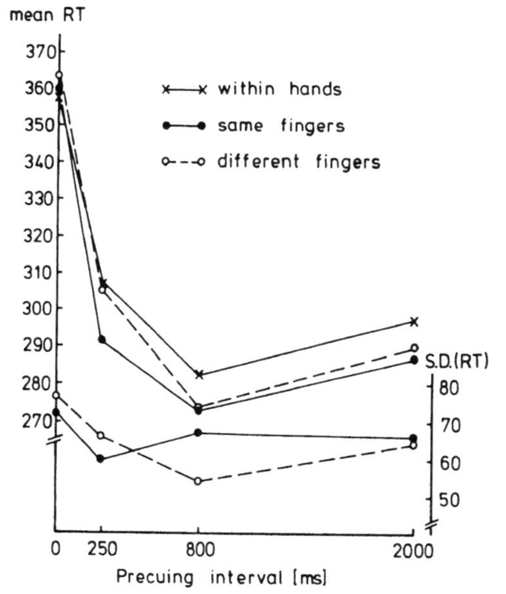

Fig.1. Mean reaction time (upper three lines, left ordinate) and mean standard deviation of reaction times (lower two lines, right ordinate) as a function of choice condition and precuing interval

Four male and four female subjects, aged 22 to 29 years, took part in a training session and four test sessions. They were paid a fixed amount of money for their participation. All subjects were right-handed according to the Edinburgh Handedness Inventory (Oldfield 1971).

Results and Discussion

Data from the four test sessions were collapsed. For each subject, finger, and choice condition, means and standard deviations of reaction times were computed. Errors were neglected in these computations. For the analysis of errors a distinction was made between those that involved a precued finger and those in which a nonprecued finger was involved.

Mean reaction times are presented in Fig. 1. In passing it should be noted that choice between hands takes less time than choice within hands, with the exception of the shortest precuing interval. An ANOVA with the factors choice within vs between hands, precuing interval, and finger reveals a significant interaction of choice within vs between hands and precuing interval ($F(3,21) = 3.6$, $p < 0.05$). This finding contrasts with other recent data from very similar experiments, in which an advantage of within-hand conditions was found (Miller 1982; Reeve and Proctor 1984), but is in line with older data from experiments in which choice conditions were blocked (Kornblum 1965; Rosenbaum and Kornblum 1982).

The most important result is the interaction between precuing interval and conditions in which same or different fingers of the left and right hand were used ($F(3,21) = 4.5$,

Table 1. Median percentages of erroneous responses with nonprecued and precued fingers

	Nonprecued fingers precuing intervals (ms)				Precued fingers precuing intervals (ms)			
	0	250	800	2000	0	250	800	2000
Same fingers	1.25	0.94	0	0	0.97	1.56	0.31	0.63
Different fingers	2.50	1.25	0.63	0	0	0.31	0	0.63
Difference	−1.25*	−0.63	−0.63(*)	0	0.97*	0.63	0	0

Wilcoxon signed−rank tests between choice conditions: *, $p < 0.05$; (*), $p < 0.10$

$p < 0.05$). At the 250−ms precuing interval there is a significant advantage (13 ms) of conditions with homologous fingers that is not present at the other precuing intervals (4, zero, and 3 ms for precuing intervals of 0, 800, and 2000 ms, respectively). This finding is in line with the hypothesis of transient effects of homologous coupling.

The second prediction of this hypothesis concerns the frequency of erroneous responses performed with nonprecued fingers. These should be more frequent in conditions with non-homologous fingers, since the finger that is not precued for one hand is homologous to the one that is precued for the other hand. Table 1 (left side) shows the median percentages. (Medians give a better description of the central tendency since the distributions are skewed. The same trends, however, are evident also in the means.) With longer precuing intervals the frequency of these errors declines to zero (Friedman test, $\chi^2(3) = 15.3$, $p < 0.01$). This "locking in" of the responses assigned to the left and right hand has been observed informally before (Heuer 1983). The decrease of the difference between conditions with homologous and nonhomologous fingers is also significant (Friedman test, $\chi^2(3) = 9.0$, $p < 0.05$). An advantage of conditions with homologous fingers is indeed only observed at the short precuing intervals.

As described above, when the responses assigned to the left and right hand differ with respect to relative timing, it is not only mean reaction time that is increased, but in addition the variability of reaction times is increased and the frequency of choice errors is reduced. These two dependent variables were also analysed for the present experiment to examine the potential effects of same vs different fingers.

The mean standard deviations are presented in Fig. 1. The interaction between same vs different fingers and precuing interval is significant ($F(3,21) = 3.2$, $p < 0.05$). At the short precuing intervals an advantage of conditions with homologous fingers can be seen, that is reversed at the longer precuing intervals. Although the advantage is small, it is reliable at the 250 ms interval. (The large difference in the opposite direction at the 800 ms interval is mainly due to a single subject, who had an extremely large variability in one of the choice conditions. Note that in this experiment there was no screening for RT outliers.)

The median percentages of choice errors are presented in Table 1 (right side). These are erroneous responses in which the wrong hand is chosen, but the correct finger –precued for that hand– is used. There is a slight trend for these errors to be more frequent in conditions which involve homologous fingers, provided that the precuing interval is short.

In general the results of the first experiment are well in line with the hypothesis of transient effects of homologous coupling during finger specification. Support for this hypothesis can also be found in the results of Reeve and Proctor (1984). Their first experiment differs from the present one mainly in that their subjects were untrained and their precuing intervals were randomized. Although their findings on choice between vs within hands are quite different from the present ones, they also observed the interaction between precuing interval and conditions with homologous and nonhomologous fingers, which is predicted by the present hypothesis.

EXPERIMENT 2

The precuing method is not without problems, especially when the precuing intervals are short (see Reeve and Proctor 1984). For example, it is generally hard to separate perceptual interpretations involving differences in precue processing from interpretations that refer to differences in associated response preparation processes. Further, the effects of a gradual shift from a 4–choice task to a 2–choice task with increasing precuing interval cannot be assessed. Instead of a lengthy discussion of the data of the first experiment with respect to these issues, a different experimental situation was tried in the second experiment, again with the aim of demonstrating transient effects of homologous coupling. This time, however, the potential confounding effects inherent to the precuing methodology were eliminated.

The basic idea was to block choice conditions and to allow subjects to know early which responses are assigned to the left and right hand. Thus subjects are in a position to specify fingers well in advance of the imperative signal. However, an attempt was made to keep their motivation low for doing so, by introducing catch–trials in which a warning signal, but not an imperative signal, was presented. With a sufficient frequency of catch–trials, the effort invested in advance specification and its maintenance would be wasted on the majority of trials, and therefore, it was hypothesized, would be saved to a larger or smaller degree. If the occurrence of catch–trials indeed succeeds in preventing advance specification to varying degrees, effects analogous to those in the first experiment should occur.

Method

Choice conditions were presented in separate blocks of trials in this experiment. In two conditions, homologous fingers were assigned to the hands (index fingers or thumbs); in the other two conditions the assignment was for nonhomologous fingers. Catch–trial frequencies were 0%, 30%, and 70% in different sessions.

The response board was the same as used by Heuer (1983, 1984a). Subjects grasped two horizontally mounted handles and placed their extended index fingers and thumbs on four contact plates in the appropriate locations. In performing a response, the finger had to be lifted for about 3 cm and then lowered.

Imperative signals were presented on an oscilloscope screen, in front of which a mask was placed with a window 6 cm wide and 2 cm high. The imperative signal was a horizontal line 2 cm long presented in the left or right half of the window. Behind the oscilloscope a monitor was placed, which was used to display knowledge of results. In addition, the fingers assigned to the left and right hand were displayed continuously throughout a block of trials.

A single trial started when the four fingers were placed on the contact plates. After 1 s had elapsed, an auditory warning signal (1000 Hz, 60 dBA) was presented by a loudspeaker for 1.5 s, which was replaced by the imperative signal except on catch−trials. On catch−trials, the response board was monitored for 800 ms to detect any responses. On regular trials, the imperative signal remained on until the end of a correct response. If an error occurred, knowledge of results was presented during the last second of the trial.

Choice conditions were presented in separate blocks of 10 warm−up and 100 test trials each. The sequence of catch−trials and regular trials was randomized as was the sequence of imperative signals. Each test session consisted of four blocks of trials, one for each choice condition. The order of choice conditions was varied across subjects according to a digram−balanced latin square. For half of the subjects there were 0% catch−trials in the first test session, 30% in the second, and 70% in the third and fourth. For the other half the order of catch−trial frequencies was reversed. In the training session that preceded the test sessions all four choice conditions and three catch−trial frequencies were presented.

Nine male and seven female subjects, aged 20 to 35 years, were paid to take part in the experiment. All but one of the subjects were right−handers according to the Edinburgh Handedness Inventory.

Results and Discussion

Trials of the four test sessions were collapsed. For each subject, choice condition, catch−trial frequency, and hand, means and standard deviations of reaction times were computed. Trials on which an error occurred or on which reaction time exceeded 800 ms or movement time 1200 ms were excluded.

Figure 2 presents the mean reaction times. The main result is the interaction between choice conditions and catch−trial frequency ($F(2,30) = 7.0$, $p < 0.01$), in addition to the main effects of same vs different fingers and catch−trial frequency. At 0% catch−trials the advantage of homologous finger conditions is 19 ms, and at 70% it increases to 40 ms. This is the interaction that is predicted by the hypothesis of transient effects of homologous coupling.

Although this interaction is in line with expectations, the results are not exactly the ideal ones that one would have liked to see. An effect of same vs different fingers is already present at

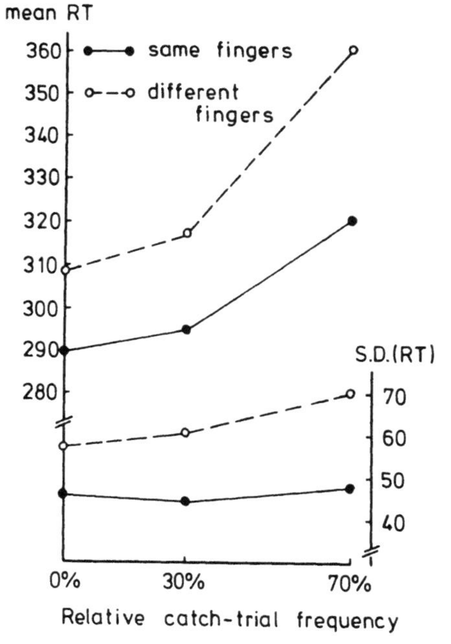

Fig.2. Mean reaction time (upper two lines, left ordinate) and mean standard deviation of reaction times (lower two lines, right ordinate) as a function of choice condition and catch–trial frequency

0% catch–trials, although it has not been found under these conditions in earlier experiments (Heuer 1982a,b,c; 1984b). It is likely that the effects of catch–trials are not strictly confined to those sessions in which they are presented, but spread slightly across the whole experiment. In the first session with 0% catch–trials, all subjects have had experience with 70% catch–trials also, at least in the initial practice session. Therefore even with 0% catch–trials advance specification of fingers may have not been as perfect as in the previous experiments in which no catch–trials at all occurred.

Table 2 (left side) presents the median frequencies of erroneous responses in which the correct hand was used. These errors are responses with the incorrect finger or responses with the correct finger, but of insufficient height. In previous experiments, in which only the latter type of errors was recorded, no difference between conditions with homologous and nonhomologous fingers was found (Heuer 1982a,b; 1984b). In this experiment it is clearly present. Moreover, it tends to increase when catch–trial frequency increases and advance specification of fingers presumably becomes less complete. This result is again in line with the hypothesis of transient effects of homologous coupling.

As in the first experiment, variability of reaction times and choice error frequencies were analysed in this experiment. Figure 2 shows the expected interaction for the mean variability of reaction times ($F(2,30) = 4.2$, $p < 0.05$). The advantage of conditions with homologous fingers increases from 12 ms at 0% catch–trials to 23 ms at 70% catch–trials. The advantage at 0% catch–trials again indicates that in this condition advance specification of fingers has not been as complete as in previous experiments without catch–trials.

Table 2. Median percentages of erroneous responses with the correct and incorrect hand

	Correct hand catch – trial frequency			Incorrect hand catch – trial frequency		
	0%	30%	70%	0%	30%	70%
Same fingers	1.02	1.08	1.67	1.50	1.07	0.83
Different fingers	3.01	2.20	3.84	1.50	0.71	0
Difference	-1.01^{**}	$-1.11^{(*)}$	-3.01^{**}	0	0	0.42

Wilcoxon signed – rank tests between choice conditions: **, $p < 0.01$; $^{(*)}$, $p < 0.10$

Mean percentages of choice errors are presented in Table 2 (right side). Although there is no convincing effect, there is at least a slight trend that an advantage of conditions with non-homologous fingers shows up when catch – trial frequency is increased.

GENERAL DISCUSSION

The results of the two experiments are quite similar in spite of considerable differences in the experimental setups. Whenever advance specification of fingers is incomplete, the triad of effects that can be reliably observed when subjects have to choose between left – hand and right – hand responses that differ in relative timing, tends also to show up when responses have to be performed with nonhomologous fingers. The increased frequency of responses with fingers that have not been assigned to the responding hand, in particular when the fingers assigned are nonhomologous, indicates that advance specification is indeed incomplete in the case of short precuing intervals or high catch – trial frequencies.

The data clearly support the hypothesis of transient effects of homologous coupling during finger specification. According to this hypothesis interference due to homologous coupling takes place during the process of finger specification only, but the final results of this process are not affected by whether a homologous or a nonhomologous finger of the other hand has been specified at the same time. Therefore the time needed to choose between a left – hand response and a right – hand response does not depend on whether homologous or nonhomologous fingers are assigned to the hands, provided that finger specification has been completed well in advance of response execution. If, however, finger specification and response execution are not separated in time, the effects of homologous coupling during finger specification will show up in reaction times and/or errors.

The hypothesis of transient effects of homologous coupling has been developed here with respect to finger specification only. But it seems that it can be generalized to other muscle groups of symmetric limbs, provided of course that the preferred mode of cooperation is indeed homologous, which is not always the case (see Heuer and Wing 1984). The most obvious prediction is that simultaneous nonhomologous movements are relatively easy if there is sufficient time to overcome the transient coupling. If, however, activations of nonhomologous muscle groups follow in quick succession, homologous coupling will show its effects: It will force the limbs into a homologous mode of cooperation (see Cohen 1971; Kelso 1984).

An obvious weakness of the concept of homologous coupling (or of coupling with respect to any other movement characteristic) is the lack of a detailed model on the nature of the coupling. A vague reference to some facilitation/inhibition is not a sufficient basis for deriving quantitative predictions. But this lack of specific predictions for some experimental situations is compensated by a broader range of phenomena to which the general approach taken here can be applied: It allows not only predictions for movements performed simultaneously, but also for choice between movements with the left and right hand and for pairs of bimanual movements in which execution of one response temporally overlaps with programming of the other (Heuer 1985).

The second issue which is addressed by the present experiments is the validity of the explanatory principle of advance specification that is widely used in the study of motor programming. The support for the hypothesis of transient effects of homologous coupling indicates that this principle should not be applied indiscriminately. For example, if a difference between responses does not lead to an increased reaction time, it is not really justified to conclude that these responses can be programmed simultaneously and independently. A slight change of experimental conditions, that induces less preprogramming in advance of the imperative signal, can lead to a reaction time difference. This suggests that differences between choice responses need not always limit the final quality of the partial program set up in advance, but rather an impairment can be restricted to the "dynamic phase" during which the partial program is actually organized.

Acknowledgements: I wish to thank Scott Kelso and two anonymous reviewers for their helpful comments on an earlier draft.

REFERENCES

Brunia CHM, Haagh SAVM, Scheirs JGM (1985) Waiting to respond: electrophysiological measurements in man during preparation for a voluntary movement. In Heuer H, Kleinbeck U, Schmidt KH (eds) Motor behavior: programming, control, and aquisition. Springer, Berlin Heidelberg New York Tokyo

Cohen L (1971) Synchronous bimanual movements performed by homologous and nonhomologous muscles. Percept Mot Skills 32: 639–644

Goodman D, Kelso JAS (1980) Are movements prepared in parts? Not under compatible (naturalized) conditions. J Exp Psychol: General 109: 475–495

Heuer H (1982a) Binary choice reaction time as a criterion of motor equivalence. Acta Psychol 50: 35–47

Heuer H (1982b) Binary choice reaction time as a criterion of motor equivalence: further evidence. Acta Psychol 50: 49–60

Heuer H (1982c) Choice between finger movements of different and identical form: the effect of relative signal frequency. Psychol Res 44: 323–342

Heuer H (1983) Reaktionszeit und Genauigkeit bei der Wahl zwischen formgleichen und formverschiedenen Bewegungen beider Hände. Psychol Beitr 25: 532–556

Heuer H (1984a) Binary choice reaction time as a function of the relationship between durations and forms of responses. J Mot Behav 16: 392–404

Heuer H (1984b) Bewegungen mit der rechten und linken Hand: Interferenzen bei der Programmierung. Psychol Beitr 26: 561–581

Heuer H (1985) Intermanual interactions during simultaneous execution and programming of finger movements. J Mot Behav 17: 335–354

Heuer H, Wing AM (1984) Doing two things at once: process limitations and interactions. In Smyth MM, Wing AM (eds) Psychology of human movement. Academic, London

Kelso JAS (1984) Phase transitions and critical behavior in human bimanual coordination. Am J Physiol 246: R1000–R1004

Kelso JAS, Southard DL, Goodman D (1979) On the coordination of two–handed movements. J Exp Psychol: Hum Percept Perform 5: 229–238

Klapp ST (1979) Doing two things at once: The role of temporal compatibility. Memory Cognition 7: 375–381

Kornblum S (1965) Response competition and/or inhibition in two–choice reaction time. Psychon Sci 2: 55–56

MacKay DG, Soderberg GA (1971) Homologous intrusions: an analogue of linguistic blends. Percept Mot Skills 32: 645–646

Miller J (1982) Discrete versus continuous stage models of human information processing: in search of partial output. J Exp Psychol: Hum Percept Perform 8: 273–296

Oldfield RC (1971) The assessment and analysis of handedness. The Edinburgh Inventory. Neuropsychologia 9: 97–113

Rabbitt PMA, Vyas SM (1970) An elementary preliminary taxonomy for some errors in laboratory choice RT tasks. In Sanders AF (ed) Attention and performance III. North-Holland, Amsterdam

Rabbitt PMA, Vyas SM, Fearnley S (1975) Programming sequences of complex responses. In Rabbitt PMA, Dornic S (eds) Attention and performance V. Academic, London

Reeve TG, Proctor RW (1984) On the advance preparation of discrete finger responses. J Exp Psychol: Hum Percept Perform 10: 541-553

Requin J (1980) Towards a psychobiology of preparation for action. In: Stelmach GE, Requin J (eds) Tutorials in motor behavior. North-Holland, Amsterdam

Rosenbaum DA (1980) Human movement initiation: specification of arm, direction, and extent. J Exp Psychol: General 109: 444-474

Rosenbaum DA (1985) Motor programming. In Heuer H, Kleinbeck U, Schmidt KH (eds) Motor behavior: programming, control, and aquisition. Springer, Berlin Heidelberg New York Tokyo

Rosenbaum DA, Kornblum S (1982) A priming method for investigating the selection of motor responses. Acta Psychol 51: 223-243

Rosenbaum DA, Saltzman E (1984) A motor-program editor. In: Prinz W, Sanders AF (eds) Cognition and motor processes. Springer, Berlin Heidelberg New York Tokyo

Schmidt RA (1980) On the theoretical status of time in motor program representations. In Stelmach GE, Requin J (eds) Tutorials in motor behavior. North-Holland, Amsterdam

Schmidt RA, Zelaznik HN, Hawkins B, Frank JS, Quinn JT (1979) Motor-output variability: a theory for the accuracy of rapid motor acts. Psychol Rev 86: 415-451

Wakelin DR (1976) The role of the response in psychological refractoriness. Acta Psychol 40: 163-175

Zelaznik HN, Shapiro DC, Carter MC (1982) The specification of digit and duration during motor programming: a new method of precuing. J Mot Behav 14: 57-68

Modulation and Maintenance of Action Patterns

Pattern Formation in Speech and Limb Movements Involving Many Degrees of Freedom*

J. A. S. Kelso

Center for Complex Systems, Florida Atlantic University and Haskins Laboratories, 270 Crown Street, New Haven, CT 06510, USA

INTRODUCTION

There are over 792 muscles and 100 joints in the human body, while, according to my eldest son's biology textbooks, the elephant's trunk contains over 40000 muscles and tendons. Thus, any activity of the human body or the elephant's trunk involves the cooperative effort of very many degrees of freedom. But what form do principles of cooperation in multivariable movements take? For some years now, my colleagues and I have viewed this question as continuous with the general issue of understanding the emergence of order and regularity in complex systems (see, e.g., Yates 1979, for defining characteristics of complexity). The core idea that we have pursued is that the collective action among multiple neuromuscular components is fundamentally task–related, and that the significant units of control and coordination are *functional* groupings of muscles and joints which we call coordinative structures or functional synergies (e.g. Fowler et al. 1980; Kelso et al. 1979; Kelso and Tuller 1984a; Kugler et al. 1980; Saltzman and Kelso 1985b; Turvey 1977). The hallmark of a coordinative structure is the temporary marshalling of several articulators into a task–specific pattern.

This notion of functional units of action, or coordinative structures differs in significant ways from conventional treatments of movement control that are based on either the neurophysiological notion of a central pattern generator or the information processing notion of a motor program. First, unlike the notion of a hard "prewired" central pattern generator, the coordinative structure construct underscores the *soft* or flexible nature of action units that are functionally specific, not anatomically specific. One of the goals of this paper is to buttress this claim using examples from recent research on the motor control of speech and limb movements. Second, contrary to the motor program formulation that relies on symbol–string manipulation familiar to computer technology, the coordinative structure construct highlights the analytic tools of qualitative (nonlinear) dynamics (e.g. Kelso et al. 1981, 1985a; Saltzman and Kelso 1985b) and the physical principles of cooperative phenomena (e.g. Kelso and Tuller 1984a, b; Kugler et al. 1980, 1982). Thus, the problem of pattern formation for skilled actions is couched as a specific aspect of the more general topic of cooperative phenomena in nonlinear, open systems (see, e.g. Haken 1975, 1977, 1983). Such systems display ordered states that are not imposed by programs, but that actively evolve from the dynamic interplay

* This work was supported in part by NIH grant NS–13617, Biomedical Research Support grant RR–05596 and contract no. N0014–83–K–0083 from the US Office of Naval Research. Comments by Bruce Kay, Kevin Munhall, Elliot Saltzman, and Betty Tuller were much appreciated.

of processes, in a so-called "self-organized" fashion. Although the present theoretical approach is in preliminary form as far as biological movements are concerned (see Kelso 1981a; Kelso and Tuller 1984a; Kugler et al. 1980, 1982), this paper attempts to convey the flavor of the approach, not only theoretically, but in terms of the kinds of experiments that it motivates. In the following sections I shall briefly address four questions drawing from our own and others' experimental work on limb movements and speech articulators involving many degrees of freedom.

1. The cooperativity question. What kind of unitary organization is formed by an ensemble of neuromuscular components?
2. The control question. What kind of control structure underlies the generation of certain movement patterns? What are the essential control parameters and how are parameter values specified?
3. The stability question. What characterizes the stability of a movement pattern, and what is the informational basis for the stability? Colloquially speaking, what holds a pattern together?
4. The change question. What are the necessary and sufficient conditions that give rise to change in an articulatory pattern?

THE COOPERATIVITY QUESTION

The Concept of Coordinative Structure

Do relatively independent articulators (muscles, joints) function as a unitary ensemble, and, if so, what kind of ensemble is it? Consider the act of speaking. Even a simple speech gesture involves cooperation among very many degrees of freedom operating at respiratory, laryngeal, and supralaryngeal levels. Yet in spite of (or perhaps because of) such a large number of neuromuscular elements, speech emerges as a coherent and organized activity. An attractive hypothesis proposed by Bernstein (1928/1967) and developed by his colleagues (e.g., Gelfand et al. 1971) is that the central nervous system, rather than controlling each degree of freedom separately, organizes them into "collectives", "linkages", or "synergies" that then behave, from the perspective of control, as a single degree of freedom.

Of course, as emphasized earlier, our notion of synergy or coordinative structure is unlike that of Sherrington (1906) or Easton (1972) (see, Kelso and Tuller 1984a) in that the collective action among multiple muscles or kinematic components is not rigid or machine-like, but fundamentally task- or function-specific. In this, Darwinian-like hypothesis, function dictates the form of cooperativity observed in an aggregate of neuromuscular components, not anatomical connections. But how might this notion be tested, and what evidence exists in its favor for complex actions?

A window into the behavior of a complex system possessing large numbers of active, interacting components can be gained by perturbing it dynamically during an activity and examining how the system reconfigures itself (e.g., with respect to response latencies, magnitudes, etc.). Thus, a group of potentially independent articulators could be said to behave in a unitary fashion if it were shown that a disruption to one (or more) members of the group was responded to by other members of the group at a site remote from the challenge. According to the concept of coordinative structure, the response of the articulatory ensemble would not be stereotypic; rather, it would be adapted quickly and precisely to accomplish the task. In general, the components of the neuromuscular system would cooperate in such a way as to preserve the performer's intent. Some evidence of so-called "remote compensation" phenomena that support a motor system design based on coordinative structures exists in both the speech and limb movement behavior literatures. These data are considered below.

Coordinative Structures in Multidegree of Freedom Movements

Although the speech literature contains a number of observations that are consistent with a coordinative structure mode of articulator organization, few experiments have been designed to test the notion explicitly. In one experiment by Folkins and Abbs (1975) the jaw was occasionally loaded during the closure movement for the initial /p/ in the utterance "a /hae paep/ again." Lip closure was attained in all cases apparently by exaggerated displacement and velocities of the lip closing gestures, particularly by the upper lip. Similarly, Folkins and Zimmermann (1982) used electrical stimulation to produce an unexpected depression of the lower lip prior to, and during, bilabial closure. Compensatory changes in the jaw and upper lip were observed to effect the bilabial closure. Abbs and colleagues (see, Abbs et al. 1984, for review) report that both autogenic (i.e., lower lip) and remote (upper lip) effects occur when a 40-g load is applied to the lower lip unexpectedly 30 ms before the onset of the phasic EMG burst in orbicularis oris inferior. They interpret these remote effects, after Houk and Rymer (1981), as evidence for open-loop, feed-forward control which is "a precise, experience-based representation of the relationship between afferent signals from one movement (from which a potential error is detected) and the motor output of a parallel synergistic movement (where the adjustment is implemented)." Autogenic compensations made by the structure perturbed are slower and thought to be under closed-loop feedback control.

Although these findings are consistent with the coordinative structure concept it is not clear whether, in fact, the patterns of articulator coupling following perturbations are in any sense standardized (as one might predict if they were completely preprogrammed or a result of fixed input-output loops) or whether they are indeed "functionally organized," i.e., directed to the stable production of the intended utterance. If the former, the pattern of response to a given perturbation should be the same regardless of the utterance. If the latter, different patterns of articulator cooperation (coordinative structures) should occur, tailored to the particular phonetic requirements.

Direct evidence that speech articulators (lip, tongue, jaw) do make *functionally specific*, near-immediate compensations to unexpected perturbations at sites remote from the locus of perturbation comes from our recent work (Kelso et al. 1982, 1984). An unexpected constant force load (5.88 N) applied during upward motion for final /b/ closure in /baeb/ revealed near-immediate changes in upper and lower lip muscles and movements (15-30 ms), but no changes in tongue muscle activity. The *same* perturbation applied during the utterance /baez/ evoked rapid and increased tongue muscle activity (genioglossus) for /z/ frication, but no active lip compensation. Although the jaw perturbation represented a threat to both utterances, no perceptible distortion of speech occurred. That a challenge to one member of a group of potentially independent articulators was met – on the very first perturbation experience – by remotely linked members of the group provides preliminary support for coordinative structures. Further anecdotal support for coordinative structures is mentioned in a review paper by Abbs and Gracco (1983). They report that for the utterance /*aba*/, upper *and* lower lips compensate when the lower lip is loaded in order to preserve bilabial closure. In contrast, for /*afa*/ which in theory does not require upper lip movement, only lower lip compensatory responses to a lower lip perturbation occur.

Analogous results emerge from recent studies of human posture (e.g. Cordo and Nashner 1982; Marsden et al. 1983). For example, in response to a perturbation applied to the thumb, which was performing a tracking task, Marsden et al. observed reactions in muscles remote from the prime mover (e.g., in pectoralis major of the same limb, in triceps of the *opposite* limb, in the opposite thumb when it served to stabilize motion, etc.). These distant reactions are much faster than typical reaction time responses; indeed they are sometimes faster (e.g., 40 ms in pectoralis) than the local, autogenetic reflex in the structure perturbed. But most interesting for the coordinative structure hypothesis is that postural responses occur only if they perform a useful *function* and they are flexibly tuned to that function. For example, postural responses in triceps disappear if the hand is not exerting a firm grip on an object. If, instead of holding a table top, the nontracking hand holds a cup of tea, the responses in triceps *reverse*, which is precisely what they have to do to prevent the tea from spilling. Marsden et al. (1983) conclude that these rapid, remote effects "constitute a distinct, and apparently new, class of motor reaction" (p. 645) that has led them to abandon an account based on stretch reflexes. Such remarks, however, reflect a strong Western bias. For example, Russian studies done in the 1960s reveal similar interactions between posture and voluntary movement (see Gelfand et al. 1971). Moreover, Bernstein (1967) refers to his published experimental work (in Russian) in the early 1920s that affords the conclusion that "movements react to changes in one single detail with a whole series of others which are sometimes very far removed from the former, both in space and in time" (p. 69).

The microscopic workings of a coordinative structure can be further explored by varying the *phase* of the jaw perturbation during bilabial consonant production. For example, recent work has asked: Does perturbing the jaw during the opening phase of the utterances /baeb/ and /baep/ induce a remote reaction in the upper lip? If the cooperativity between oral structures is functionally based, remote effects are predicted only when a jaw perturbation occurs in the closing phase (i.e., during the transition out of the vowel into the final consonant), when the upper lip is actively involved in producing consonantal closure. On the other hand, if the form of interarticulator coupling is in any sense rigid, remote reactions should be seen regardless of when the jaw is perturbed. In fact, the data support the former hypothesis. Remote reactions in the upper lip were observed only when the jaw (Vatikiotis–Bateson and Kelso 1984; Kelso et al. 1984) or lower lip (Munhall and Kelso 1985) was perturbed during the closing phase of motion, that is, when the reactions were necessary to preserve the identity of the spoken utterance.

The phase–specific patterning in speech shares a likeness to recent work in other motor systems. For examples, in cat locomotion (cf. Forssberg 1982, for review), when light touch or weak electrical shock is applied to a cat's paw during the flexion phase of the locomotor cycle, an abrupt withdrawal response occurs as if the cat were trying to lift its leg over an obstacle. When the *same* stimulus is applied during the stance phase of the cycle, the flexion response (which would make the animal fall over) is inhibited, and the cat responds with added extension (Forssberg et al. 1975). This "stumble corrective reaction" is present in intact and spinal animals and, like speech compensation, occurs remarkably quickly. The earliest flexor burst in response to a tactile stimulus applied during the swing phase, for example, occurs with a latency of 10 ms. Just as the foregoing data on articulatory reactions to perturbation appear specific to the spoken utterance, so also do the data on cat locomotion reveal reactions that are nonstereotypic and functionally suited to the phase–dependent requirements of locomotion.

In summary, the evidence presented in this section in support of task–specific action units poses a challenge not only to the neuroscientist, but to anyone who seeks to understand the relation between an organism's structure and its function. The adaptive reactions discussed here could certainly be *described* as reflexive because of their speed. Their mutability, on the other hand, speaks against any hypothesis about fixed reflex connections or rigidly constructed servomechanisms. Similarly, it is not parsimonious to assume that the computation is pre-programmed in such a way that the articulatory ensemble produces precisely those movements that accomplish the task. The problem is exacerbated when unexpected environmental challenges are introduced whose dimensions (e.g., magnitude, duration, locus) are potentially manifold. The main message that emerges is that the multiple components or the motor system are "softly" assembled and flexible in function, not machinelike and rigid – in either the hard–wired language of central pattern generators of the hard–algorithmed language of computers, which are the source of the motor program idea.

THE CONTROL QUESTION

What are the essential control structures that govern the patterning of articulator motion in space and time? Though this question is of much interest to many in the field of motor control in general, here the movements of speech articulators will be the primary focus. However, the kinematic relationships that we shall identify and focus upon are not unique to speech at all, a fact that is quite appealing in that it suggests a common vocabulary might exist to describe the underlying control structure of speech and other actions.

Obviously there are many surface features of a movement that one might propose as significant candidates for controlled variables. What then, fashions the constraints on the choices one makes? Is the selection among controlled variables really like a multiple choice exam (cf. Stein 1982)? Or, might a "deep structure" for motor control exist, that can be recognized in the face of much surface variability? And, if so, on what principle(s) is it based? In the following, the idea is developed that a *dynamic* control regime governs movement patterns.

After Maxwell (1877), dynamics can be viewed as the simplest and most abstract description of the motion of a system. The relations among, and the values of, dynamic parameters (e.g., mass, stiffness, damping) can produce a wide variety of kinematic consequences (e.g., position, velocity). Thus, kinematics provides a surface description of the movements of a system which are generated from a given type of dynamical organization. Note that the dynamics referred to here is not to be interpreted as local and concrete, or to be equated with pure biomechanics. Rather the branch of dynamics emphasized here, nonlinear dynamics, is concerned with the underlying, abstract basis of *forms of motion* or pattern formation in complex, multidegree of freedom systems (e.g., Abraham and Shaw 1982; Haken 1983). These forms of motion are specified, roughly, by the qualitative shapes observed in phase portraits of a system's behavior (see below). For example, the muscles, joints, and neuronal structures that cooperate to produce a walking pattern involve literally thousands of degrees of freedom, but the pattern itself represents a low dimensional form – a cyclical motion of the limbs – which can be operated by low dimensional control (see Garfinkel 1983). In fact, changes in gait in the decerebrate cat can be manipulated experimentally by a single parameter – the intensity of electrical stimulation delivered to the midbrain (Shik et al. 1966).

Such low dimensional forms are called *attractors* and represent the asymptotic stable behavior of a whole family of trajectories. As a simple example, a damped mass spring system can have many trajectories depending on its initial conditions and its parameter values (mass, stiffness, damping). Such a system is called a *point attractor*, a generic dynamical category that reflects the fact that all trajectories converge to an asymptotic, static equilibrium state. Importantly, however, a *multi*degree of freedom system whose trajectories likewise converge to a single rest position can also be described as a point attractor. Thus, a point attractor is a low dimensional description of a potentially high dimensional state space and exhibits the property of equifinality – the tendency to achieve an equilibrium position regardless of initial conditions. Though the language and the concepts of nonlinear dynamics may be unfamiliar (but see Kelso and Kay 1986, for a tutorial), the intent here is to show that one can apply this framework (combined with a quantitative treatment of articulator trajectories) to the analysis of speech production and other biological activities.

The advantages of a dynamical approach to control are several (see Kelso and Kay 1986, for details; also Saltzman, this volume). Among these, hinted at above are: (a) Generativity, an invariant dynamic structure can give rise to much surface kinematic variability; (b) no explicit representation or pointwise control of the system's planned trajectory need exist in a dynamical system; (c) different dynamic regimes (e.g., point attractor, periodic attractor) can serve to categorize different tasks (see Kelso and Tuller 1984b; Saltzman, this volume; Saltzman and Kelso 1986). For example, recent work in the motor control field – especially on voluntary limb and finger movements – indicates that discrete and rhythmic movements can be modeled as a damped mass–spring, point attractor (e.g., Bizzi et al. 1976; Cooke 1980; Fel'dman 1966; Houk 1978; Kelso 1977; Kelso and Holt 1980; Schmidt and McGown 1980) or limit cycle, periodic attractor system, respectively (Fel'dman 1980; Kelso et al. 1981). These control structures are characterized by sets of invariant dynamic parameters (e.g., damping, stiffness, and equilibrium length), and kinematic variations (e.g., position, velocity, acceleration over time) can be viewed as *consequences* of these underlying patterns of dynamic control parameters. A final, related advantage is that the abstract task level of description and the description of muscle–joint properties are entirely commensurate. That is, a dynamical description applies at all levels. The problem becomes one of relating dynamics that operate on different time scales.

In complex movements like speech, however, we seldom make direct measurements of the dynamic parameters themselves, e.g., the mass, damping, and stiffness values for an organization of neuromuscular elements. In our ongoing work, we measure and compute articulatory kinematics during the production of simple syllables and use the relations among these kinematic variables to infer the underlying functionally defined dynamic control regimes. One main paradigm involves reiterant speech, in which subjects are required to substitute a simple syllable (e.g., /bɑ/ or /mɑ/) for the real syllable in an utterance, yet still maintain the utterance's normal prosodic structure. The benefit of the reiterant technique for production studies is that the removal of segmental factors (i.e., the different consonants and vowels of real speech), besides having minimal effects on the timing/metrical pattern, allows one to measure movements of those supralaryngeal articulators that are consistently active over the entire utterance, in this case the lips and jaw involved in /bɑ/ or /mɑ/.

Kelso et al. (1985a) employed a phase plane analysis (a continuous plot of articulator position vs velocity) of lip and jaw movement trajectories followed by a quantitative kinematic analysis of opening and closing gestures. Several interesting kinematic results were obtained (see Fig. 1). First, largely unimodal velocity patterns of jaw and lips occurred for opening and closing gestures at both slow and fast speaking rates (Fig. 1A); second, a given gesture's peak velocity (Vp) covaried with its displacement (d). Regression analysis of the data showed not only a strong relation between Vp and d, but also that the slope of the relation changed depending on the reiterant syllable's stress and rate. As shown in Fig. 1, shorter amplitude motions corresponding to unstressed gestures and faster speaking rates had steeper slopes than stressed gestures spoken at a normal rate.

The impressive scaling relation between Vp and d is not unique to speech where it has been reported before, often as an incidental result (e.g., Kent and Moll 1975; Sussman et al. 1973). An inventory of other activities, ranging from natural reaching movements to tongue movements (see Kelso and Kay 1986, for review; and Viviani, this volume) to infant kicking (Thelen et al. 1985) shows the same relationship. *Thus, this lawful regularity is observed not only in different material structures, but in activities involving multiple degrees of freedom.*

What kind of dynamic control structure could give rise to such kinematic relations? Consider the relationship "ut tensio sic vis; that is the power of any spring is in the same proportion with the tension thereof" (Hooke 1678, cited in Timoschenko 1953). By spring, Hooke meant any "springy body" and by "tension", what we would now call "extension" or more generally, strain. This linear relationship is called Hooke's law ($F = -kx$), where F is the restoring force, k is a proportionality constant representing spring stiffness, and x is displacement. The elementary equation of motion can be derived from Newton's second law, $F = m\ddot{x}$. That is, $F = -kx = m\ddot{x}$; therefore, $m\ddot{x} + kx = 0$, where m is mass and $\ddot{x}$ is acceleration.

This last equation describes the motion of a simple harmonic oscillator with a given mass and stiffness and no damping. On the phase portrait, all concentric trajectories of the oscillator have the same shape with the same periodicity for a given set of dynamic parameters. Note that any changes in initial conditions $(x,\dot{x})$ are precisely accommodated by changes in peak

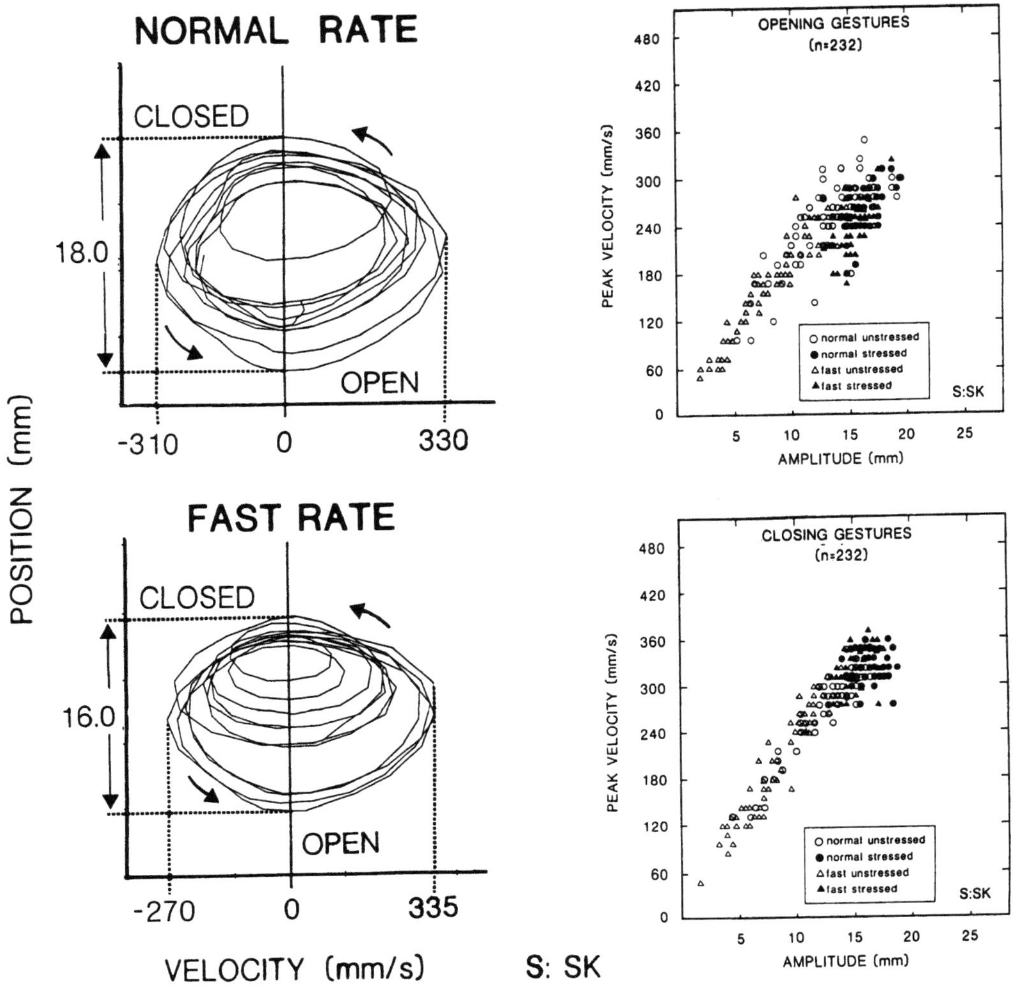

Fig. 1. Left: Phase plane trajectories of lower lip plus jaw (i.e., from a sensor placed on the lower lip) for reiterant speech spoken at a normal (top) and fast (bottom) rate with /ba/ as the reiterant syllable. Right: Scatter plot of peak velocity vs displacement (lower lip plus jaw) of a subject's opening gestures associated with the consonant–vowel portion of the syllable (top) and closing gestures associated with the vowel–consonant portion of the syllable. The legend specifies conditions (from Kelso et al. 1985a)

velocity. Thus, the Vp/d scaling relationship is *specified* by this particular dynamical system, since $\omega_0 A$ is the peak velocity of simple harmonic motion, and the slope of $\omega_0 A$ vs A is ω_0 (where A is cycle amplitude and ω_0 [= $(k/m)^{1/2}$] is the angular frequency of motion). Assuming constant mass, the slope of the Vp/d relationship is proportional to $k^{1/2}$. Changing stiffness changes the eccentricity of the phase plane trajectories – which is what Kelso et al. (1985) observed – and increases the slope of the peak velocity–displacement relation (see also Cooke 1980; Ostry and Munhall 1985). Time or duration per se plays no explicit role here as a controlled variable. Rather, a spatiotemporal pattern arises as a consequence of a dynamic regime in which, at worst, only two system parameters, stiffness and rest length, are specified according to task requirements. Movement certainly *evolves* in time, but time is not directly controlled or metered out by a central executive or timekeeper in this scenario (cf. Schmidt 1980, 1982).

This simple model strongly suggests that the stiffness or elasticity of the system is an important control parameter for skilled actions. In concluding this section, the potential (and more generalized) theoretical significance of this claim is addressed. To do this, we need to develop briefly a perspective based on elasticity theory (see Landau and Lifshitz 1981; Love 1927; Timoschenko 1953).

The most general form of Hooke's law, i.e., beyond a simple force–displacement description, is that over a wide range of applied stresses, the measured strain increases in the same proportion. The proportionality linking stresses to strains is the elastic constant, k. Thus, Hooke's law is fundamentally a description of *elastic deformation processes*. This generalization, though entirely consistent with recent work demonstrating stiffness or impedance control (e.g., Hogan 1985) offers a very different image for movement control. It characterizes movement fundamentally as *form*: solid bodies (limbs, jaws, tongues) can be made to change their size and shape, that is, their *configuration*, by the application of suitable forces (stresses). In this view, any new configuration is expressed by the specification of strains. Note that displacement is only a *measure*, often on a single plane of motion, of strain or deformation. Strains themselves are changes in the *relative* positions (or configuration) of a body. They usually, require a tensorial description (e.g., Love 1927). In Kelso et al. (1985a), changes in movement duration and displacement that occurred when speaking rate and stress changed, were characterized as consequences of the dynamic parameters of stiffness and equilibrium position. This formulation can now be recast into an equivalent, but more conceptually meaningful form, one that affords insights into the regulation of *multiple* muscles during action, not simply an agonist–antagonist pair (e.g., Bizzi et al. 1982; Cooke 1980).

When an effector system, say the jaw–lip complex, moves from one configuration to another, the system in general does some work. A way to envisage the system specification of equilibrium position and stiffness is to express the work done as a potential strain or energy function. The latter specifies the macroscopic relation between stresses and strains. In Fig. 2A, a linear force–displacement relation is mapped onto a strain–energy surface in which the potential energy is a quadratic function of the strain components (in this case simply displacement). The corresponding phase portrait is also shown. For comparison purposes, the case in which stiffness changes nonlinearly as a function of displacement, the so–called "soft" spring (cf. Jordan and Smith 1977; Kelso et al. 1983) is illustrated in Fig. 2B.

It is apparent from Fig. 2 that the amount of potential energy is proportional to displacement (or more generally, configuration) and that the slope of the force–displacement function specifies stiffness. In this view, the system's "endpoints" or "targets" correspond to minima of potential energy functions whose gradients define spring force. As Kugler et al. (1980) emphasize, to produce a movement is to effect a change in the underlying geometry of the dynamics, captured as a potential field. Some years ago this language was not common in the field of motor control. However, it is interesting to quote Greene and Boylls's (1984) assessment of trends in the field –

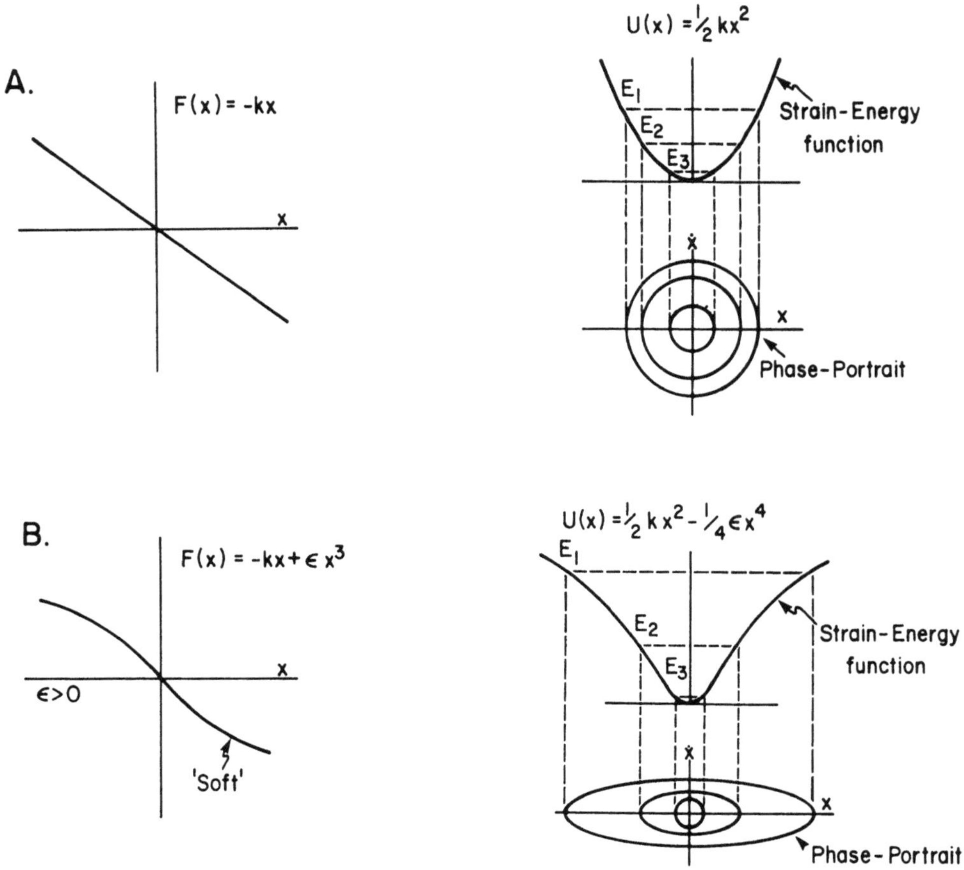

Fig. 2. A. Left: A graph illustrating Hooke's law. The deviation of the force (F) from linearity is symmetrical about the equilibrium position (x = 0). Right: The potential (or strain–) energy function corresponding to the linear force–displacement relation and its associated phase diagram in $x,\dot{x}$ coordinates. Three phase paths are drawn, corresponding to the three values of total energy, E, indicated by dotted lines in the potential function. B. The form of the above relationships when the force is less than the linear term alone and the system is said to possess a "soft" nonlinearity

post–Bernstein – "that bear watching. ... [It] seems likely that the theory of impedance or endpoint control will soon be recast in terms of potential functions (with endpoints as extrema of such functions to be 'sought', gradient fashion, by the state of the skeletomuscular system)" (p. xxiii). See Kugler et al. (1980, pp. 34–40) and Hogan (1981) for applications to robotic motion. Recently, Hogan (1984) has elaborated this framework for the trajectories of multijoint movements. Successive target locations are specified by means of a time–varying potential field with stable equilibria at the "target" locations.

There are two main points that arise from the perspective advanced here. First, because we are dealing with potential energy functions, only scalar quantities are involved. Several advantages for control accrue immediately. Since energy is a scalar quantity, unlike force (which is a vector), energy is invariant under coordinate transformations. Thus, the coordinate system can be chosen to simplify the problem (see Marion 1970). Also, it is often impossible to define exactly what the forces are (e.g., in a multimuscle system), whereas it is often possible to express the kinetic and potential energies. The latter are intrinsic to the system under study, whereas the standard force description places its emphasis on an outside agency acting *on* a body. Relatedly, because scalar potentials may be superimposed, the overall effect of multiple muscle activity can be obtained by addition of the potential functions (Marion 1970). This characterization may offer considerable advantages for a compact description of control in multidegree of freedom movements.

But the second main feature of the present perspective is that the potential or strain–energy function (Love 1927) can be properly conceived as an *elastic field.* (As an aside, Asada (1982) has recently demonstrated how elastic fields can be used for planning stable grasp in a robot manipulator.) The notion that movement involves deformation of an elastic field may ground one of Bernstein's most interesting intuitions, namely that movement is a morphological object. The ubiquity of the peak velocity–displacement relation, then, may offer a window into processes that form and deform the configuration of the body. If correct, an ancient theme for "static" forms – that a few simple rules can fashion come very intricate products (e.g., Gould 1980; Stevens 1974) – may apply equally well to the forms of patterned motion that interest us here.

THE STABILITY QUESTION

It is now well established that as an articulator ensemble performs its task at different speeds and forces, the relative timing of muscle contractions and/or articulator motions is preserved invariantly (e.g., Boylls 1975; Kelso et al. 1979; Kugler et al. 1980; Schmidt 1982; Shapiro 1978; Shapiro et al. 1981, for reviews). Such results rest on the assumption that the central program determines when pulses of muscular force are applied to the limbs and their durations and relative sizes. Thus, according to Schmidt (1980) the determination of *time* (emphasis his) of contractions and relaxations appears to be directly controlled. Many view an identified invariant as indicating a relevant control parameter. Our position has been exactly the opposite (e.g., Fowler and Turvey 1978; Kelso et al. 1979) – namely, that invariance represents a system constraint, a "freezing" of degrees of freedom, that is, an invariance tells the investigator what does *not* have to be directly controlled.

In the previous sections, it was argued that though muscles contract and relax and though movements flow in time, a movement's temporal structure may be a consequence of the system's dynamic parameterization. Here I want to show how stable *relative timing* among gestures may be understood without recourse to an extrinsically imposed timing program (see Kelso and Tuller 1985a, 1985b). But if timing is not controlled extrinsically in such a fashion, what processes might underlie the observed temporal stability? How in a complex system of articulators, does a given gesture/articulator "know" when it should be activated in relation to other gestures/articulators? With respect to our relative timing data in speech, for example, what information is needed for the upper lip (a remote, nonmechanically linked articulator associated with a consonantal gesture) to move in appropriate temporal relation to the vocalic movement cycle of the jaw? As we shall see, different views of relative timing emerge when the articulator motions are examined in different coordinative spaces.

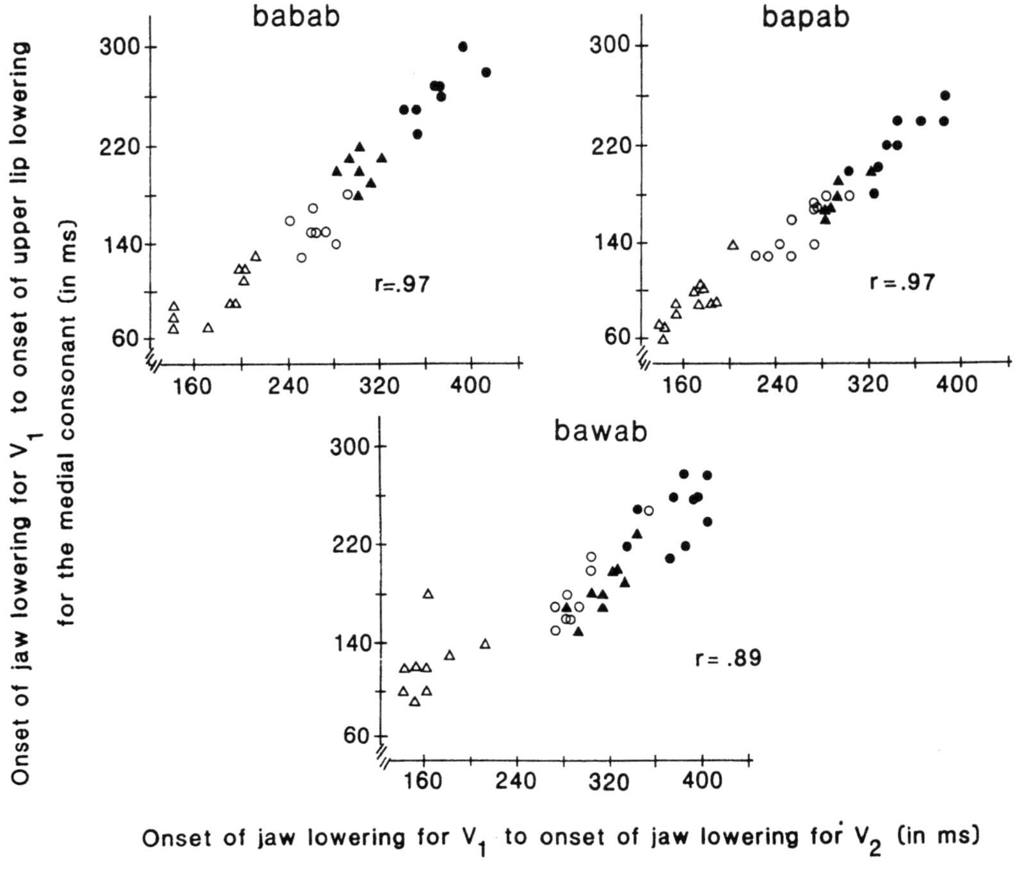

Fig. 3. Timing of upper lip lowering for medial consonant articulation as a function of vowel-to-vowel period for one subject's production of the indicated utterances. Each point represents a single token of the utterance: (•), primary stress on the first syllable spoken at a conversational rate; (○), primary stress on the second syllable (conversational rate); (filled triangle) and (open triangle), primary stress on the first and second syllables, respectively, spoken at a faster rate (from Tuller and Kelso 1984)

Consider first a very simple, but paradigmatic case in which the delay (in ms) of onset of upper lip motion for a medial consonant is measured relative to the interval (in ms) between onsets of jaw motion for flanking vowels. Figure 3, taken from Tuller and Kelso (1984), plots these events for one of four speakers who produced the utterances /babab/, /bapab/, and /bawab/, at two speaking rates and with emphatic stress placed on either the first or second syllable. The data for all four subjects were very similar. This figure shows that over changes in speaking rate and stress, the measured intervals change considerably, as do the magnitude of the events themselves, but the function relating these events is linear. That is, the metrics (amplitude, velocity, duration) of the events change, but the relative timing does not. Note that this is a strictly temporal description relating discrete movement events. Like most, if not all of the work on relative timing, measurements are confined to the onsets and offsets of articulator movement (see, e.g., Schmidt 1982).

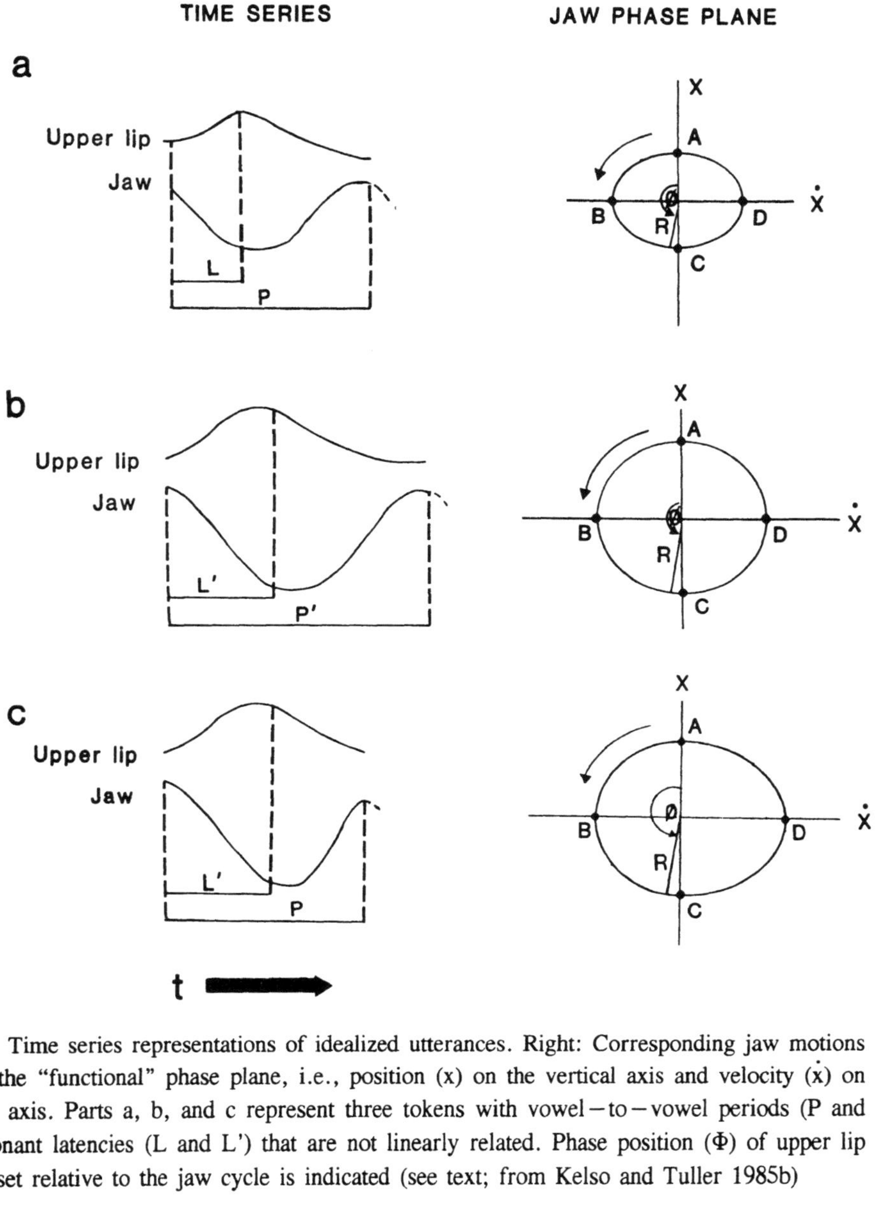

Fig. 4. Left: Time series representations of idealized utterances. Right: Corresponding jaw motions displayed on the "functional" phase plane, i.e., position (x) on the vertical axis and velocity ($\dot{x}$) on the horizontal axis. Parts a, b, and c represent three tokens with vowel–to–vowel periods (P and P') and consonant latencies (L and L') that are not linearly related. Phase position (Φ) of upper lip movement onset relative to the jaw cycle is indicated (see text; from Kelso and Tuller 1985b)

A very different view of articulatory "timing" emerges when a reanalysis of the movements using phase plane trajectories is employed (Kelso and Tuller 1985a). Figure 4 illustrates the mapping from time domain to phase plane trajectories. On the left, hypothetical jaw and upper lip motions (position as a function of time) are shown for an unstressed /bab/ (top left) and a stressed /bab/ (bottom left). On the right are shown the corresponding idealized phase plane trajectories. In this figure we have reversed the typical orientation of the phase plane so that displacement is shown on the vertical axis and velocity on the horizontal axis. Thus, downward movements of the jaw are displayed as downward movements of the phase path. The vertical crosshair indicates all points of zero velocity and the horizontal crosshair indicates zero position (midway between minimum and maximum displacement). As the jaw moves from its highest to its lowest point (from A to C) velocity increases to a local maximum (B), then decreases to zero when the jaw changes direction of movement (C). Similarly, as the jaw is raised from the low vowel /ɑ/ into the following consonant constriction, velocity peaks approximately midway through the gesture (D) then returns to zero (A). It is useful to transform the Cartesian position–velocity coordinates into equivalent polar coordinates, namely, a phase angle, $\phi = \tan^{-1} [\dot{x}/x]$ and a radial amplitude, $R = [x^2 + \dot{x}^2]^{1/2}$. The phase angle is a key concept in the reanalysis of interarticulator timing because it signifies position on a cycle of states.

Notice in Fig. 4 that the phase plane trajectory preserves some important differences between stressed and unstressed syllables. For example, maximum displacement of the jaw for the unstressed vowel is less than displacement for the stressed vowel and maximum articulator velocity differs noticeably between these two orbits. In contrast, note that the different *durations* taken to traverse the orbit as a function of stress are not represented in this description. Time, although implicit and recoverable from the phase plane description, does not appear explicitly. Jaw cycles of different durations are characterized as single orbits on the plane and they are topologically equivalent.

Now one can pose the question of how the upper lip "knows" when to begin its movement for the medial consonant by asking *where* on the cycle of jaw phase angles the lip motion for medial consonant production begins. One possibility is that upper lip motion begins at the same phase angle of the jaw across different jaw motion trajectories (that is, across rate and stress). In other words, the information for timing of a remote articulator, such as the upper lip, would not be time itself, nor absolute position of another articulator (e.g., the jaw), but rather a relationship defined over the position–velocity *state* (or, in polar coordinates, the phase angle) of the other articulator. It is also important to recognize that the motion need not be perfectly sinusoidal in order to apply a phase angle analysis. In fact, the jaw motions actually observed are usually not sinusoidal; the displacements at zero velocity are affected by the stress and rate characteristics of the surrounding vowels. For this reason, we normalize each jaw cycle's amplitude and peak velocity to unity.

When the original Tuller and Kelso (1984) data were reanalyzed in this fashion, the result was that phase angle was indeed constant across both rate and stress variations. The complete statistical analysis is presented in Kelso et al. (1986). The mean phase position of the upper lip relative to the jaw was found to be constant and the standard error of the mean tiny. It should be emphasized that a critical phase angle description in no way entails, or is predicted by, the relative timing results. Instead, it constitutes an alternative description of the data set. For example, two utterances that have *identical* vowel–to–vowel periods and consonant latencies can nonetheless show very different phase positions for upper lip movement onset relative to the cycle of jaw states. Specifically, the phase angle analysis incorporates the full space–time trajectory of motion; the relative timing analysis ignores trajectory, once movement has begun.

There are at least two empirical advantages of this result over others' and our relative timing description. First, in the relative timing analysis, the overall correlations across rate and stress conditions are very high, but the within–condition slopes tend to vary somewhat. In the phase analysis, on the other hand, the mean phase angle is the same across conditions. Second, although the relative timing scenario is described by two parameters, a slope and an intercept (Fig. 3), the phase description requires only a single parameter (phase angle). Thus, if nothing else, the phase description is more parsimonious.

The phase angle conceptualization also has a number of theoretical advantages over our original relative timing analysis. First, once articulatory motions are represented geometrically on the phase plane, duration is normalized across stress and speaking rate. Strictly speaking, the system's *topology* is unaffected by durational changes. Second, neither absolute nor relative durations have to be extrinsically monitored or controlled in this formulation. There is no need to posit a *timing* program. This fact potentially provides a grounding for, and a principled analysis of so-called intrinsic timing theories of speech production (e.g., Fowler et al. 1980; see also Kelso and Tuller 1985a). The present view is bolstered indirectly by demonstrations in the articulatory structures themselves of afferent bases for phase angle information (e.g., position and velocity sensitivities of muscle spindle and joint structures), but not for time-keeping information (e.g., time receptors; cf. Kelso 1978). It might well be the case that certain critical phase angles provide information for orchestrating the temporal flow of activity among articulators (beyond those considered here) and/or vocal tract configurations. (Note that the statement that lip motion starts at a particular phase angle of the jaw is equivalent to saying that it occurs when the rate of dilation of the jaw (i.e., $\dot{x}/x$) reaches a particular value. Lee's work (see this volume for references) similarly shows how the inverse of the rate of dilation of the optic image of a surface specifies time-to-contact τ (t) with that surface. The critical phase angle may be the *proprioceptive flow field* analogue of Lee's optic flow field variable so crucial to the visual guidance of action.) Such phase angles would serve as natural, i.e., dynamically specified, information sources for guaranteeing the stability of coordination in the face of scalar (metrical) changes. As in a candle (which provides a metric for time by a change in its *length*) or a water clock (where the metric is number of drops), the units of time for speech production might be defined entirely in terms of the *state* variables of the system. Thus, according to the present analysis, it is gestural phase angle (a space-time description) not gestural time (a purely temporal description) that captures the stable cooperative relations among articulators. This essential parameter, phase angle, will take on added significance in the final section.

THE CHANGE QUESTION

The previous sections focused on the stability characteristics within and among coordinative structures – a description of what remains stable in articulatory ensembles as the metrics of the activity are systematically scaled. Here the other side of the coin is addressed: How do new (or different) forms of spatiotemporal behavior come about? Although the invariance aspect of coordinative structures has been emphasized, it is nevertheless clear that such organizations are not strictly invariant, but change over time according to different time scales. Over the relatively long time span of early childhood, skills are acquired: the forms of motion that emerge must, in turn, be adapted to slow changes in body morphology that accompany growth. As adults, we can learn new skills (within limits), such as tennis and juggling, given sufficient practice. And finally, certain activities involve coordinative structures whose forms change swiftly and dramatically within the performance of a particular skill, as with gait transitions in locomotion (e.g., Hoyt and Taylor 1981). It is this faster kind of change that I want to address here. The reasons are as follows. First, one can design experiments to examine the necessary and sufficient conditions that may underlie such rapid changes in organization (see below). In contrast, slower kinds of change that occur with learning and development often require longitudinal studies and intervening variables can play a significant role. Second, it seems possible that fast and slow changes in perception – action systems follow similar kinds of principles, except that the time scales are very different. Just as evolution may occur in qualitative "jumps" (see, Eldredge and Gould 1972), so also may skill learning and development. What constitutes a jump at one time scale, however, can be nearly continuous or quasistatic in another. Nevertheless, principles of change in spatiotemporal behavior may provide a test field for comparing the motor program/central pattern generator account of motor control with the "movement as cooperative phenomenon" approach promoted here. A fundamental prediction of this approach is that movement patterns, like other cooperative phenomena (see, e.g., Haken 1975; Prigogine 1980) exhibit qualitatively new modes of organization when certain parameters are scaled past critical bounds. Unlike the motor program construct, however, no *a priori* prescription exists before the new mode of organization appears (see, Kelso 1981a; Kugler et al. 1980).

The concept of *mode* is quite crucial here. Modes are macroscopic descriptors for collective behavior in systems with many degrees of freedom. Modal descriptions are distinct from those at a microscopic level. For example, an oscillating string made up of 10^{22} atoms is described by "macro" quantities like wavelength and amplitude that are entirely different from the atomistic description (see, Haken 1977). Analogously, in certain biological activities the relative phase among movement components serves as a macroscopic description of the spatiotemporal order, say, among the limbs during the act of locomoting, or the articulators during speech. Thus, particular phasings among the legs of a quadruped correspond to particular modes or locomotory gaits. A microscopic description, on the other hand, requires minimally an identification of the ensemble's neuromuscular elements, their membrane and synaptic properties, and all the connections among them. Though it is commonplace for the neuroscientist to talk of neural circuits controlling behavior, it has proved difficult – even in the simplest neural networks – to relate specific patterns of electrical activity to behavioral action. Indeed, if one were to experimentally manipulate the parameters of a central pattern generator, one would be confronted (by some limited estimates) with a space that contains 46 parameters (Bullock 1976). Clearly some other principles – continuous perhaps with the treatment of cooperative phenomena in other natural systems – are needed to guide the selection of relevant parameters.

As Haken et al. (1985) note, this problem of relating neuronal events to global behavioral patterns – say, abrupt changes and other characteristic indices of a movement – is reminiscent of problems faced by physicists 50 years ago (and in many cases today as well). Even though the microscopic properties of atoms were thought to be theoretically understood, it still proved difficult to derive the system's macroscopic behavior from its microscopic features. In the field of synergetics, for example, which deals with the formation of order in open, nonequilibrium systems (e.g., Haken 1975) it has been shown that the behavior of complex systems can be successfully modeled by means of a few macroscopic quantities, – called *order parameters*, – in those situations where the system's behavior changes qualitatively.

Elsewhere, we have presented numerous examples, – drawn largely from Haken and Prigogine's work, – of dissipative or synergetic structures in physics, chemistry, and biology (Kelso and Tuller 1984a; see also, Kelso et al. 1980 and Kugler et al. 1980, for empirical and theoretical treatment of such structures in the realm of action systems). The mechanisms common to all these systems is that the values of one or more order parameters become unstable and undergo sudden discontinuous changes when control parameters are scaled (usually under experimental manipulation). The observed bifurcation results from the competition, as is were, between the "forces" or inputs that are systematically scaled (e.g., by increasing the velocity of a treadmill and forcing an animal to move faster), and the "forces" holding the system together (e.g., the order parameter describing, say, a synergistic modal pattern or locomotory gait). Thus, under the influences of continuous scaling, a given mode may suddenly become dominant, and capture or *slave* (in Haken's terms) the other modes. The significant, and *universal feature of such critical behavior is that around transition regions, where stability is lost, the behavior of the system is governed by the order parameters alone*. This implies a tremendous reduction in the degrees of freedom since the behavior of all the subsystems is now governed by a single order parameter.

These kinds of sharp, discontinuous behaviors are omnipresent in the action system when system–sensitive parameters are appropriately scaled, e.g., in voluntary limb movements (Kelso 1981b, 1984), speech (Kelso and Tuller 1984a), locomotion (e.g., Hoyt and Taylor 1981; Kugler et al. 1980) and posture (e.g., Nashner and McCollum 1985; Saltzman and Kelso 1985). For example, in recent work on bimanual activities, Kelso (1981b, 1984) had subjects move their right and left hands together at a comfortable rate in both an out–of–phase (180° phase difference) and in–phase (0° phase difference) modal pattern, and either with or without an added frictional resistance. The preferred frequencies and amplitudes of each hand were measured under the two resistance conditions. Subjects then attempted to perform the out–of–phase rhythmic movement at steadily increasing frequencies. Of special interest was the critical frequency at which the out–of–phase movements could no longer be sustained, and the rhythmic organization abruptly became in–phase. Though this critical phase transition frequency was different for subjects, when expressed in units of each subject's preferred frequency, the same dimensionless number was obtained. As in many physical and biological systems new "modes" or spatiotemporal orderings were observed when the system was scaled beyond equilibrium. Continuous scaling on frequency in Kelso's experiments resulted in the initial out–of–phase modal pattern (or phase relation) becoming unstable, until, at a critical point, bifurcation occurred and a different modal pattern appeared. Though not given a bifurcation interpretation, similar results have been obtained by Cohen (1971), Mackenzie and Patla (1983), and Baldissera et al. (1982).

Recently, Haken et al. (1985) have modeled these bimanual phase transitions, using some of the central concepts and mathematical tools of synergetics and nonlinear oscillator theory. Using *relative phase* as an order parameter, they first specified a potential function corresponding to the layout of modal attractor states (i.e., the stable in–phase and out–of–phase patterns), and showed how that layout was altered as a control parameter (driving frequency) was scaled. (Note: There are several criteria for the identification of an order parameter. A main one is that the order parameter changes much more slowly than the subsystems it is said to govern. Relative phase fits this criterion well. Remember, it is the phasing structure of many different activities that remains invariant across scalar transformations. Thus, in the bimanual experiments, relative phase changes much more slowly than the kinematic variables describing the motion of each hand.) From the behavior of the potential function Haken et al. then derived the equations of motion for each hand, and the nonlinear coupling between the hands. Analytic derivations and consequent numerical simulation revealed that if the system was "prepared" in the out-of-phase mode (i.e., by instruction to the subject), and driving frequency was increased slowly, the oscillation remained in that mode until the solution of the coupled equation of motion became unstable. At this point, a jump occurred and the only stable stationary solution produced by the system corresponded to the in-phase mode (see Haken et al. 1985, for more details). Ongoing empirical and theoretical work (Kelso and Scholz 1985; Schöner et al. 1986) has revealed that the nonlinear coupling strength as well as fluctuations (both intrinsically generated due to noise in system parameters and extrinsically generated due to an added random forcing function) play an important role in effecting the modal transitions between the hands.

Although it is tempting to ascribe transitions in phasing among the limbs to "switches" or (in the case of gait) a "gait selection process" (Gallistel 1980), such an account possesses a Kiplingesque "just so" quality. To assign a phenomenon, switching (an abrupt shift in spatiotemporal order) to a device or a mechanism that is said to perform the duty of explaining the phenomenon, is a questionable strategy at best. Yet modal shifts in coordination are often "explained" in this fashion, e.g., by motor programs (cf. Schmidt 1982, p. 316). The synergetic framework offered here asks instead: What are the necessary and sufficient conditions giving rise to order in biological activities? It is antithetical to views that try to account for complex behaviors by devices that embody (or represent) these behaviors. A principled account of new spatiotemporal patterns should not rest on the introduction of special mechanisms, even when such "mechanisms" are borrowed from current computer technology.

EPILOGUE (After Kelso and Tuller 1984a)

Unlike machines that are designed by people to exhibit special structures and functions, the structures and functions discussed here develop in a self-organized fashion.* Often a new mode emerges when a random event occurs in an unstable region of the system's parameter space and the fluctuation becomes amplified. Such is the case, one suspects, in the gait of a quadruped or in the bimanual experiments described here (see, Kelso and Scholz 1985, for a more complete treatment of critical fluctuations in the bimanual case). Near the unstable region, where it is energetically expensive to maintain a given mode, a small change in speed produces dramatic effects – a new mode arises. Literally, a phase transition occurs.

Throughout the present paper the emphasis has been on similarities, in terms of dynamic behavior, exhibited by articulatory systems which vary widely in their material composition. Common to all of them is their intrinsically nonlinear and dissipative nature, and the fact that they possess many degrees of freedom. These are features that the perception-action system shares with many other

* In fact, neuroscience is beginning to talk this way. A recent report has described systematic changes in topographic maps of sensorimotor cortex that occur due to finger ablation and cortical tissue removal, as evidence that the brain "has embedded processes ... that make it self-organizing ... ." And that "the dominant view of the nervous system [as] a machine with static properties ... [is] incorrect" (Fox 1984, quoting Merzenich and colleagues' work). Times, it seems, are achanging.

natural systems. The focus in this paper has been on the discovery and elaboration of principles that embrace cooperative phenomena, regardless of any particular structural embodiment. From such principles it may be possible to generate an account of the emergence and stability of movement patterns without hermeneutic devices that prescribe such patterns.

REFERENCES

Abbs JH, Gracco VL (1983) Sensorimotor actions in the control of multimovement speech gestures. Trends Neurosci 6: 393–395

Abbs JH, Gracco VL, Cole KJ (1984) Control of multimovement coordination: Sensorimotor mechanisms in speech motor programming. J Mot Behav 16: 195–232

Abraham RH, Shaw CD (1982) Dynamics – the geometry of behavior. Aerial, Santa Cruz

Asada H (1982) A geometrical representation of manipulator dynamics and its application to arm design. In: Book WJ (ed) Robotics research and advanced applications. American Society of Mechanical Engineers, New York, pp 1–8

Baldissera F, Cavallari P, Civaschi P (1982) Preferential coupling between voluntary movements of ipsilateral limbs. Neurosci Lett 34: 95–100

Bernstein NA (1967) The coordination and regulation of movements. Pergamon, London

Bizzi E, Polit A, Morasso P (1976) Mechanisms underlying achievement of final head position. J Neurophysiol 39: 435–444

Bizzi E, Accornero N, Chapple W, Hogan N (1982) Arm trajectory formation in monkeys. Exp Brain Res 46: 139–143

Boylls CC (1975) A theory of cerebellar function with applications to locomotion. II. The relation of anterior lobe climbing fiber function to locomotor behavior in the cat. COINS technical report 76–1. University of Massachusetts, Department of Computer and Information Science, Amherst

Bullock TH (1976) In search of principles of neural integration. In: Fentress JD (ed) Simple networks and behavior. Sinauer, Sunderland, pp 52–60

Cohen L (1971) Synchronous bimanual movements performed by homologous and non–homologous muscles. Percep Mot Skills 32: 639–644

Cooke JD (1980) The organization of simple, skilled movements. In: Stelmach GE, Requin J (eds) Tutorials in motor behavior. North Holland, Amsterdam

Cordo PJ, Nashner LM (1982) Properties of postural adjustments associated with rapid arm movements. J Neurophysiol 47: 287–302

Easton TA (1972) On the normal use of reflexes. Am Sci 60: 591 – 599

Eldredge N, Gould SJ (1972) Punctuated equilibria: an alternative to physical gradualism. In: Schopf TJM (ed) Models in paleobiology. Freeman, San Francisco, pp 82 – 115

Fel'dman AG (1966) Functional tuning of the nervous system with control of movement or maintenance of a steady posture. III. Mechanographic analysis of execution by man of the simplest motor tasks. Biophysics 11: 766 – 775

Fel'dman AG (1980) Superposition of motor programs. I. Rhythmic forearm movements in man. Neurosci 5: 81 – 90

Folkins JW, Abbs JH (1975) Lip and jaw motor control during speech: responses to resistive loading of the jaw. J Speech Hear Res 18: 207 – 220

Folkins JW, Zimmermann GN (1982) Lip and jaw interaction during speech: responses to perturbation of lower – lip movement prior to bilabial closure. J Acoust Soc Am 71: 1225 – 1233

Forssberg H (1982) Spinal locomotion function and descending control. In: Sjölund B, Björkland A (eds) Brainstem control of spinal mechanisms. Ferström, New York

Forssberg H, Grillner S, Rossignol S (1975) Phase dependent reflex reversal during walking in chronic spinal cats. Brain Res 55: 247 – 304

Fowler CA, Turvey MT (1978) Skill acquisition: an event approach with special reference to searching for the optimum of a function of several variables. In: Stelmach G (ed) Information processing in motor control and learning. Academic, New York

Fowler CA, Rubin P, Remez RE, Turvey MT (1980) Implications for speech production of a general theory of action. In: Butterworth B (ed) Language production. Academic, New York

Fox JL (1984) The brain's dynamic way of keeping in touch. Science 225: 820 – 821

Gallistel CR (1980) The organization of action: a new synthesis. Erlbaum, New York

Garfinkel A (1983) A mathematics for physiology. Am J Physiol 245: R455 – R466

Gelfand IM, Gurfinkel VS, Fomin SV, Tsetlin ML (eds) (1971) Models of the structural – functional organization of certain biological systems. MIT, Cambridge

Gould SJ (1980) The evolutionary biology of constraint. Daedalus 109: 39 – 52

Greene PH, Boylls CC Jr (1984) Introduction: Bernstein's significance today. In: Whiting HTA (ed) Human motor actions: Bernstein reassessed. North – Holland, Amsterdam

Haken H (1975) Cooperative phenomena in systems far from thermal equilibrium and in nonphysical systems. Rev Mod Phys 47: 67 – 121

Haken H (1977) Synergetics: an introduction. Nonequilibrium phase transitions and self-organization in physics, chemistry, and biology. Springer, Berlin, Heidelberg, New York

Haken H (1983) Advanced synergetics: instability hierarchies of self-organizing systems and devices. Springer, Berlin, Heidelberg, New York

Haken H, Kelso JAS, Bunz H (1985) A theoretical model of phase transitions in human hand movements. Biol Cybern 51: 347-356

Hogan N (1981) Impedance control of a robotic manipulator. In: Winter annual meeting. American Society of Mechanical Engineers, Washington DC

Hogan N (1985) Impedance control: an approach to manipulation. Part II: implementation. J Dynamic Syst Meas Control 107: 1-24

Houk JD (1978) Participation of reflex mechanisms and reaction time processes in compensatory adjustments to mechanical disturbances. In: Desmedt J (ed) Cerebral motor control in man: long loop mechanisms. Karger, Basel

Houk JC, Rymer W (1981) Neural control of muscle length and tension. In: Brooks VB (ed) Handbook of physiology, section 1, vol. II, motor control, part 1. American Physiology Society, Bethesda, pp 257-323

Hoyt DF, Taylor CR (1981) Gait and the energetics of locomotion in horses. Nature 292: 239-240

Jordan DW, Smith P (1977) Nonlinear ordinary differential equations. Clarendon, Oxford

Kelso JAS (1977) Motor control mechanisms underlying human movement reproduction. J Exp Psychol: Hum Percept Perform 3: 529-543

Kelso JAS (1978) Joint receptors do not provide a satisfactory basis for motor timing and positioning. Psychol Rev 85: 474-481

Kelso JAS (1981a) Contrasting perspectives on order and regulation in movement. In: Long J, Baddeley A (eds) Attention and performance (IX). Erlbaum, Hillsdale

Kelso JAS (1981b) On the oscillatory basis of movement. Bull Psychonom Soc 18: 63

Kelso JAS (1984) Phase transitions and critical behavior in human bimanual coordination. Am J Physiol 246: R1000-R1004

Kelso JAS, Holt KG (1980) Exploring a vibratory systems analysis of human movement production. J Neurophys 43: 1183-1196

Kelso JAS, Holt KG, Kugler PN, Turvey MT (1980) On the concept of coordinative structures as dissipative structures: II. Empirical lines of convergence. In: Stelmach G, Requin J (eds) Tutorials in motor behavior. North-Holland, Amsterdam, pp 49-70

Kelso JAS, Holt KG, Rubin P, Kugler PN (1981) Patterns of human interlimb coordination emerge from the properties of nonlinear limit cycle oscillatory processes: Theory and data. J Motor Behav 13: 226–261

Kelso JAS, Kay BA (1986) Information and control: A macroscopic analysis of perception–action coupling. In: Heuer H, Sanders AF (eds) Perspectives on perception and action. Erlbaum, Hillsdale

Kelso JAS, Putnam CA, Goodman D (1983) On the space–time structure of human interlimb coordination. Q J Exp Psychol 35A: 347–376

Kelso JAS, Saltzman EL, Tuller B (1986) The dynamical perspective on speech production: data and theory. J Phonet

Kelso JAS, Scholz JP (1985) Cooperative phenomena in biological motion. In: Haken H (ed) Complex systems: operational approaches in neurobiology, physics, and computers. Springer, Berlin, Heidelberg, New York, Tokyo

Kelso JAS, Southard DL, Goodman D (1979) On the nature of human interlimb coordination. Science 203: 1029–1031

Kelso JAS, Tuller B (1984a) A dynamical basis for action systems. In: Gazzaniga MS (ed) Handbook of cognitive neuroscience. Plenum, New York, pp 321–356

Kelso JAS, Tuller B (1984b) Converging sources of evidence for common dynamical principles in speech and limb coordination. Am J Physiol 15: R928–R935

Kelso JAS, Tuller B (1985a) Intrinsic time in speech production. J Acoust Soc Am [Suppl 1] 77: S53

Kelso JAS, Tuller B (1985b) Intrinsic time in speech production: theory, methodology, and preliminary observations. In: Keller E, Gopnik M (eds) Motor and sensory processes of language. Erlbaum, Hillsdale

Kelso JAS, Tuller B, Fowler CA (1982) The functional specificity of articulatory control and coordination. J Acoust Soc Am 72: S103

Kelso JAS, Tuller B, Vatikiotis–Bateson E, Fowler CA (1984) Functionally specific articulatory cooperation following jaw perturbations during speech: evidence for coordinative structures. J Exp Psychol: Hum Percept Perform 10: 812–832

Kelso JAS, Vatikiotis–Bateson E, Saltzman EL, Kay BA (1985a) A qualitative dynamic analysis of reiterant speech production: phase portraits, kinematics, and dynamic modeling. J Acoust Soc Am 77: 266–280

Kent RD, Moll K (1975) Articulatory timing in selected consonant sequences. Brain Lang 2: 304–323

Kugler PN, Kelso JAS, Turvey MT (1980) On the concept of coordinative structures as dissipative structures: I. Theoretical lines of convergence. In: Stelmach GE, Requin J (eds) Tutorials in motor behavior. North-Holland, Amsterdam, pp 3-47

Kugler PN, Kelso JAS, Turvey MT (1982) On the control and coordination of naturally developing systems. In: Kelso JAS, Clark JE (eds) The development of movement control and coordination. Wiley, Chichester, pp 5-78

Landau LD, Lifshitz EM (1981) Theory of elasticity. Pergamon, Oxford

Love AEH (1927) A treatise on the mathematical theory of elasticity. Dover, New York (1944)

Mackenzie CL, Patla AE (1983) Breakdown in bimanual finger tapping as a function of orientation and phasing. Soc Neurosci (abstract)

Marion JB (1970) Classical dynamics of particles and systems. Academic, New York

Marsden CD, Merton PA, Morton HB (1983) Rapid postural reactions to mechanical displacement of the hand in man. In: Desmedt JE (ed) Motor control mechanisms in health and disease. Raven, New York, pp 645-659

Maxwell JC (1877) Matter and motion. Dover, New York

Munhall K, Kelso JAS (1985) Phase-dependent sensitivity to perturbation reveals the nature of speech coordinative structures. J Acoust Soc Am 78: S38

Nashner LM, McCollum G (1985) The organization of human postural movements: a formal basis and experimental synthesis. Behav Brain Sci 8

Ostry DJ, Munhall K (1985) Control of rate and duration in speech. J Acoust Soc Am 77: 640-648

Prigogine I (1980) From being to becoming: time and complexity in the physical sciences. Freeman, San Francisco

Saltzman E, Kelso JAS (1985) Synergies: stabilities, instabilities, and modes. Behav Brain Sci 8: 161-163

Saltzman E, Kelso JAS (1986) Skilled actions: a task dynamic approach. Psychol Rev

Schmidt RA (1980) On the theoretical status of time in motor program representations. In: Stelmach GE, Requin J (eds) Tutorials in motor behavior. North-Holland, Amsterdam

Schmidt RA (1982) Motor control and learning: a behavioral emphasis. Human Kinetics, Champaign

Schmidt RA, McGown C (1980) Terminal accuracy of unexpectedly loaded rapid movements: evidence for a mass–spring mechanism in programming. J Mot Behav 12: 149–161

Schöner G, Haken H, Kelso JAS (1986) A stochastic theory of phase transitions in human hand movement. Biol Cybernet 53: 442–452

Shapiro DC (1978) The learning of generalized motor programs. PhD dissertation, University of Southern California

Shapiro DC, Zernicke RF, Gregor RJ, Diestal JD (1981) Evidence for generalized motor programs using gait–pattern analysis. J Mot Behav 13: 33–47

Sherrington CS (1906) The integrative action of the nervous system. Constable, London

Shik ML, Severin FV, Orlovskii GN (1966) Control of walking and running by means of electrical stimulation of the mid–brain. Biofizyka 11: 659–666

Stein RB (1982) What muscle variables does the central nervous system control? Behav Brain Sci 5: 535–577

Stevens PS (1974) Patterns in nature. Little Brown, Boston

Sussman HM, MacNeilage PF, Hanson RJ (1973) Labial and mandibular dynamics during the production of bilabial consonants: preliminary observations. J Speech Hear Res 16: 397–420

Thelen E, Skala KD, Kelso JAS (1985) Spontaneous kicking in very young infants: evidence for a dynamic bilateral system. Hask Lab Stat Rep Speech Res SR–81: 305–313, Developmental Psychology (in press)

Timoschenko SP (1953) History of strength of materials. Dover, New York (1983)

Tuller B, Kelso JAS (1984) The timing of articulatory gestures: evidence for relational invariance. J Acoust Soc Am 76: 1030–1036

Turvey MT (1977) Preliminaries to a theory of action with reference to vision. In: Shaw R, Bransford J (eds) Perceiving, acting and knowing: toward an ecological psychology. Erlbaum, Hillsdale

Vatikiotis–Bateson E, Kelso JAS (1984) Remote and autogenic articulatory adaptation to jaw perturbations during speech: more on functional synergies. J Acoust Soc Am 76: S23–S24

Yates FE (1979) Physical biology: a basis for modeling living systems. J Cybern Infor Sci 2: 57–70

Task Dynamic Coordination of the Speech Articulators: A Preliminary Model*

E. Saltzman

Haskins Laboratories, 270 Crown Street, New Haven, CT 06510, USA

INTRODUCTION

It is perhaps a truism that skilled actions of the limbs and speech articulators are goal directed. It is equally true, however, that such actions are performed by effector systems that are indifferent to the goals of would-be performers. An effector system is the set of limb segments or speech articulators used in a given action; a terminal device or end-effector is the part or a controlled effector system that is directly related to the goal of a performed action. Thus, in a reaching task, the fingers define the terminal device and the arm and hand comprise the effector system; in a "cup-to-mouth" task, the grasped cup is the terminal device and the combination of hand and arm constitutes the effector system; in a steady-state vowel production task, the tongue body is the terminal device and the jaw and tongue comprise the effector system. During skilled actions, the numerous degrees of freedom defined by the muscles and joints of such effector systems must be harnessed functionally in a manner specific to the task or goal at hand.

In addition to a skill's goal directedness, it is also clear that ordinary actions (such as walking or talking) or extraordinary actions (such as ballet or operatic singing) are never performed twice in exactly the same way. Yet observers and students of such activities seem to share the intuition that there is a task-specific commonality or invariance that underlies the separate task performances. In the present paper, a theoretical approach to these dual issues of contextual variation and task-specific invariance in skilled actions is described. This approach is called *task dynamics* (Saltzman and Kelso 1983a), and promises to provide within a single framework a parsimonious account of both variable and invariant aspects of well-learned, skilled actions. Saltzman and Kelso (1983a) describe how a mathematical, task-dynamic model can be applied to tasks involving relatively simple arm movements in the horizontal or sagittal planes (e.g., reaching discretely and cyclically, transporting cup-to-mouth, and crank-turning). The present paper describes how task-dynamic modeling is being extended by this author and his colleagues at Haskins Laboratories to the coordination and regulation of the speech articulators during linguistically meaningful tasks (cf. Browman and Goldstein 1985; Browman et al. 1984; Kelso et al. 1985).

There are (at least) two signature properties of skilled actions – *trajectory shaping* and *immediate compensation* – that must be accounted for by a theory of coordination and control. Trajectory

* This research was supported in part by NIH grant NS-13617 and contract no. N0014-83-K-0083 from the US Office of Naval Research.

shaping refers to the tendency of end-effector trajectories to display forms that are characteristic of the demands of performed tasks. For example, it has been demonstrated in several laboratories that in planar reaching tasks using the shoulder and elbow joints, the hand moves in a quasistraight line toward the target (e.g., Bizzi and Abend 1982; Bizzi et al. 1981; Morasso 1981; Soechting and Lacquaniti 1981; Wadman et al. 1980; see also Hollerbach and Atkeson, this volume). Similarly, in cup-to-mouth tasks, the grasped cup must maintain a spillage-preventing horizontal orientation while en route from table to mouth.

The second characteristic of skilled gestures, immediate compensation, refers to the task-specific flexibility of action systems in reorganizing themselves when faced with unexpected disturbances or perturbations. Thus, compensation for the perturbation of a given effector during a movement trajectory is achieved by readjusting the activity over the entire system in order to achieve the task goal (e.g., Bernstein 1967; Marsden et al. 1983; Nashner and McCollum 1985). Further, these readjustments appear to occur automatically without the need to detect the disturbance explicitly, replan a new movement, and execute the new movement plan. Kelso et al. (1984) have demonstrated such behavior in the speech articulators (jaw, upper and lower lip, tongue body) when subjects produced the utterances /baeb/ or /baez/ across a series of trials in which the jaw was occasionally and unpredictably tugged downward while moving upward to the final /b/ or /z/ constriction (see also, Abbs and Gracco 1983; Folkins and Abbs 1975). The system's response to the jaw perturbation was measured by observing the motions of the jaw and upper and lower lips as well as the electromyographic (EMG) activities of the orbicularis oris superior (upper lip), orbicularis oris inferior (lower lip), and genioglossus (tongue body) muscles. The investigators found relatively "immediate" task-specific compensation (i.e., 20-30 ms from onset of jaw pull to onset of compensatory response) in remote articulators to jaw perturbation. For /baeb/ (in which final lip closure is crucial) they found increased upper lip activity (motion and EMG) relative to the unperturbed control trials, but normal tongue activity; for /baez/ (in which final tongue-palate constriction is important) they found increased tongue activity relative to controls, but normal upper lip motion. The speed of these task-specific patterns suggests that compensation does not occur according to traditionally defined "intentional" reaction time processes, but rather according to an automatic, "reflexive" type of organization. However, such an organization is not defined in a hard-wired input/output manner. Instead, these data imply the existence of a selective pattern of coupling or gating among the component articulators that is specific to the utterance produced. Such compensatory behavior represents the classic phenomenon of motor equivalence (Hebb 1949; Lashley 1930), according to which a system will find alternate routes to a given goal if an initially intended route is unexpectedly blocked.

What type of coordinative processes could generate, in a task-specific manner, both characteristic trajectory patterns for unperturbed movements and spontaneous, compensatory behaviors for perturbed movements? The task-dynamic model for effector systems having many articulatory degrees of freedom was developed in an effort to deal with these issues (Saltzman and Kelso 1983a; see also, Boylls and Greene 1984, for related discussions of task-specific dynamics). The model is labeled task dynamic since: (a) it deals with the performance of well-learned skilled movements or gestures designed to accomplish real-world tasks; and (b) it is defined with respect to the dynamics that underlie a given action's kinematics. Note that kinematics refers to a gesture's observable spatiotemporal properties (e.g., its position, velocity, and acceleration trajectories over time), while dynamics refers to the pattern of the underlying field of forces that gives rise to these kinematics. The task-dynamic approach extends and elaborates the view that the functional units of

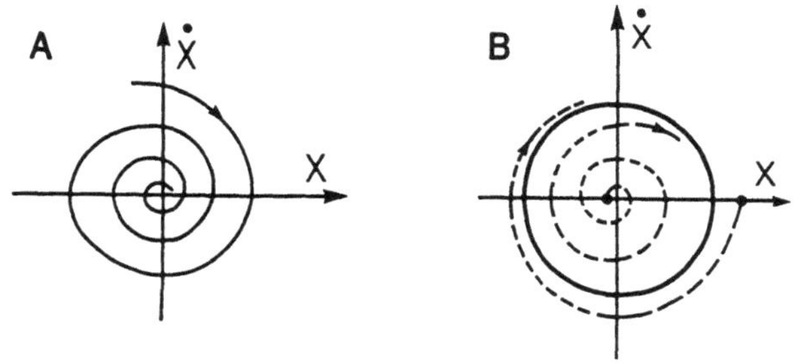

Fig. 1. Representative phase plane $(x, \dot{x})$ trajectories for point attractor (A) and periodic attractor (B) systems. Examples of motion equations are: A. $m\ddot{x} + b\dot{x} + kx = 0$, where m is mass, b is damping, and k is stiffness. B. $m\ddot{x} + b\dot{x} + kx = f(x,\dot{x})$, where $f(x,\dot{x})$ is a nonlinear damping (i.e., escapement) term, and all other coefficients are as in A

action (or *coordinative structures*; e.g., Easton 1972; Fowler 1977; Kelso et al. 1979; Turvey 1977) underlying the performance of a given gesture may be identified with abstractly defined, task-specific control regimes whose dynamic parameters (e.g., stiffness, damping, rest position) remain constant over the course of the gesture (cf. Fitch and Turvey 1978; Kelso et al. 1980; Kugler et al. 1980, 1982; Saltzman and Kelso 1985). In the task-dynamic model, the control regime that governs the performance of a particular gesture or task is defined functionally as an abstract (*task space*) dynamic system that is effector-independent, i.e., it does not explicitly incorporate the particular end-effectors directly involved in performing the task. It is hypothesized that a common task-space description underlies the functional equivalence of different effector systems for the performance of a given task, e.g., writing one's signature using a pencil held in the hand or between the teeth. Relatedly, qualitative differences between tasks are captured by corresponding *topological* distinctions among task-space dynamical systems (see also, Arbib 1984, for a related discussion of the relation between task and controller structures.)

For example, gestures involving a hand's discrete motion to a single spatial target and repetitive cyclic motion between two such targets are characterized by *point attractor* and *periodic attractor* dynamic regimes, respectively (cf. Abraham and Shaw 1982). The behaviors of these two types of dynamical systems may be represented in the phase plane (i.e., where system velocity is plotted vs position) as illustrated in Fig. 1, along with examples of corresponding equations of motion. Figure 1A shows a point attractor regime characterized by an (underdamped) mass-spring equation of motion. This system displays *point stability* or *equifinality*, in that it will asymptotically attain the equilibrium position, x_0, regardless of initial conditions for x and $\dot{x}$ and despite any transient perturbations encountered during its motion trajectory. Figure 1B shows a periodic attractor regime with a stable cyclic orbit (i.e., *limit cycle*) that is approached asymptotically by all trajectories (except those starting exactly at x_0) regardless of transiently introduced perturbations. The value of specifying a system's behavior in terms of topologically defined attractors is that such attractors provide task-specific, low dimensional descriptions for movement systems with many degrees of freedom, and promise to provide an elegant notational scheme for capturing the dynamical invariance across different effector systems that are observed to perform identical tasks. Distinct topologies correspond, therefore, to distinct patterns of task-dynamic parameters (e.g., damping and stiffness coefficients), and have been labeled the *organizational invariants* for skilled actions of different types (Fowler and Turvey 1978; Saltzman and Kelso 1983a, b). Such patterns denote functions that are preserved invariantly over changes in the parameter's specific values. In the task–dynamic model, the values of these *tuning* parameters (e.g., Greene 1972; Saltzman and Kelso 1983a, b) are determined according to factors such as the rate or amplitude of movement, and are defined to be constant over the course of a given gesture.

The task–dynamic model is able to account for the phenomena of trajectory shaping and immediate compensation without the need for explicit trajectory planning or replanning (see Saltzman and Kelso 1983a, for further details). Note that defining invariant patterns of dynamic parameters at the level of articulatory degrees of freedom (e.g., stiffness and damping parameters at the joints of an arm) will not suffice to generate these behaviors. Constant articulatory–dynamic parameters will not suffice to generate the quasistraight–line hand trajectories seen in planar reaching tasks (Delatizky 1982; Hollerbach 1982); rather, such trajectory shapes must result from task–specific patterns of change in these parameters during the reaching gestures. Similarly, the immediate compensation data for speech described above (Kelso et al. 1984) could not be generated by a system with a constant *rest configuration* parameter (i.e., a vector whose components are constant rest positions for the lips and jaw). As shown in these data, when sustained perturbations were introduced during articulatory closing gestures, the system "automatically" achieved the same constriction goals as for unperturbed gestures, but with different final or rest configurations. Thus, both trajectory shaping and immediate compensation behaviors appear to result from the way that dynamic parameters at the articulatory level are constrained to change during a gesture in a context–dependent manner. In the task–dynamic model, such patterns of constraint originate in corresponding invariant patterns of dynamic parameters at the task–space level of description.

Example 1. Planar Reaching, Three Joints

Using, for illustrative purposes, a discrete reaching task in the horizontal plane with angular motion at the shoulder, elbow, and wrist joints, the operation of a given task–dynamic regime may be understood in the following way. First, the functional aspects of a reaching gesture are specified in a two–dimensional task space as an invariant point attractor (e.g., a two–dimensional damped mass–spring system; see Fig. 2A). These dynamics give rise to an evolving pattern of state–dependent "forces" exerted on an effector–independent terminal device (i.e., a *task mass*). In the task space, the reach target defines the origin of a Cartesian coordinate system, with axis t_1 ("reach" axis) defined along a line from the initial position of the task mass to the target, and axis t_2 ("normal" axis) defined normal to t_1. The equations of motion for this task–dynamic regime are described in matrix notation as follows:

$$\mathbf{M}_T\ddot{\mathbf{t}} + \mathbf{B}_T\dot{\mathbf{t}} + \mathbf{K}_T\mathbf{t} = \mathbf{0}, \text{ where} \tag{1}$$

$$\mathbf{M}_T = \begin{bmatrix} m_T & 0 \\ 0 & m_T \end{bmatrix}; \ \mathbf{B}_T = \begin{bmatrix} b_{T1} & 0 \\ 0 & b_{T2} \end{bmatrix};$$

$$\mathbf{K}_T = \begin{bmatrix} k_{T1} & 0 \\ 0 & k_{T2} \end{bmatrix};$$

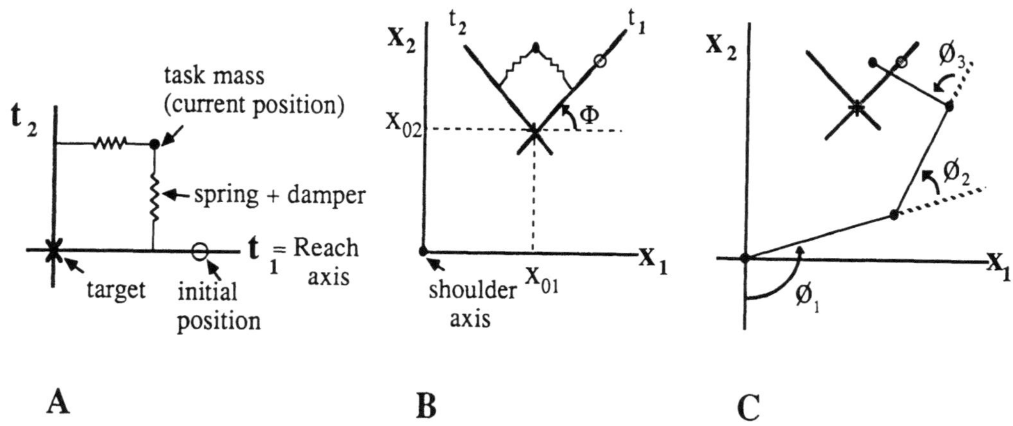

Fig. 2. Discrete reaching: A. Task space (**t**). B. Shoulder space (**x**). Task space is located and oriented in shoulder-centered reference frame via $\mathbf{x}_0$ and Φ, respectively. C. Model articulator space ($\boldsymbol{\phi}$); ϕ's denote joint angles

where m_T is the task-mass coefficient; b_{T1}, b_{T2} are damping coefficients; and k_{T1}, k_{T2} are stiffness coefficients.

Equation (1) describes a linear, uncoupled set of task-space equations, whose terms are defined in units of force, and whose dynamic parameters (i.e., M_T, B_T, K_T) are constant. In Fig. 2A, the corresponding damping and stiffness elements are represented in lumped form by the squiggles in the lines connecting the task mass to axes t_1 and t_2.

Second, the task mass is identified with the relevant "virtual" end-effector (e.g., a virtual fingertip), and the task-space dynamic system is transformed kinematically into a two-dimensional body-space system (x_1, x_2; shoulder space) governing movements of the virtual end-effector (see Fig. 2B). Thus, the task space is located and oriented in body-space coordinates according to the tuning parameters $\mathbf{x}_0$ (the body-space position vector of the task-space origin) and Φ (the orientation angle between task axis t_1 and body axis x_1), respectively. The resulting set of linear body-space equations of motion for the task's terminal device are defined in matrix form as follows (in these and the following equations, a superscript T denotes the vector or matrix *transpose* operation):

$$M_B\ddot{\mathbf{x}} + B_B\dot{\mathbf{x}} + K_B\,\Delta\mathbf{x} = \mathbf{0} \qquad (2)$$

where $M_B = M_TR$, with M_T being the task-space mass matrix, and R the rotation transformation matrix with elements $r_{ij}(\Phi)$ converting task-space variables into body-space form; $B_B = B_TR$, where B_T is the task-space damping matrix; $K_B = K_TR$, where K_T is the task-space stiffness matrix; and $\Delta\mathbf{x}=\mathbf{x}-\mathbf{x}_0$, where $\mathbf{x}=(x_1,\ x_2)^T$, the current body-space position vector of the terminal device.

Note that Eq. (2), unlike Eq. (1), represents a set of body-space equations that are (usually) coupled due to the rotation transformation (i.e., the off-diagonal matrix elements are generally non-zero). However, as with the task-space equations, the terms of Eq. (2) are defined in force units and the resultant set of body-space dynamic parameters is constant.

Third, the body-space dynamic system is transformed into a three-dimensional "model" articulator space where the moving segments (upper arm, forearm, and hand) have lengths, but are massless (see Fig. 2C). Like the transformation from task space to body space, this transformation is a strictly kinematic one (since the segments have no mass) and involves only the substitution of variables defined in one coordinate system for variables defined in another coordinate system. As illustrated in Fig. 2C, this corresponds to expressing body-space variables ($\mathbf{x}$, $\dot{\mathbf{x}}$, $\ddot{\mathbf{x}}$) as functions of an arm model's kinematic variables ($\boldsymbol{\phi}$, $\dot{\boldsymbol{\phi}}$, $\ddot{\boldsymbol{\phi}}$; where $\boldsymbol{\phi} = [\phi_1, \phi_2, \phi_3]^T$, and ϕ_1 is the shoulder angle defined relative to axis x_2, ϕ_2 is the elbow angle defined relative to the upper arm segment, ϕ_3 is the wrist angle defined relative to the forearm segment), and the arm's proximal (shoulder) and distal (fingertip) ends are attached to the body-space origin and the terminal device/task mass, respectively. The body-space variables of Eq. (2) are transformed into the joint-angle variables of the massless arm model using the following kinematic relationships:

$$\mathbf{x} = \mathbf{x}\,(\boldsymbol{\phi}) \tag{3a}$$

$$\dot{\mathbf{x}} = \mathrm{J}\,(\boldsymbol{\phi})\,\dot{\boldsymbol{\phi}} \tag{3b}$$

$$\begin{aligned} \ddot{\mathbf{x}} &= \mathrm{J}\,(\boldsymbol{\phi})\,\ddot{\boldsymbol{\phi}} + (\mathrm{dJ}\,(\boldsymbol{\phi})/\mathrm{dt})\dot{\boldsymbol{\phi}} \\ &= \mathrm{J}\,(\boldsymbol{\phi})\ddot{\boldsymbol{\phi}} + \mathrm{V}(\boldsymbol{\phi})\,\dot{\boldsymbol{\phi}}_p \end{aligned} \tag{3c}$$

where $\mathbf{x}(\boldsymbol{\phi}) = (x_1(\boldsymbol{\phi}), x_2(\boldsymbol{\phi}))^T$, the current body-space position vector of the terminal device expressed as a function of the current model arm configuration; $\dot{\boldsymbol{\phi}}_p = [\dot{\phi}^2_1, \dot{\phi}_1\dot{\phi}_2, \dot{\phi}_1\dot{\phi}_3, \dot{\phi}^2_2, \dot{\phi}_2\dot{\phi}_3, \dot{\phi}^2_3]^T$ is the current model arm joint velocity product vector; $\mathrm{J}(\boldsymbol{\phi})$ is the *Jacobian* transformation matrix whose elements J_{ij} are partial derivatives, $\partial x_i/\partial \phi_j$, evaluated at the current $\boldsymbol{\phi}$; and $\mathrm{V}(\boldsymbol{\phi})$ is a matrix resulting from rearranging the terms of the expression $(\mathrm{dJ}(\boldsymbol{\phi})/\mathrm{dt})\,\dot{\boldsymbol{\phi}}$ in order to segregate the joint velocity products into a single vector $\dot{\boldsymbol{\phi}}_p$.

Using the kinematic relationships in Eq. (3), the model effector system's equation of motion is as follows:

$$\mathrm{M_B J}\ddot{\boldsymbol{\phi}} + \mathrm{B_B J}\dot{\boldsymbol{\phi}} + \mathrm{K_B}\Delta\mathbf{x}(\boldsymbol{\phi}) = -\mathrm{M_B V}\dot{\boldsymbol{\phi}}_p \tag{4}$$

where $\mathrm{M_B}$, $\mathrm{B_B}$, $\mathrm{K_B}$ are the same constant matrices used in Eq. (2); and $\Delta\mathbf{x}(\boldsymbol{\phi}) = \mathrm{x}(\boldsymbol{\phi}) - \mathbf{x}_0$, where $\mathbf{x}_0$ is the same constant vector used in Eq. (2). It should be noted that since $\Delta\mathbf{x}$ in Eq. (2) and (4) is *not* assumed to be "small," a differential approximation $\mathbf{dx} = \mathrm{J}(\boldsymbol{\phi})\,\mathbf{d}\boldsymbol{\phi}$ is not justified and, therefore, Eq. (3a) was used instead for the kinematic displacement transformation into model arm variables.

The terms of Eq. (4) are still defined in units of force, not torque, and may be rewritten in units of angular acceleration:

$$\ddot{\boldsymbol{\phi}} + \mathrm{J^*M_B}^{-1}\mathrm{B_B J}\dot{\boldsymbol{\phi}} + \mathrm{J^*M_B}^{-1}\mathrm{K_B}\Delta\mathbf{x}(\boldsymbol{\phi}) + \mathrm{J^*V}\dot{\boldsymbol{\phi}}_p = \mathbf{0} \tag{5}$$

where J* is a weighted Jacobian pseudoinverse (e.g., Benati et al. 1980; Klein and Huang 1983; Whitney 1972) that is used because there are a greater number of model articulator variables than spatial variables for this task. Hence, the model effector system is *redundant* (e.g., Saltzman 1979), the *inverse kinematic* transform from spatial to model articulator coordinates is indeterminate, and the Jacobian inverse (J^{-1}) cannot be defined. More specifically, $J^* = A^{-1}J^T(JA^{-1}J^T)^{-1}$, where A is a positive definite *articulatory weighting* matrix whose elements are constant during a given gesture. Using J* provides a unique, optimal least squares solution for the differential transformation from body–space to model articulator variables that is weighted according to the pattern of elements in the A–matrix. In current modeling, the A–matrix is defined to be of diagonal form, and a given set of articulator weights will constrain motion of an articulator in direct proportion to the magnitude of the corresponding weighting element. Hence, different articulator weighting patterns are associated with different patterns of relative angular motions of the three joints for the same task–space motion of the task mass (or body–space motion of the virtual fingertip). For example, one weighting pattern might correspond to predominant shoulder motion, while a second weighting pattern might correspond to predominant elbow motion for the same task– or body–space trajectory of the terminal device. In this sense, elements of the A–matrices used in the associated J*'s define a further set of tuning parameters for the model effector system's equation of motion (Eq. 5).

The task–dynamic model allows one to define for the discrete reach (as well as other tasks) an invariant task–space dynamic regime that: (a) is specified by a constant set of task–dynamic parameters, and (b) constrains in a context–dependent way the evolving pattern of changes in the model arm's articulatory–dynamic parameters (i.e., stiffness, damping, and equilibrium positions of shoulder, elbow, and wrist joints) during the course of the gesture. Thus, one may interpret the task–specific, coherent movements of the model effector system as resulting from the way that instantaneous task–space "forces" acting on the associated terminal device are distributed across the model arm's articulatory degrees of freedom during the course of the planar reach. At any given instant during this gesture, the partitioning is based on two factors:

1. The task–specific, constant set of task space (Eq. 1), body space (Eq. 2), and model articulator space (Eq. 4 and 5) dynamic parameters
2. The current values of elements in the posturally dependent transformation matrices (i.e., the J and J^T matrices in Eq. 4 and 5) that relate motions of the articulators at their current configuration to corresponding body–space motions of the virtual fingertip. Because these elements are nonlinear functions of the current arm model posture $\boldsymbol{\phi}$, the elements of the matrix products in Eq. (4) and (5) (i.e., the coefficients that define articulatory–dynamic parameters) are also dependent on the evolving configuration of the arm model.

The final step in the task–dynamic approach is to exploit algebraic relations between the model arm's dynamic regime and the physical and control parameters of the "real" (biological, robotic, or prosthetic) arm in order to specify patterns of control parameters over time for the real arm. Saltzman and Kelso (1983a) discuss two related methods for specifying these controls. Both methods are applicable to the control and coordination of artificial linkage systems (e.g., robotic or prosthetic devices), although one offers a more biologically plausible style of control than the other (see also Hogan and Cotter 1982). The aim of both methods, however, is to make the real arm behave identically or near–identically to the model arm. Further, the essence of the task–dynamic approach lies in its account of the coordinated movement patterns that arise in a task–specific and posturally conditioned form in the model effector system. Consequently, for the purposes of the present paper, further discussion will focus on behavioral phenomena in the model articulators only. The interested reader is referred to Saltzman and Kelso (1983a), however, for details concerning the hypothesized relationships between control processes of the model and real effector systems.

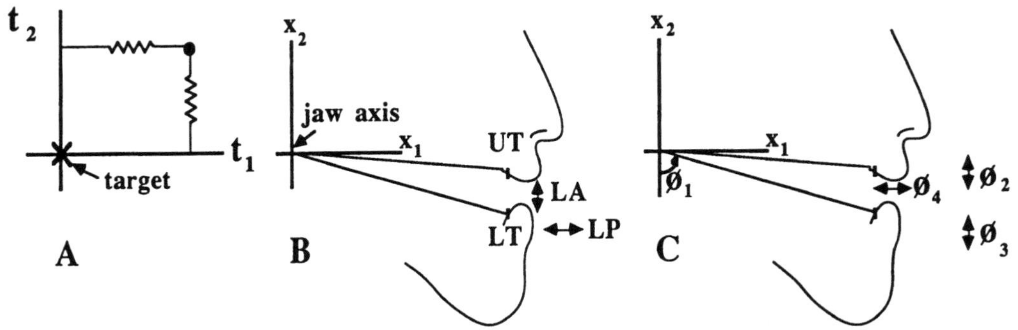

Fig. 3. Bilabial tasks: A. Task space (**t**). Closed circle denotes current system configuration. Squiggles denote each axis' dynamics in lumped forms. B. Jaw space (**x**). Local tract variables (LP, lip protrusion; LA, lip aperture) are expressed in jaw coordinates. UT and LT denote positions of upper and lower front teeth, respectively. C. Model articulator space ($\boldsymbol{\phi}$). ϕ's denote articulator variables

TASK DYNAMICS AND SPEECH: BILABIAL GESTURES

The task–dynamic approach has been extended in a preliminary way to speech gestures in order to explore the hypothesis that speech production involves task–specific, dynamically specified coordination of the articulators.

Example 2. Discrete Bilabial Closure, Unperturbed Gestures

As with the limb tasks described earlier, the first step in generating simulated movements of the speech articulators is to specify the functional aspects of these gestures with reference to the movements of an effector–independent terminal device (i.e., an idealized vocal tract constriction). This is done in a two–dimensional task space whose axes represent constriction location (t_1) and constriction degree (t_2), and the topological structure of the control regime for each task–space variable is specified according to the qualitative characteristics of the given speech task. Thus, for example, discrete and repetitive speech gestures will have point attractor and limit cycle regimes, respectively, along each axis. At the task–space level, then, the control regime is an abstract one in that the constriction being controlled is independent of any particular effector system, and can refer, for example, to either a bilabial constriction produced by the lips and jaw or to a tongue–palate constriction produced by the tongue and jaw. Since simulations to date have focused on bilabial gestures, we will begin by examining a discrete bilabial closure task involving (uncoupled) point attractor dynamics along each task axis (see Fig. 3A). The task–space equation of motion is expressed as follows:

$$M_T\ddot{\mathbf{t}} + B_T\dot{\mathbf{t}} + K_T\mathbf{t} = \mathbf{0} \tag{6}$$

where $M_T = \begin{bmatrix} m_{T1} & 0 \\ 0 & m_{T2} \end{bmatrix}$; $B_T = \begin{bmatrix} b_{T1} & 0 \\ 0 & b_{T2} \end{bmatrix}$; and $K_T = \begin{bmatrix} k_{T1} & 0 \\ 0 & k_{T2} \end{bmatrix}$

In Eq. (6) m_{T1}, m_{T2} are inertial coefficients; b_{T1}, b_{T2} are damping coefficients; and k_{T1}, k_{T2} are stiffness coefficients.

The forms of these task–space dynamics and corresponding equation of motion are identical to those for the discrete limb reaching task described earlier (Fig. 2A; Eq. 1). This identity highlights the fact that functional equivalence among tasks does not depend on the specific effector systems involved, but only on the topological equivalence of dynamical regimes in the task spaces. The two main differences between the limb and speech examples are: (a) the task–space axes for the bilabial task do not share a common task mass, but rather are characterized by their own inertial coefficients m_{T1} and m_{T2} (compare Eq. 2 and 6); and (b) the axes for the bilabial task are not differentiated into distinct "reach" and "normal" axes as they were in the limb reaching task. Finally, as in the reaching example, movements along the task axes do not influence one another, since the corresponding equations of motion are defined to be uncoupled.

The next step in modeling the bilabial closure is to transform the task–space system kinematically into a two–dimensional body–space system (x_1, x_2) defined in the midsagittal plane of the vocal tract and centered on the jaw's rotation axis (see Fig. 3B). In contrast to the task–space regime, the body–space dynamics are effector–specific, in that they refer to the movement of a "virtual" terminal device (i.e., the bilabial constriction) of the effector system defined by the lips and jaw. The result of transforming from task–space (t_1, t_2) to jaw–space (x_1, x_2) coordinates, then, is to define a two–dimensional set of motion equations, one for each axis of jaw space. As with the task–space equation, the jaw–space equation has the same form as its corresponding shoulder–space reaching equation (Eq. 2). The jaw–space equation is as follows:

$$M_B\ddot{\mathbf{x}} + B_B\dot{\mathbf{x}} + K_B\Delta\mathbf{x} = \mathbf{0} \tag{7}$$

where $\mathbf{x}$, $\dot{\mathbf{x}}$, $\ddot{\mathbf{x}} = (x_1, x_2)^T$ and its derivatives with respect to time; $\Delta\mathbf{x} = \mathbf{x} - \mathbf{x}_0$, where $\mathbf{x}_0$ is the target vector for lip protrusion (x_{01}) and lip aperture (x_{02}); $M_B = M_T$, $B_B = B_T$, and $K_B = K_T$, since no rotation is involved in the transformation from task– to jaw–space coordinates.

Equation (7) contains a constant set of dynamic parameters, and governs the movements for the bilabial constriction along the dimensions of lip aperture (LA) and lip protrusion (LP). Lip aperture and protrusion are labeled *local tract variables*, and represent the effector–specific body–space versions of the effector–independent task–space variables of constriction degree and location, respectively. Lip aperture is defined by the vertical distance between the upper and lower lips, and lip protrusion by the horizontal distances in the anterior–posterior direction of the upper and lower lips from the upper and lower teeth, respectively. It should be noted that upper and lower lip protrusion movements are not independent in this formulation, but have been constrained to be equal in the model for purposes of simplicity. Consequently, like constriction location in task space, lip protrusion in body space constitutes only a single degree of freedom. Finally, it should be noted that the control regimes for each jaw–space coordinate are independent, since their corresponding equations of motion are uncoupled. This is due to the fact that lip aperture and protrusion are defined parallel to the x_2 and x_1 jaw–space axes, respectively. Note, however, that such non-interacting dynamics are not usually found at the body–space level of description. For example, with movements of the tongue body orthogonal and tangential to the (curved) palate, the set of uncoupled task−space equations would be transformed into a set of (generally) coupled jaw−space equations.

The last step in modeling the closure is to transform kinematically the two−dimensional jaw−space regime into the coordinates of a four−dimensional model articulator space. The model articulators are moving segments that have lengths but are massless (see Fig. 3C), and are defined with reference to the simplified articulatory degrees of freedom adopted in the Haskins Laboratories software articulatory speech synthesizer (Rubin et al. 1981). For bilabial gestures, the articulator set associated with lip aperture includes rotation of the jaw (ϕ_1), and vertical displacements of the upper lip (ϕ_2) and lower lip (ϕ_3) relative to the upper and lower front teeth, respectively; for lip protrusion, the articulator set includes yoked horizontal displacements in the anterior−posterior direction of the upper and lower lips (ϕ_4) relative to the upper and lower front teeth, respectively. Expressed in units of linear acceleration, the model articulator equation has the same form as Eq. 5 and is expressed as follows (the angular acceleration terms in the jaw's motion equation have been multiplied by a unit scaling factor to ensure dimensional homogeneity along all articulatory degrees of freedom):

$$\ddot{\boldsymbol{\phi}} + J^*M_B^{-1}B_BJ\dot{\boldsymbol{\phi}} + J^*M_B^{-1}K_B\Delta\mathbf{x}(\boldsymbol{\phi}) + J^*V\dot{\boldsymbol{\phi}}_p = \mathbf{0} \tag{8}$$

where M_B, B_B, K_B are the same constant matrices used in Eq. (7); and $\Delta\mathbf{x}(\boldsymbol{\phi}) = \mathbf{x}(\boldsymbol{\phi}) - \mathbf{x}_0$, where $\mathbf{x}(\boldsymbol{\phi})$ is expressed as a function of model articulator variables, and $\mathbf{x}_0$ is the same constant vector used in Eq. (7); the elements of the Jacobian matrix (J, and hence also V and J*) reflect the geometrical relationship among motions of the (simplified) model speech articulators (4 degrees of freedom) and motions of the corresponding local tract variables (2 degrees of freedom); and A is the articulator weighting matrix, a component of the pseudoinverse J*. A's elements reflect task−specific constraints on the relative motions of the articulators during the closing gesture.

Given a fixed set of tuning parameters (i.e., M_T, B_T, K_T, $\mathbf{x}_0$, and A) and a set of initial conditions ($\boldsymbol{\phi}_I$, $\dot{\boldsymbol{\phi}}_I$, and hence a corresponding $\mathbf{x}_I$ and $\dot{\mathbf{x}}_I$) Eq. 8 will generate a pattern of coordinated motion in the model speech articulators that will achieve the task goals specified for the local tract variables. For an initial configuration ($\boldsymbol{\phi}_I$) corresponding to open and relatively unprotruded lips, and with an initial velocity vector of zero, the coordinated articulator movements will reflect the evolving task−specific motions of the local tract variables en route to their specified targets ($\mathbf{x}_0$), with motion characteristics (e.g., speed, degree of overshoot, etc.) specified by the pattern of M_T, B_T, and K_T parameters. Assuming the system is not perturbed during its motion trajectory, the relative extents of movement for the articulators associated with lip aperture (i.e., ϕ_1, ϕ_2, ϕ_3 in Fig. 3C) will be specified by the relative values of articulator weights in the associated articulatory weighting matrix, A. Figure 4 (configurations A and B) illustrates an unperturbed movement from an initially open and relatively unprotruded configuration (Fig. 4A) to a closed and relatively protruded final configuration (Fig. 4B). Since the articulators associated with lip aperture were weighted equally in the corresponding A−matrix, the extents of motion for these articulators were equal over the course of the gesture.

Example 3. Immediate Compensation, Bilabial Closure, Perturbed Gestures

Previous dynamical accounts of coordinated actions performed by the limbs and speech articulators have posited that invariant sets of dynamic parameters could be defined at the level of articulatory

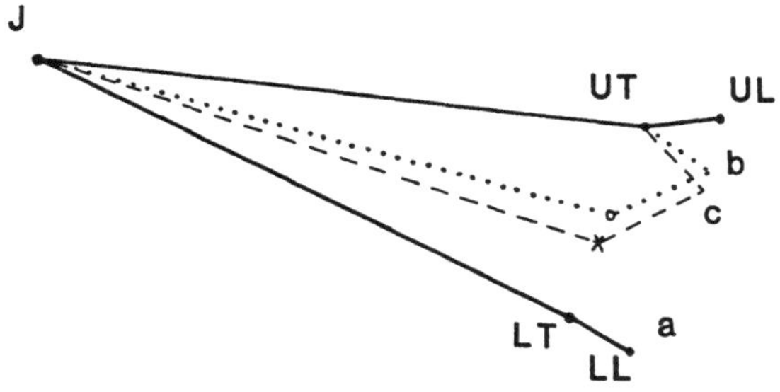

Fig. 4. Simulated articulator configurations for bilabial closure task. A. Initial configuration (solid lines). B. Final configuration, unperturbed trajectory (dotted lines). C. Final configuration, perturbed trajectory (broken lines). Note that closure occurs lower in jaw space in C than in B. J, jaw axis; UT, upper teeth; UL, upper lip; LT, lower teeth; LL, lower lip

degrees of freedom (e.g., Cooke 1980; Fel'dman 1966; Fowler 1977; Kelso 1977; Polit and Bizzi 1978). Thus, for example, discrete targeting tasks of the elbow joint were modeled as damped mass – spring systems (having point attractor dynamics) where the target angle was specified by the value of the rest angle dynamic parameter. As discussed earlier, this approach implies that the task of reaching a bilabial closure target for speech is specified according to a corresponding rest – configuration parameter for the articulators. However, recent work (Abbs and Gracco 1983; Folkins and Abbs 1975; Kelso et al. 1984) has shown that this formulation must be modified. In particular, the Kelso et al. (1984) study demonstrated that if the jaw is retarded en route to a bilabial closure target for /b/, then the closure is still attained and the final articulatory configuration for the perturbed movement is different from the final configuration for unperturbed movements. Significantly, the upper lip compensation is absent if the jaw is perturbed en route to an alveolar closure target for /z/. These results show that an invariant dynamic description of a movement does not apply at the articulator level, since the articulatory – dynamic parameters must be able to change according to a movement's context in an utterance – specific (i.e., /b/ vs /z/) manner. Furthermore, the speed of these compensatory behaviors suggests that they must occur "automatically" without reference to traditional stimulus – response reaction – time correction procedures.

The task – dynamic model handles such immediate compensation as follows. Bilabial closing gestures are simulated as discrete movements toward target constrictions, using point attractor dynamics for the local tract variables of lip aperture and protrusion (see Eq. 7 above). When the simulated jaw is "frozen" in place during the closing gesture at the level of the model effector system, the main qualitative features of the perturbation data are captured, in that: (a) compensation is immediate in the upper and lower lips to the jaw perturbation, i.e., the system does not require reparameterization in order to compensate; and (b) the target bilabial closure is reached (although with different final articulator configurations and, hence, different jaw – space locations for the closure) for both perturbed (Fig. 4C) and unperturbed (Fig. 4B) "trials."

Example 4. Cyclic Bilabial Motion, Unperturbed Gestures

The point attractor task – space (and local tract variable) topology that was used in the discrete bilabial closure task is inappropriate, however, for generating cyclic bilabial gestures, e.g., a

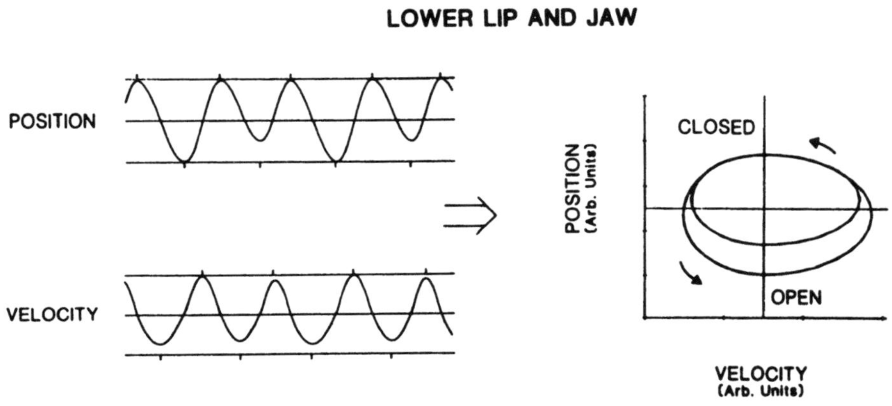

Fig. 5. Simulated trajectories for lower lip height (i.e., jaw and lower lip) in the time domain (left) and phase plane (right) for a repetitive sequence of /ma/'s with alternating stress (from Kelso et al. 1985)

sequence of repeated /ma/'s as in "...mamama...." The task–dynamic model has been used to simulate a repetitive gestural sequence that is characterized by an alternating stress pattern, e.g., "...mamamama...," where the underlining denotes the pattern of stress (Browman et al. 1984). Mass–spring dynamics were specified for the local tract variables of lip aperture and protrusion in order to generate sustained cyclic motions of the model articulators. (Mass–spring and limit cycle dynamics can produce near–identical motion trajectories in the absence of perturbations to the system. Since the modeled cyclic bilabial gestures were unperturbed by design, (undamped) mass–spring dynamics were adequate for these purposes. The model is presently being extended, however, to include limit cycle dynamics at the task–space level, in order to explore the simulated effects of perturbations introduced during cyclic speech tasks.) Focusing on lip aperture, the parameters of rest position and stiffness were estimated from articulatory movement data collected in an experiment on reiterantly produced speech (Kelso et al. 1985). In reiterant speech, talkers substitute a given syllable (e.g., /ma/) for the real syllables in an utterance while maintaining the utterance's normal stress pattern (e.g., the sentence "When the sunlight strikes raindrops in the air" becomes "ma ma ma ma ma ma ma ma ma ma"). The lip aperture parameters for the task–dynamic simulation were estimated using the average amplitudes and frequencies of the articulatory data obtained for the stressed and unstressed syllables spoken reiterantly at a given rate. Figure 5 illustrates the resultant cyclic trajectories for lower lip height, both in the time domain and the phase plane. For a given simulated cyclic gesture (closure–to–closure), the equilibrium position was set only once because, in the data, the jaw–lip complex returned roughly to the same position at closure for each syllable. The values for the equilibrium positions in temporally adjacent cycles alternated in value, however, since stressed syllables were found to involve greater movement amplitudes than unstressed syllables. Additionally, because closing gestures were faster than opening gestures in these data, two values of stiffness were specified within each cycle: one at the start of the opening gesture and another at the start of the closing gesture. The set of task–dynamic parameters were invariant, therefore, over the course of a given opening or closing gesture.

CONCLUSION

The task – dynamic model is able to generate coordinated movement patterns for the model articulators in both discrete and cyclic unperturbed (bilabial) utterances. Additionally, for discrete bilabial closing gestures it provides task – specific patterns of compensatory responses to jaw perturbations that are qualitatively similar to those observed experimentally. Finally, Browman et al. (1984) have used sets of simulated articulator trajectories from an alternating stress, repetitive, bilabial speech task as inputs to the Haskins Laboratories articulatory speech synthesizer (Rubin et al. 1981; see also Example 2 above) with promising acoustic and perceptual results. Note that, although these simulated utterances involve a simple stress pattern and segmental structure, the task – dynamic approach to articulatory speech synthesis could certainly be used to generate more complex utterances on a gesture – by – gesture basis. The elegance of the procedure would still be maintained, however, since utterance – specific and contextually variable patterns of articulator trajectories and compensatory responses would still emerge automatically as implicit consequences of task space control regimes that are invariant within a given speech gesture. There is no need to invoke either explicit trajectory planning or replanning procedures on a timeframe – to – timeframe basis within the gesture. My colleagues and I are encouraged by these preliminary results, and are currently engaged in extending the task – dynamic model to account for phenomena such as coarticulation (e.g., Harris 1984; Kent and Minifie 1977) and relative timing (e.g., Kent et al. 1974; Tuller et al. 1982, 1983) among serially ordered speech gestures.

REFERENCES

Abbs JH, Gracco VL (1983) Sensorimotor actions in the control of multimovement speech gestures. Trends Neurosci 6: 393 – 395

Abraham RH, Shaw CD (1982) Dynamics – the geometry of behavior. Aerial, Santa Cruz

Arbib MA (1984) From synergies and embryos to motor schemas. In: Whiting HTA (ed) Human motor actions: Bernstein reassessed. North – Holland, Amsterdam

Benati M, Gaglio S, Morasso P, Tagliasco V, Zaccaria R (1980) Anthropomorphic robotics. II. Analysis of manipulator dynamics and the output motor impedance. Biol Cybern 38: 141 – 150

Bernstein NA (1967) The coordination and regulation of movements. Pergamon, London

Bizzi E, Abend W (1982) Posture control and trajectory formation in single and multiple joint arm movements. In: Desmedt JE (ed) Brain and spinal mechanisms of movement control in man: new developments and clinical applications. Raven, New York

Bizzi E, Accornero N, Chapple W, Hogan N (1981) Processes underlying arm trajectory formation. In: Ajmone – Marsan C, Pompeiano O (eds) Brain mechanisms of perceptual awareness and purposeful behavior. IBRO monograph series. Raven, New York, pp 311 – 318

Boylls CC Jr, Greene PH (1984) Introduction: Bernstein's significance today. In: Whiting HTA (ed) Human motor actions: Bernstein reassessed. North – Holland, Amsterdam

Browman CP, Goldstein LM (1985) Dynamic modeling of phonetic structure. In: Fromkin V (ed) Phonetic linguistics. Academic, New York

Browman CP, Goldstein L, Kelso JAS, Rubin P, Saltzman E (1984) Articulatory synthesis from underlying dynamics. J Acoust Soc Am 75: S22–S23 (abstract)

Cooke JD (1980) The organization of simple, skilled movements. In: Stelmach GE, Requin J (eds) Tutorials in motor behavior. North–Holland, Amsterdam

Delatizky J (1982) Final position control in simulated planar horizontal arm movements. PhD thesis, Department of Electrical Engineering and Computer Science, MIT

Easton TA (1972) On the normal use of reflexes. Am Sci 60: 591–599

Fel'dman AG (1966) Functional tuning of the nervous system with control of movement or maintenance of a steady posture. III. Mechanographic analysis of execution by man of the simplest motor tasks. Biophysics 11: 766–775

Fitch H, Turvey MT (1978) On the control of activity: some remarks from an ecological point of view. In: Landers D, Christina R (eds) Psychology of motor behavior and sports. Human Kinetics, Urbana

Folkins JW, Abbs JH (1975) Lip and jaw motor control during speech: responses to resistive loading of the jaw. J Speech Hear Res 18: 207–220

Fowler C (1977) Timing control in speech production. Indiana University Linguistics Club, Bloomington

Fowler CA, Turvey MT (1978) Skill acquisition: an event approach with special reference to searching for the optimum of a function of several variables. In: Stelmach G (ed) Information processing in motor control and learning. Academic, New York

Greene PH (1972) Problems of organization of motor systems. Prog Theoret Biol 2: 303–338

Harris KS (1984) Coarticulation as a component of articulatory descriptions. In: Daniloff RG (ed) Articulation assessment and treatment issues. College Hill, San Diego

Hebb DO (1949) The organization of behavior. Wiley, New York

Hogan N, Cotter SL (1982) Cartesian impedance control of a nonlinear manipulator. In: Book WJ (ed) Robotics research and advanced applications. ASME, New York, pp 121–128

Hollerbach JM (1982) Computers, brains, and the control of movement. Trends Neurosci 5: 189–192

Kelso JAS (1977) Motor control mechanisms underlying human movement reproduction. J Exp Psychol: Hum Percept Perform 3: 529–543

Kelso JAS, Southard DL, Goodman D (1979) On the nature of human interlimb coordination. Science 203: 1029–1031

Kelso JAS, Holt KG, Kugler PN, Turvey MT (1980) On the concept of coordinative structures as dissipative structures: II. Empirical lines of convergence. In: Stelmach G, Requin J (eds) Tutorials in motor behavior. North–Holland, Amsterdam, pp 49–70

Kelso JAS, Tuller B, V–Bateson E, Fowler CA (1984) Functionally specific articulatory cooperation following jaw perturbations during speech: evidence for coordinative structures. J Exp Psychol: Hum Percept Perform 10: 812–832

Kelso JAS, V–Bateson E, Saltzman EL, Kay BA (1985) A qualitative dynamic analysis of reiterant speech production: phase portraits, kinematics, and dynamic modeling. J Acoust Soc Am 77: 266–280

Kent RD, Minifie FD (1977) Coarticulation in recent speech production models. J Phon 5: 115–133

Kent RD, Carney PJ, Severeid LR (1974) Velar movement and timing: evaluation of a model for binary control. J Speech Hear Res 17: 470–488

Klein CA, Huang C–H (1983) Review of pseudoinverse control for use with kinematically redundant manipulators. IEEE Trans Syst Man Cyber SMC–13: 245–250

Kugler PN, Kelso JAS, Turvey MT (1980) On the concept of coordinative structures as dissipative structures: I. Theoretical lines of convergence. In: Stelmach GE, Requin J (eds) Tutorials in motor behavior. North–Holland, Amsterdam, pp 3–47

Kugler PN, Kelso JAS, Turvey MT (1982) On the control and coordination of naturally developing systems. In: Kelso JAS, Clark JE (eds) The development of movement control and coordination. Wiley, Chichester, pp 5–78

Lashley KS (1930) Basic neural mechanisms in behavior. Psychol Rev 37: 1–24

Marsden CD, Merton PA, Morton HB (1983) Rapid postural reactions to mechanical displacement of the hand in man. In: Desmedt JE (ed) Motor control mechanisms in health and disease. Raven, New York

Morasso P (1981) Spatial control of arm movements. Exp Brain Res 42: 223–227

Nashner LM, McCollum G (1985) The organization of human postural movements: a formal basis and experimental synthesis. Behav Brain Sci 8: 135–172

Polit A, Bizzi E (1978) Processes controlling arm movements in monkeys. Science 201: 1235–1237

Rubin P, Baer T, Mermelstein P (1981) An articulatory synthesizer for perceptual research. J Acoust Soc Am 70: 321–328

Saltzman E (1979) Levels of sensorimotor representation. J Math Psychol 20: 91–163

Saltzman E, Kelso JAS (1983a) Skilled actions: a task dynamic approach. Hask Lab Status Rep Speech Res SR–76: 3–50

Saltzman E, Kelso JAS (1983b) Toward a dynamical account of motor memory and control. In: Magill R (ed) Memory and control of action. North–Holland, Amsterdam, pp 17–38

Saltzman E, Kelso JAS (1985) Synergies: stabilities, instabilities, and modes. Behav Brain Sci 8: 161–163

Soechting JE, Lacquaniti F (1981) Invariant characteristics of a pointing movement in man. J Neurosci 1: 710–720

Tuller B, Kelso JAS, Harris KS (1982) Interarticular phasing as an index of temporal regularity in speech. J Exp Psychol: Hum Percept Perform 8: 460–472

Tuller B, Kelso JAS, Harris KS (1983) Further evidence for the role of relative timing in speech: a reply to Berry. J Exp Psychol: Hum Percept Perform 9: 829–833

Turvey MT (1977) Preliminaries to a theory of action with reference to vision. In: Shaw R, Bransford J (eds) Perceiving, acting and knowing: toward an ecological psychology. Erlbaum, Hillsdale

Wadman WJ, Denier van der Gon JJ, Derksen RJA (1980) Muscle activation patterns for fast goal–directed arm movements. J Human Movement Stud 6: 19–37

Whitney DE (1972) The mathematics of coordinated control of prosthetic arms and manipulators. ASME J Dyn Sys Measurement Control 94: 303–309

Understanding Human Movement*

P. Morasso and R. Zaccaria

Dipartimento di Informatica, Sistemistica e Telematica, Universita di Genova, Via Opera Pia 11 A, 16145 Genova, Italy

INTRODUCTION

One of the main topics of Artificial Intelligence is presently the representation of knowledge. This means, in a broad sense, to define concepts and rules able to capture the essential complexity of a given problem domain which escapes a direct approach to understanding. Thus, a knowledge representation (KR) approach emphasizes the goals and the functions of a system more than its specific mechanisms, and in so doing it generalizes and abstracts our understanding of the system, because it is likely that different mechanisms can be found which can implement the same function and goal. Abstraction and generality can be considered weak or strong aspects, according to the point of view: weak for the neuroscientist, who can find only suggestive power on a KR model, but strong for a designer of skilled machines, who may rely on a KR model while developing a robot with anthropomorphic skills without being limited to the antropomorphic hardware.

The richness of human skills cannot be understood if we do not take adequately into account the deeply seated complementarity between a performing human (or, in general, an actor) and its surrounding: the "richness" of an actor's skill must match the "diversity" of its surrounding. The actor/environment relationship may be characterized by a "golden tetrahedron", whose vertices are "action", "perception", "cognition", "environment": each vertex of the tetrahedron is strongly linked to all the others by fundamental flows of information and/or energy.

Important theoretic contributions have been provided by several authors on different aspects of the golden tetrahedron, such as the concept of motor synergy (Bernstein 1967) or of motor schema (Piaget 1971; Schmidt 1975), the emphasis on the "ecological" nature of skills (Gibson 1979; Turvey and Kugler 1984), the recognition of the linkage between the perceptual structures and motor control (Arbib 1981); nevertheless, this kind of knowledge remains essentially at the conceptual level and it does not provide the computational tools for performing, accordingly, "general" experiments, either in the real or the simulated world.

* This work has been partially supported by the Center of Bioengineering and Anthropomorphic Robotics of the National Research Council, Italy, and by an Esprit Project.

Plenty of relevant methodological tools do exist, which deal with general spatial–temporal–logic topics at the base of action complexity. Some of these tools are:

1. Computational geometry, 3D computer graphics, solid modeling
2. Geometric reasoning, task planning
3. Object–oriented programming, message passing, actors
4. Logic programming, expert systems
5. Analogic computation, dynamics, compliance
6. Qualitative reassoning, naive physics

However, such tools are being developed in different and largely unrelated areas and one of the main problems which need to be faced is to interface and integrate them in the common framework of action.

Taking all of this into account, this paper summarizes some experimental findings about human movement, then discusses some general aspects about action and finally describes a language, Nem (Morasso et al. 1985), for the representation and animation of humanoids in a scene, which is meant to provide a software laboratory for experimenting with action.

REVIEW OF EXPERIMENTAL FINDINGS

The complexity of the motion of the human body has attracted the attention of artists, in the first place, and of scientists. Perhaps Marey (1873) and Muybridge (1901) can be chosen to mark the borderline indicating transition from the artistic to the scientific approach to human movement. However, immediately after the beginning of the scientific study of human motion, it became apparent that the eidetic beauty of the human body had to be reduced to something more formal, such as a stick figure (see the cyclogrammetric studies of gait by Bernstein 1967).

In mathematical terms, a stick figure can be represented as an ensemble of kinematic chains, i.e., as a structure of joints and links. The complexity of such a system, on the other hand, has directed the mainstream of research on human movement to simpler subsystems of the body, characterized by "few" degrees of freedom. The overall stick figure model has received attention mainly in static configurations, for example in the field of computer graphics representation of the human body (Badler 1983) or in the fields of anthropometry and ergonomy, focusing on the range of permissible and/or comfortable movements which constrain the human stick figure (Korein 1985).

A stereotyped spatiotemporal structure has been found to characterize simple rotations of different parts of the body, such as the eyes (Morasso et al. 1973), the head (Bizzi et al. 1976), or the forearm (Polit and Bizzi 1982). That is, these rotations are characterized by a smooth acceleration phase, followed by a smooth deceleration phase in a type of structure which has been shown to have a "minimum–jerk" (Hollerbach and Flash 1982).

The joint level characterization of human movements, however, is inappropriate for describing more complex movements, such as cursive script handwriting, as it was first suggested by

Bernstein, when he formulated his principle of "equal simplicity." The suggestion by Bernstein has been confirmed by subsequent investigations which showed the invariant characteristics of hand movements performed in different parts of the workspace, with different amplitudes, and with different orientations (Viviani and Terzuolo 1980; Morasso 1981; Abend et al. 1982; Lacquaniti and Soechting 1982). Also these movements have bell-shaped, minimum-jerk velocity profiles and such a characterization implies a constraint on the shape and speed of the trajectory. In particular, the regular segmentation of the velocity and its coupling with the corresponding curvature profile suggest association of each velocity peak with a primitive minimum-jerk motion and decomposition, accordingly, of the overall trajectory as a chain of primitive motions, smoothed by a partial time overlap (Morasso and Mussa Ivaldi 1982; Morasso 1983).

With regard to dynamics, it must first be pointed out that remarkable inertial changes due, for example, to loads, have been reported to influence the spatiotemporal characteristics of movements very little (Bizzi et al. 1978). Furthermore, the viscous-elastic properties of the muscles have been shown to provide an analogic automatic feedback for the compensation of disturbances (Feld'man 1966; Bizzi et al. 1978), and their action can be represented by means of a stiffness tensor (Mason 1981; Hogan 1982; Mussa Ivaldi et al. 1984). As a logical consequence of these findings, it has been suggested that each posture is the equilibrium point of the elastic forces due to the different muscles and the dynamics of the movements are counteracted by shifting such a point along an appropriate trajectory (McKeon et al. 1984).

The stiffness tensor may be expressed more intuitively in geometric terms by means of stiffness ellipses (in the plane, or stiffness ellipsoids, in the space) which specify the relation between vectors of displacement from an equilibrium point and the corresponding restoring-force vectors, in such a way making explicit the preferential directions of compliance. The shape and orientation of the stiffness ellipses have been found to vary regularly in the workspace, therefore providing the possibility to formulate in geometric terms the compliance aspects of action.

In general, the picture which emerges from the experimental findings briefly reported in this review is that motor control problems may be divided into two gross categories:

1. Energy driven processes which solve dynamic aspects of action in an implicit/analogic way, mostly relying on the mechanical properties of muscles and on the complexity of their geometric arrangement.

2. Spatiotemporal processes, which allow significant relations between an actor and the surrounding environment to be expressed explicitly and the corresponding trajectory formation problems to be solved.

ACTORS, ACTIONS, SCENES

The goals of an action may be quite diverse, even discarding nonmotoric purposes. Let us consider some examples:

1. Reaching or approaching something or someone or himself, according to a specific "trajectory" and with respect to a specific geometric framework

2. Hitting, i.e., impacting, something or someone or himself, according to selective directional characteristics of the impact, i.e., according to a specific compliance

3. Catching and grasping something or someone or himself, adapting the actor's self to the geometry and the physics of the grasped entity and so modifying the topology and the metrics of the actors's self

4. Pressing/pushing/pulling/twisting, i.e., applying a force field to something or someone or himself, according to a planned direction and intensity

By means of the four gross categories of general skills outlined above (which correspond to generalized forms of positioning, impacting, application of force vectors, and grasping), it is possible to compose an unbounded variety of actions, characterized by a highly structured network of specifications. Geometric, kinematic, and dynamic aspects are all tightly interrelated for any skill. In general, however, it is convenient to distinguish, for an actor, two different levels of "knowledge" of the particular aspect of a skill: (a) a qualitative/symbolic/explicit level, and (b) an implicit/analogic/quantitative level (Funt 1983).

For example, a kind of knowledge of level (a) is the so-called "naive physics" (De Kleer 1975), i.e., a set of rules for judging about stability, forces, constraints, impacts, etc. in the environment and at the actor/environment interface. This kind of knowledge, since it must provide explicit expectations of probable events due to imprecisely known causes, can only be qualitative and then it is naturally suitable for storage in a data base of "rules-of-thumb".

On the contrary, however, coordination among the many degrees of freedom which are involved, for example, in generating planned trajectories of the hand, can only be the result of some kind of multidimensional signal processing which transforms (with adequate accuracy) trajectory plans into actual trajectories. Such transformations are likely to be performed, in part, by the mechanical properties of muscles (McKeon et al. 1984) and, in part, by specialized signal processors which implement inverse kinematics (Benati et al. 1982) or inverse dynamics processors (Hollerbach 1980): What the degree of mix is among these computational modules is still an open question, but, whatever the answer, they represent an implicit/analogic computational layer.

Let us define an action as a purposeful change in the relation between an actor and his surrounding, which may consist of objects and/or other actors. Let us put aside, for the moment, the "purpose" of an action, which has to do with the nongeometric aspect of action planning, and let us consider the mutual "change" between the actor and his surrounding, which characterizes the specific motoric performance.

It is important to realize that such a change is always multidimensional and highly redundant: every significant geometric feature of the actor "changes" his geometric relation relative to any other geometric feature of the actor himself and of all other objects and actors in the scene.

Let us now associate with each feature a local system of coordinates (a frame). The action can then be considered as a stream of variations of some of the mutual relations between local coordinate systems, which results from the implementation of a motor plan.

The description of the elements of a scene in terms of local coordinate systems, linked to features, needs obviously to be constrained in such a way that the actions do not violate the integrity of actors and objects. Furthermore, it is convenient to group together frames which refer to the same structure (an actor, an object, or a part thereof).

As a consequence, a scene can be represented as a "forest" of frames which changes over time, and by visiting the forest it is possible to express the geometric relations which exist between any two features of the scene. Some of these geometric relations can be interpreted as "motor" and some as "proprioceptive"; whereas, others can be interpreted as "visual" or, more in general, "exteroceptive".

Furthermore, it is worth noting that ensembles of frames can only represent, in a direct way, the "skeletal" structure of actors or objects; if smoother and more detailed representations are needed, it would be necessary to associate appropriate "shape formation" attributes with frames and frame ensembles. This points out the problem of interfacing an action-oriented system with solid modeling concepts (Requicha 1980; Binford 1982) and with techniques of path planning and obstacle avoidance (Lozano Perez 1982).

It is important to note that the point of view about action discussed in this section is not limited to the so-called motor program concept and it does not imply a neglect of the importance of "feedback" in motor control. On the contrary, both centrally planned and feedback-driven movements fit into the same general framework, which stresses the geometrical complexity and abstractness of action.

NEM: A LANGUAGE FOR REPRESENTING ACTORS, ACTIONS, SCENES

As any programming language, Nem is characterized by "data types" and "instructions". The main data type in Nem is the frame, which corresponds internally to an orthogonal system of reference and is represented by means of an homogeneous 4 x 4 matrix (Foley and Van Dam 1983). Homogeneous matrices express the translation/rotation of the given frame with respect to an ancestor frame and such a relation is denoted by the operator "@". For example, a tetrahedron can be defined by the following frame structure

```
@ env FRAME tetrahedron
@ tetrahedron FRAME vertex_1 vertex_2 vertex_3 vertex_4
```

where env is the parent frame of the scene, tetrahedron is a frame associated with some reference point of the tetrahedron (such as its centroid), and a frame is associated with each vertex of the tetrahedron (i.e., with each significant feature of the shape).

Limbs can be defined as chains of frames. For example,

@ scapula LIMB shoulder elbow wrist hand

corresponds to the following chain of specifications:

```
@ scapula    FRAME shoulder
@ shoulder   FRAME elbow
@ elbow      FRAME wrist
@ wrist      FRAME hand
```

Furthermore, prefixes can be used to denote distinct limbs with a similar frame structure, such as a left or right arm. In particular, the following frame structure can represent a humanoid skeleton:

```
@ env        FRAME   humanoid
@ humanoid   LIMB    pelvis torso neck
@ pelvis     LIMB    'l' ilium hip knee ankle foot
@ pelvis     LIMB    'r' ilium hip knee ankle foot
@ torso      LIMB    'l' scapula shoulder elbow wrist hand
@ torso      LIMB    'r' scapula shoulder elbow wrist hand
@ neck       FRAME   head
@ head       FRAME   mouth nose forehead
@ head       FRAME   'l' eye ear
@ head       FRAME   'r' eye ear
```

The definition of a frame structure and the corresponding initialization can be embedded into a definition block, which can be stored in a data base of structures and recalled when needed, e.g.:

```
define_human(humanoid,size_parameters)
  @ env FRAME humanoid
          .
          .
          .
  {
   initialize the humanoid frames according
   to the size_parameters
  }
```

(see Fig. 1 for the visualization of a humanoid frame structure).

The same frame structure can be "instanced" several times, i.e., different "instances" of the same structure can be generated with different parameters. For example, if two humanoids, "Adam" and "Eve", are required in an action, the corresponding frame structures can be instanced as follows:

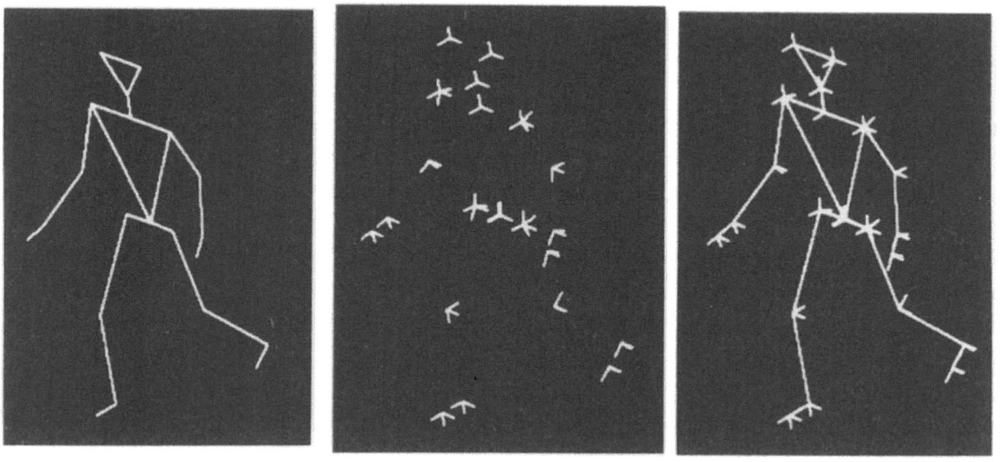

Fig. 1. Humanoid frame structure: the stick figure on the left is generated by displaying the vectors which join the (selected set of) frames, the center figure displays only the frames of the humanoid frame structure, and the right figure shows the frame structure together with the stick figure

```
define_human(Adam'humanoid,parameter_set_A)
define_human(Eve'humanoid,parameter_set_B)
```

and if, during the action, we are interested in the geometric relation between, say, the right hand of Eve and the mouth of Adam, we can express it by means of a frame:

```
FRAME dontdo
dontdo=Adam'mouth @ Eve'r'hand
```

Other data types in Nem are triple (arrays) and scalars. For dealing with the triples it is possible to use vector algebra operators (sum, subtraction, vector product, scaling, scalar product) and it is possible to assemble/disassemble FRAMEs in terms of TRIPLEs by means of another set of operators. In any case, the main requirement which motivated the development of such a FAN (frame algebra notation) was to provide the language with a powerful formalism for expressing general geometric relations among any two features of the (simulated) world, whether features of an actor or of an object.

By itself, however, FAN does not capture another important geometric aspect which distinguishes, for example, a collapsed marionette from a correct human posture. In general, this has to do with rules for partitioning the configurational space of a frame structure into admissible and nonadmissible regions, and with the definition of some metric which allows to measure the "distance" of a posture from the configuration boundary in order to plan well-formed movements. In Nem, this kind of important geometric modeling aspect can be embedded either in motion procedures (making them aware of the motion limits and providing them with local correction rules) and/or in movement-planning algorithms. (Note that movement planning in face of motion limits is conceptually akin to another classic robotic planning problem, i.e., path planning in face of obstacles; see, for example, Lozano Perez (1982), for a treatment of this aspect.)

An additional data feature of Nem, that we need to mention, is the event, which allows the synchronization of different motions, or processes, or actors.

Primitive motion operators are defined at two levels: (a) joint level and (b) limb level. At the joint level, there are "primitives" for rotating a frame F with respect to its ancestor A around a given axis: "bend", "fold", "rotate" use the standard Euler axes (yaw, pitch, roll), and "turn" can use an arbitrary axis, i.e., the screw axis of Rodrigues. Another joint–level primitive is "screw" which expresses the kind of movement performed by an arm when the hand is kept fixed and the arm is rotated around the wrist–shoulder axis: this is the combination, in general, of a "turn" of F around an axis which joins F to its ancestor–ancestor and of a "turn" of that frame around the same axis, but with opposite sign.

At the limb level (or manipulator level, as it is called in the robotic literatur) there are two primitives: (a) one primitive ("move") specifies that the end effector of a limb has to be moved toward a target frame, carrying along the rest of the limb, i.e., solving the inverse kinematic problem (Benati et al. 1982), and (b) another primitive ("fix") specifies that an end effector has to keep its current relation with respect to a target frame until further notice, i.e., until an event occurs.

Primitive motions are executed in a single "time slot". Composite motions can be defined by combining different primitive or composite motions by means of three different motions execution paradigms: (a) sequential execution (e.g., motion_1 ; motion_2 ; motion_3 ; ...); (b) parallel execution (e.g., motion_1 & motion_2 & motion_3 & ...); (c) guarded execution (e.g., WHILE condition DO action; WAIT condition THEN action; PERFORM action_1 EXCEPT condition THEN action_2;).

By means of the sequential paradigm, for example, it is possible to implement motion procedures with a bell–shaped velocity profile which can be used as basic spatiotemporal units. Furthermore, by using the parallel and guarded paradigms it is possible to overlap in time two or more spatiotemporal units, thus, providing an unlimited capability of trajectory formation even with a simple set of basic movements. Complex activation patterns can be generated by mixing the three composition paradigms and we like to point out the subtle complexity which results from the fact that simultaneously active processes may affect the same frame structure.

Figure 2 shows an example in which the spatiotemporal units are straight segments of the hand with a bell–shaped velocity profile; the algorithm which is used has been proposed (Morasso and Mussa Ivaldi 1982) for modeling the cursive script handwriting of humans: a sequence of spatiotemporal units (called "strokes") is chained to form a polygon surrounding the desired trace and then the strokes are executed with a partial time overlap between one stroke and the next one. It is important to point out the generality of this representation, because the same program remains valid for different sizes of the script, for different orientations of the plan in which the script is lying, and for different locations on such a plane.

In general, the movement composition mechanisms described above allow the natural expression of a hierarchical, distributed approach to motor control (see also Hinton 1984 for a distributed motor program that reaches and balances) which is conceptually akin to the object–oriented programming style emerging in computer science (Goldberg 1983); in such a

Fig. 2. Handwriting Nem program

framework, "central programming" and "peripheral feedback" are fundamentally and unavoidably intermingled.

Figure 3 (left) shows a walking humanoid. This action is expressed by means of a combination of joint- and space-level primitives, as well as sequential, parallel, and guarded executions, and it was implemented, at the highest level, as a process which moves the center of gravity of the humanoid along a given trajectory and with a given speed. The other actions are synchronized, implicitly, by means of events. For example, during the stance phase, the standing foot is "fixed" until a limit geometric condition is met at the ankle joint and then

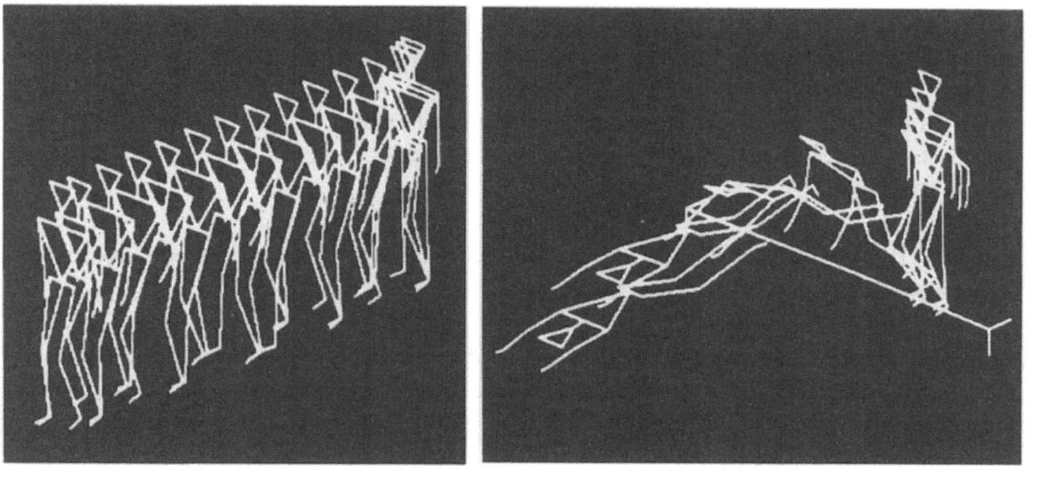

Fig. 3. Walking and diving Nem program

swing action is started; this action, in turn, can be nicely represented as the concurrent execution of a "move" of the swinging foot toward a virtual forward target (defined with respect to the humanoid body) and an overall flexion of the leg, resulting in a complex curved trajectory of the foot without really having to specify the shape of such a trajectory and/or the detailed timing of the leg joints. Furthermore, the planned motion of the center of gravity is slightly modified in an automatic–implicit way by the action of other concurrent processes, such as the extension of the supporting ankle during the final part of the stance phase and the pelvic rotation around a vertical axis (Inman et al. 1981). The role played by "events" in the Nem walking program is an example of the fundamental interaction between central control and peripheral feedback which can be expressed in the language.

Figure 3 (right) shows a diving humanoid; here it is worth noting the action of gravity, which determines, concurrently with everything else, the motion of the center of gravity.

In general, dynamics is not explicitly represented in Nem, one reason being that the viscous–elastic properties of muscles seem likely to handle it implicitly (at least for the most part), removing it from the level of action representation and planning. However, dynamic aspects can be integrated in Nem programs in terms of concurrent processes which may modify the planned action and which can be monitored by motor control modules, in order to tune control parameters.

The purpose of Nem is to provide a simulated environment in which it is possible, by means of software experiments, to gain insight into the complexity of the geometric and sequencing aspects of movement, in order to devise better experiments which may ask the right questions about movement planning and execution.

REFERENCES

Abend W, Bizzi E, Morasso P (1982) Human arm trajectory formation. Brain 42: 331–338

Arbib MA (1981) Perceptual structures and distributed motor control. In: Brook VB (ed) Handbook of physiology – The motor system II. Motor control. American Physiological Society, Bethesda, MD

Badler N (ed) (1983) Modeling the human body for animation. IEEE Computer Graphics and Applications 2: 1–120 (Special issue, November)

Benati M, Gaglio S, Morasso P, Tagliasco V (1980) Anthropomorphic robotics. I: Representing mechanical complexity. II: Analysis of manipulator dynamics and the output motor impedance. Biol Cybern 38: 125–150

Benati M, Morasso P, Tagliasco V (1982) The inverse kinematic problem for anthropomorphic manipulator arms. Trans ASME J Dynam Systems Measurement Control 104: 110–113

Bernstein NA (1967) The coordination and regulation of movements. Pergamon, New York

Binford TO (1982) Survey of model based image analysis systems. Int J Robotic Res 1: 18–64

Bizzi E, Polit A, Morasso P (1976) Mechanisms underlying achievement of final head position. J Neurophysiol 39: 435–444

Bizzi E, Dev P, Morasso P, Polit A (1978) Effect of load disturbances during centrally initiated movements. J Neurophysiol 41: 542–556

DeKleer J (1975) Qualitative and quantitative knowledge in classical mechanics. Artificial Intelligence Laboratory Technical Report 352, MIT Press, Cambridge, MA

Fel'dman AG (1966) Functional tuning of the nervous system with control of movement or maintenance of a steady posture. II. Controllable parameters of the muscles. Biophysics 11: 565–578 (translated from Biofizika 11: 498–508)

Foley JD, Van Dam A (1983) Fundamentals of interactive computer graphics. Addison–Wesley, Reading, Mass.

Funt BV (1983) Analogical modes of reasoning and process modeling. IEEE Computer 16: 10: 99–104

Gibson JJ (1979) The ecological approach to visual perception. Hougton Mifflin, Boston

Goldberg A, Robson D (1983) Smalltalk–80: the language and its implementation. Addison–Wesley, Reading, Mass.

Hinton G (1984) Some computational solutions to Bernstein's problems. In: Whiting HTA (ed) Human motor action – Bernstein reassessed. Elsevier, Amsterdam, pp 373–412

Hogan N (1982) Mechanical impedance control in assistive devices and manipulators. Proceedings of the American control conference. Arlington, Virginia

Hollerbach JM (1980) A recursive lagrangian formulation of manipulator dynamics and a comparative study of dynamics formulation complexity. IEEE Trans Systems, Man, Cybern SMC 11: 730–736

Hollerbach JM, Flash T (1982) Dynamic interaction between limb segments during planar arm movements. Biol Cybern 44: 67–77

Inman VT, Ralston HJ, Todd F (1981) Human walking. Williams and Wilkins, Baltimore

Korein J (1985) A geometric investigation of reach. MIT Press, Cambridge, Mass.

Lacquaniti F, Soechting JF (1982) Coordination of arm and wrist motion during a reaching task. J Neurosc 2: 399–408

Lozano Perez T (1982) Task planning. In: Brady M et al. (eds) Robot motion: planning and control. MIT Press, Cambridge, MA

Marey EJ (1873) La machine animale, locomotion terrestre et aèrienne. Paris

Mason MT (1981) Compliance and force control for computer controlled manipulators. IEEE Trans Systems, Man, Cybern SMC 11: 418–431

McKeon B, Hogan N, Bizzi E (1984) Effect of temporary path constraint during planar arm movements. Abstract of the 14th annual conference of the society for neuroscience, October 10–15, Anaheim, CA, p 337

Morasso P (1981) Spatial control of arm movements. Exp Brain Res 42: 233–237

Morasso P (1983) Three dimensional arm trajectory. Biol Cybern 48: 187–194

Morasso P, Mussa Ivaldi FA (1982) Trajectory formation and handwriting: a computational model. Biol Cybern 45: 131–142

Morasso P, Bizzi E, Dichgans J (1973) Adjustment of saccade characteristics during eye movements. Exp Brain Res 16: 548–562

Morasso P, Troiano E, Zaccaria R (1985) Representing motor knowledge. In Bibel W, Pethoff B (eds) Artificial intelligence – methodology, systems, applications. Elsevier, Amsterdam

Mussa Ivaldi FA, Hogan N, Bizzi E (1984) Invariant features of hand postural stiffness. Abstract of the 14th Annual conference of the society for neuroscience, October 10–15, Anaheim, CA, p 337

Muybridge E (1901) The human figure in motion. Capman and Hall, London

Piaget J (1971) Biology and knowledge: an essay on the relation between organic regulations and cognitive processes. Edinburgh University Press, Edinburgh

Polit A, Bizzi E (1982) Characteristics of motor programs underlying arm movements in monkeys. J Neurophysiol 42: 183–194

Requicha AG (1980) Representation of rigid solids: theory, methods and systems. ACM Computing Surveys 12: 437–475

Schmidt RA (1975) A schema theory of discrete motor skill learning. Psychol Rev 82: 225–260

Turvey MT, Kugler PN (1984) An ecological approch to action and perception. In Whiting HTA (ed) Human motor action – Bernstein reassessed. Elsevier, Amsterdam, pp 373–412

Viviani P, Terzuolo CA (1980) Space–time invariance in learned motor skills. In: Stelmach GE, Requin J (eds) Tutorials in motor behavior, Elsevier, Amsterdam, pp 525–533

Opposition Space as a Structuring Concept for the Analysis of Skilled Hand Movements*

T. Iberall[1], G. Bingham[2], and M. A. Arbib[1]

[1] Laboratory for Perceptual Robotics, Department of Computer and Information Science, University of Massachusetts, MA 01003, USA
[2] Haskins Laboratories, 270 Crown Street, New Haven, CT 06510, USA

INTRODUCTION

The human hand is a highly articulated device which, as an extension of the arm, provides a mechanism for complex interactions with the environment. A large variety of skilled behaviors involving the human hand can be observed ranging from subtle and precise manipulations to smooth and forceful exertions, involving over 20 degrees of freedom (*dofs*) (see Tubiana 1981; Kapandji 1982). How is this large and complex space of possibilities for movement restricted so that the uniquely useful configurations of the system observed in actual practice appear in appropriate circumstances? (see Whiting 1984, and Turvey et al. 1978 for discussions of the synergies approach to the degrees of freedom problem originally described by N. Bernstein.)

We are taking a task analysis approach, which is similar to studies done in other multiarticular systems. In analyzing the task of handwriting, Hollerbach (1981) describes two functional *dofs* (writing direction and letter height), and suggests a controller containing coupled oscilators. In posture control, McCollum et al. 1984, Nashner and McCollum 1985) find specific parameters for a controller relevant to the task of maintaining balance, such as ankle torque, relative amplitude between two muscle synergies, and a timing parameter between hip torque and braking. In speech production, researchers (Abbs et al. 1984; Kelso, this volume; Kelso et al. 1982) find the covarying of the upper and lower lips within the task (production of specific words) an important parameter, rather than the absolute position of the lips. Of the wide range of movements in the repertoire of the hand, we are studying the *grasping* and *picking up* of objects. An example of this type of task is grasping a mug (Arbib et al. 1985). A person, reaching for a visually perceived mug, preshapes the hand into a shape suitable for gripping the handle. This preshaping movement is coordinated with the timing of the reach which carries the hand to the location of the object to be grasped (Jeannerod 1981, 1984; Wing and Fraser 1983). At an appropriate point during the reach, the preshaped hand closes around the handle, and stably grips it for lifting.

In this paper, we describe a functionally effective space of kinematic possibility which is sufficient for manipulation tasks, given the basic link segment structure of the hand and its dynamic properties. Researchers in the robotics domain, in looking at manipulation tasks, have analyzed these requirements for simple two-fingered hands (Fearing 1983; Mason 1982), and dextrous three- or four-fingered hands (Salisbury 1982, Wolter et al. 1985). The task of controlling functionally

* Preparation of this paper was supported in part by grant ECS-8108818 from NSF.

Experimental Brain Research Series 15

effective activity has two mutually constraining component parts. The first is the sources of constraint which reduce the complexity of the space of kinematic possibility; the second is finding a suitable control system to manage behavior in that space.

We have defined the *virtual finger* (Arbib et al. 1985) as a group of fingers (perhaps even including the palm) which act as a single functional unit in a structure representing the task. The hand in preshaping organizes according to an appropriate arrangement of virtual fingers. In the mug task, three independent functional units operate cooperatively to effect a stable grasp, but the way in which the actual fingers are incorporated into the three virtual fingers – the thumb, the fingers that go through the handle, and the fingers that go beneath the handle – varies with handle size. We suggested a coordinated control program involving the activation of *schemas* (Arbib 1981), which organize and control the virtual fingers using dynamic visual and tactile input.

In this paper, we introduce the concept of *opposition space*. For a given manual task, this is the area within the coordinates of the hand where opposing forces can be exerted between virtual finger surfaces in effecting a stable grasp. At first, we only consider static stable postures of the grasping hand, irrespective of whether or not the arm is moving. (We recognize that this approach can achieve only a first approximation to the repertoire of human hand grasping behaviors, since active manipulation of objects involves at least transitions between postures.) We will show that regions within the broad capability of the hand correspond to specific functional capabilities of particular hand postures. Virtual finger formation is then shown to be channelled by the selection of a region (an opposition space) for a given task. Finally, we map task constraints into virtual finger configurations needed to perform the task, and show that schemas as units of motor control can be defined to control virtual fingers. Since schemas organize and control virtual fingers by taking advantage of the architectural limitations of the hand and additional dynamic properties of the prehensile actuators, only a limited but useful number of postures and kinematic forms will occur.

CONSTRAINTS IMPOSED BY TASKS

The hand, as it reaches to grasp an object, preshapes into a configuration suitable for the anticipated interaction (Jeannerod 1981). What determines the shape that the hand is taking? To gain some intuition, we observed two tasks that differed in how an object was to be used. Grasping activity was videotaped and subsequently studied in slow motion with stop action.

Description of Two Similar Tasks

Both tasks required that the same object be picked up from a table by a seated participant. The object was a medium-sized cylinder standing on its base. The first task (Fig. 1a–c) was to grasp the cylinder, pick it up, and set it onto a circle on which it could just fit (the "place cylinder" task). The cylinder, providing two opposing surfaces which were pinched between the thumb (virtual finger 1, or VF1) and finger pads (VF2), was lifted (Fig. 1b) and carefully guided onto the circle (Fig. 1c). The grasp, called a precision grasp by Napier (1956), or palmer prehension by Schlesinger (1919), allows the object to be translated or rotated in the fingers and thumb so that fitting the cylinder onto the circle could be more easily and accurately accomplished. Such fine

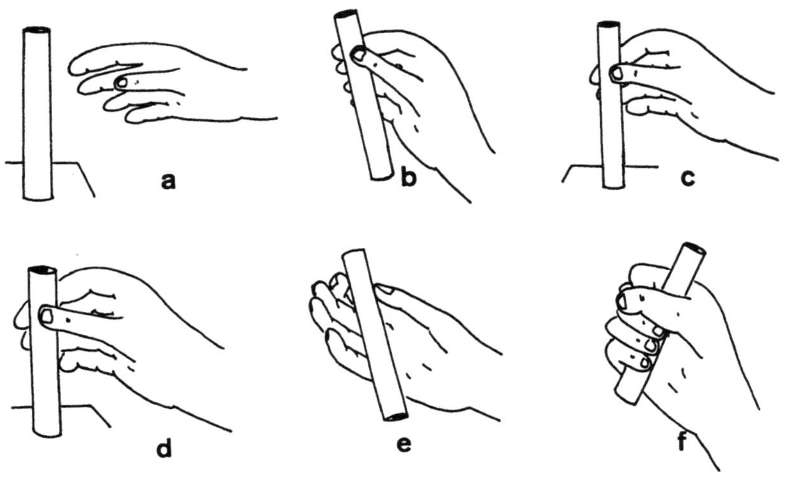

Fig. 1. Two grasping tasks. a. Preshape hand for place cylinder task. b. Grasp and lift cylinder. c. Place cylinder onto circle. d. Grasp and lift cylinder for shake cylinder task. e. Transition of cylinder from initial grasp to a more firm grasp. f. Shake cylinder

control is better effected by the fingers, a low mass system, than by the arm, with its greater structural mass.

The second task (Fig. 1d–f) was to grasp the cylinder, pick it up, and shake it vigorously two or three times along its long axis (the "shake cylinder" task). The cylinder was initially grasped (Fig. 1d) using the same precision grasp between thumb (VF1) and finger pads (VF2) as described above. However, once the object was securely lifted in the hand, a transition in grasping postures occurred (Fig. 1e). Using an axis between thumb and index, the object was allowed to be rotated by gravity into the palm where first the ring and little fingers and then the long and index fingers and thumb wrapped it in a secure grip against the palm (Fig. 1f). This grasp, called a power grasp by Napier (1956) or cylindrical prehension by Schlesinger (1919), provides a strong and stable grasp to counteract the forces that developed during the shaking task. The cylinder body again provided opposing surfaces to grasp, but this time the palm (VF1) and digits (VF2) were used. The thumb, pressing against VF2, secures the object.

Task Constraints

The particular preshape and grip configuration used during the task relate to the functions that are to be performed during the task. A task determines a set of functional requirements that must be satisfied by the postures that the hand assumes. A generic functional requirement of all prehensile lifting tasks is to avoid dropping the object. In the place cylinder task, the goal is to guide the cylinder onto the circle, which adds the functional requirement that the axis of the cylinder be matched with a normal to the circle at its center. A functional requirement of the second task is to not let the object move in the grip, specifically to avoid translations of the object transverse to the hand due to the rapid oscillations.

The hand and the manipulated object each have a large variety of physical properties. Out of these sets, the functional requirements point to a subset of physical properties relevant to the task. These subsets (from the hand and from the object), when taken in relation to one another, constitute the *physical constraints* on the activity. For example, to lift an object, the hand must supply a force which cancels the force due to gravity, and the posture taken on by the hand must be appropriate for the application of that force. To avoid dropping an object, properties such as weight, size, surface shape, and friction determine potential surfaces for opposition where these properties are scaled relative to strength, surface friction, and size properties of the hand. Potential surfaces are further constrained by the orientation of the object relative to its environment (i.e., which surfaces are available for grasping).

Thus, the chosen hand shape, which we call *opposition space*, as exhibited in the preshape, is preparatory for a grasp which satisfies the anticipated physical constraints. These physical constraints will be looked at in the next section. We then go on to explore how a controller might be designed, which can map functional requirements onto physical constraints, successfully resolving the issue of conflicting requirements.

OPPOSITION SPACE

Conditions for a Stable Grasp

A rigid object can be both translated along and rotated about any of the three Cartesian coordinate directions to any degree of combination. Thus, specifying the behavior of a rigid object over time is a matter of determining its manner of variation with respect to each of this six *dofs*. (Only a single rigid body will have six *dofs*. An object such as a pair of scissors has an additional *dofs* corresponding to its hinge. Opposition space as developed in this paper concerns only manipulations of single rigid bodies.) If the object is to be manipulated in some manner while in the grasp, then the object must vary along a specified set of *dofs* in a controlled fashion. Variation must remain within specified bounds.

Fearing (1983) defines a grasp to be stable if the object is in equilibrium, if all forces on the object are within an angle relative to the surface normal (as related to the coefficient of friction), and if motion can be prevented by increasing the grasping forces without moving the fingers. The conditions for equilibrium in any static grasp are that all forces and moments sum to zero (Williams and Lissner 1962; McLean and Nelson 1978; Fearing 1983):

$$\sum_i \vec{F}_i = 0$$

$$\sum_i \vec{r}_i \cdot \vec{F}_i = 0$$

Thus, each force tending to make the object translate in some direction must be cancelled by an equal force in the opposite direction. If forces are represented using vector notation, and if we restrict consideration to co−planar forces, each force vector of a certain length and in a certain direction must be cancelled by a co−linear force vector of equal length and opposite direction.

Successful equilibrium for an unmoving hand and object requires a force oppositely directed to gravity equal to the weight of the object. This force might be supplied simply by the hand's forming a support surface normal to gravity, or else by two or more hand surfaces making contact with the object. If the surface of the hand were frictionless, problems would arise from the fact that the human hand is subject to a low amplitude "physiological" tremor of about 10 Hz (Marsden et al. 1969). However, the hand surface has useful frictional qualities, which can be used to prevent objects from moving in a stable grasp. Coulomb's law states that the tangential force of friction is constrained to be no greater than the product of the normal force with the coefficient of static friction, or:

$$F_T < \mu_s F_N$$

This relationship, first conjectured by Leonardo da Vinci, and later described by Amontons (1699), only holds for materials where the deformation between them is plastic in nature (Comaish and Bottoms 1971; Bowden and Tabor 1950). It defines an angle of friction $\Phi_s = \tan^{-1}\mu_s$, such that, if the grasping force is applied to the surface at an angle larger than this angle, the object will slip (Fearing 1983; Mason 1982). If the object does slip, but if the motion can still be prevented by increasing the normal force, the grasp is still stable (Fearing 1983; Hanafusa and Asada 1982). For small microslips in human precision grasping, Johannson and Westling (1984) have shown that this is the case.

Surface Properties of the Hand and Friction Force

The surfaces of the palm and the pads of the digits have three properties which contribute greatly to the stability of grasps (Glicenstein and Dardour 1981). First, these surfaces are padded (Quilliam 1978), giving them a good measure of elastic compliance. This compliance tends to increase the amount of friction between the hand surface and the surfaces of objects grasped, and automatically supplies a force sufficient to counteract small amplitude perturbations of the object (Hanafusa and Asada 1982; Moore 1972; Fearing 1983). Second, these surfaces of the hand contain small ridges; and third, the pad surfaces are self-lubricating through small pores located along the tops of these epidermal ridges. The lubricant is greasy, having good adhesive qualities at low shear velocities, enhanced by the hills and valleys of the ridged surface which extends the total shearing surface area (Moore 1972, 1975). At high shear velocities, friction is reduced, thus minimizing wear and tear of the hand surface. Altogether, the surfaces of the palm, thumb, and fingers have frictional qualities quite conducive to stable grasping of objects.

Comaish and Bottoms (1971) analyzed the coefficient of friction between skin and various materials in order to determine whether it obeys Coulomb's law. They found that the frictional force is not proportional to the load, and that, for some materials, the static coefficient of friction increased with the area. While their data showed a proportional response for loads over 100 g for *in vitro* skin, the data for *in vivo* skin was definitely nonlinear. Skin has been shown to be viscoelastic (Wilkes et al. 1973), meaning that it is capable of both deformation and flow in response to an applied load. Thus, living skin does not obey Coulomb s law.

Westling and Johannson (1984, Johannson and Westling 1984) looked at the effects of varying the surface material of objects as subjects picked them up. They found that subjects maintained the applied force (the grip force) slightly above the slip force, the force below which the object would drop out of the hand. In experimental conditions where subjects could not actually see the surface, they noted that subjects varied their grip force, not to the surface, but to the friction conditions. They looked at the influence of frictional conditions from previous objects against the skin. If the previous trial required the subject to grasp a texture such as silk, with a low coefficient of friction, the subject would grasp a texture with a higher coefficient of friction (such as sandpaper) harder than normal. In one experiment, subjects washed their hands after about 15 trials of lifting the same object, and the results indicated that the grip force adapted not to the surface, but to the friction conditions (more grip force was used right after washing).

Force Balance Between Parts of the Hand

Equilibrium requires that the force applied to the object by the hand to create friction be cancelled by an equal and opposite force. The opposite force can be supplied by additional hand surfaces impinging on surfaces of the object generally parallel to (for curved surfaces, parallel planes tangent to) the initial object surface. Parallelism can be within a tolerance, such as the angle of friction (Fearing 1983; Mason 1982), allowed by the frictional properties of the hand–object contact surface. For example, when grasping a cup with sloping sides, the component of force pressing the cup upwards is counteracted not only by gravity, but by friction as well. Generally parallel oppositely facing object surfaces are referred to as *opposable surfaces*. *Opposition* occurs when equal and oppositely directed, co–linear force vectors (within a tolerance) act (along the *opposition axis*) to freeze an object's translational degree of freedom relative to the hand. Additional *dofs* may be controlled by friction at opposable surfaces.

Moment Balance

Whenever an opposition axis fails to pass through a vertical line through the center of mass of an object, a moment (or torque) around that axis will be created. In addition, to the extent that the vertical line through the center of mass is at some perpendicular distance from a plane normal to an opposition axis and bisecting it, there will be a torque tending to twist that axis of opposition.

Moments around an opposition axis can be cancelled in either of two ways short of repositioning opposition axes. Requisite cancelling torques can be provided by friction between hand and object surfaces. Thus, by squeezing an object harder, a larger contact force (normal to the surface) is created, increasing the amount of friction. However, since the effective lever arms are typically short for the friction force, such lever arm length being constrained by finger surface breadth, this source of torque quickly requires large amounts of normal force generated in proportion to the required torque. Since this can be very tiring, the second way of cancelling undesired torques is more desirable. Either the fingers or the palm positioned along the surface of the object may provide force not in opposition to force exerted by other parts of the hand, but in opposition to moments. The amount of force required will decrease with an increase in the distance from the opposition axis constituting the axis of the moment.

Moments tending to twist the axis of opposition likewise can be cancelled either by increasing friction between the object and hand surfaces or by increasing the number of fingers contacting the

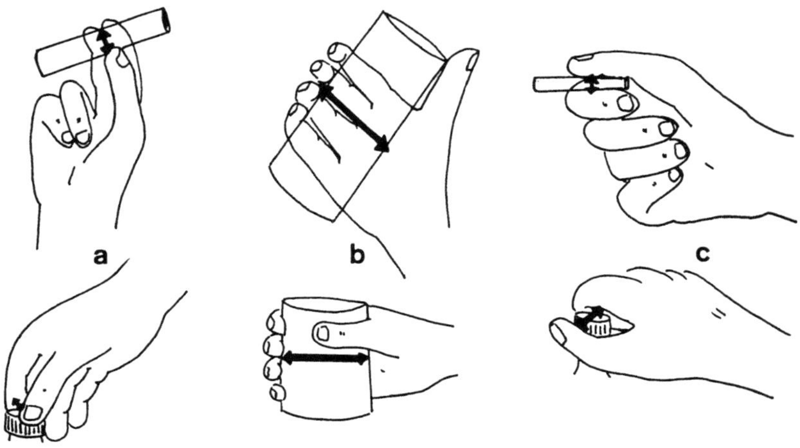

Fig. 2. Basic oppositions of the human hand. a. Pad opposition. b. Palm opposition. c. Side opposition

object. A long cylindrical object, such as a pen, could be held horizontally (wrist pronated and dorsiflexed) in a precision grasp between the thumb pad and index finger; however, if the chosen opposition axis is not near the object's center of mass, a twisting moment arises. With the help of additional fingers and a new opposition axis, this moment can be cancelled without increasing the amount of active force and without knowing the exact location of the center of mass. (This issue extends into the grasping of circular objects, such as jar lids, where again twisting moments are cancelled by using more than two fingers.)

OPPOSITION

Current terminology in the hand literature (Napier 1956; Schlesinger 1919; Landsmeer 1962; Long et al. 1970; Lyons 1985) lacks the flexibility for describing how the hand preshapes in order to bring to bear functionally effective forces for grasping an object within a task context. Grasps tend to be composites of power and precision requirements (Napier 1956). A more useful approach is to focus on the types of opposition available in the hand.

The first, *pad opposition*, is opposition between the pads of one or more fingers and the thumb pad (see Fig. 2a), along an axis roughly parallel to the palm. Opposition at the digit pads offers greater flexibility in finely controlled manipulations of an object at the expense of stability and maximum force. An object held by the digit pads can be translated in a direction roughly normal to the palm by finger and thumb flexion and extension. The object can be translated in a direction laterally across the palm by sliding and climbing along the object with the fingers and thumb. There is also a slight translation possible in the distal–proximal direction. The object can be rotated in two directions by rolling it between oppositely translating finger and thumb surfaces. Finally, the object can be rotated about the axis of opposition between thumb and finger by modulating the torque created by friction. If the opposition axis runs through the center of mass of the object, an additional torque has to be supplied manually. If the opposition axis is off from the center of mass, then gravity can be enlisted to effect the rotation. Pad opposition is strongest and most stable in the direction parallel to the axis of opposition. Forces ensuring equilibrium on the other *dofs* may be supplied by friction for both translations and rotations or by support for rotations (e.g., opposition to gravity–induced torques). Typically the weakest *dof* most subject to destabilization is rotation around the axis of opposition. In sum, flexibility regarding the manipulation of an object along its *dofs* can be had in pad opposition at the expense of a lack of stability along those same *dofs*.

The second basic type of opposition, *palm opposition* between the digits and the palm, is one in which flexibility is sacrificed in favor of stability. Essentially the object is fixed in hand coordinates (see Fig. 2b), along an axis roughly normal to the palm of the hand. Greater stability is achieved in this opposition by a combination of three factors (Thomine 1981). First, the larger surface area of palm provides increased friction force. Second, the more proximal segments of the digits are able to exert greater forces (Hazelton et al. 1975; Chao et al. 1976). Finally, the structural extent of the palm (and fingers) is used to passively cancel torques around the axes lying in the surface. Thus, a large measure of stability is provided, but flexibility must be derived strictly from motions of the wrist or more proximal joints. The thumb in palm opposition, while not part of the opposition itself, adds to its stability in two ways: moving it into a position facing the fingers (i.e., what is called thumb opposition) helps the fingers in opposition to the palm, while, at the same time, increases the surface area of the palm in contact with the object.

A third type of opposition, *side opposition*, between the thumb pad and the side of the index finger, or between the sides of the fingers, is a compromise between flexibility and stability (see Fig. 2c). Its opposition axis occurs primarily along a transverse axis. The thumb, due to the orientation of its articular surfaces, has its pad oriented to the sides of the other digits, giving side opposition using the thumb the extra frictional component. Side opposition between fingers is stronger if the object is held proximally in the fingers (with less control), and weaker more distally (with more available *dofs*). This distal opposition is similar to pad opposition, without the friction available at the pads. When an object is held in side opposition, the object is aligned in the pronation–supination rotational axis of the wrist. By using pronation–supination instead of flexion–extension or adduction–abduction at the wrist, the strongest possible force available at the fingers can be maintained (Hazelton et al. 1975; Chao et al. 1976), with the whole muscular system of the arm able to be used as well. This is especially useful when the thumb is opposing the side of the index finger, due to its strength and frictional qualities.

The shape the hand takes on during the grasp reflects the use of one or more of these oppositions. As was seen in Fig. 1, the place cylinder task used pad opposition. The shake cylinder task started with pad opposition, but changed to palm opposition, with an intermediate step of opposition to the torque arising while switching. If we look at the mug grasp (Fig. 3a), we see that both palm opposition and side opposition are used. The backs of the fingers are also opposing the torque created by the mug not being supported above or around its center of mass. Other examples of composite oppositions are seen in Fig. 3. In Fig. 3b, a pen cap is taken off the pen using two oppositions: the fingers oppose the palm to hold the body of the pen, and the thumb opposes the side of the index finger to hold the cap. In Fig. 3c, the grasp which Napier (1956) calls a power grip is seen as the combination of two oppositions: palm opposition and side opposition. Grasping different size screwdrivers (Fig. 3d–f) demonstrates how oppositions are affected by object properties: for a large screwdriver (Fig. 3d), palm opposition is seen; for a medium screwdriver (Fig. 3e), side opposition is added; for a small screwdriver (Fig. 3f), pad opposition is seen

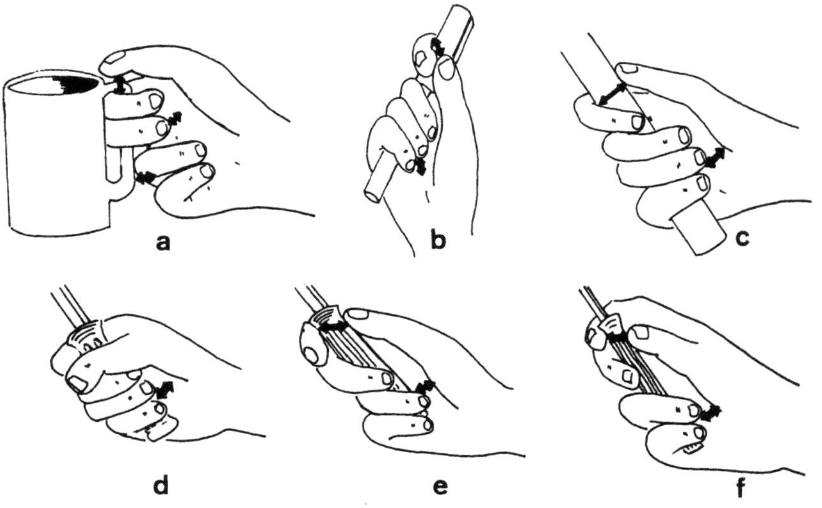

Fig. 3. Composite use of oppositions in grasps. a. Grasping mug. b. Taking cap off pen. c. Holding hammer. d. Grasping large screwdriver. e. Grasping medium screwdriver. f. Grasping small screwdriver

between the thumb and the index and long fingers, while the other two fingers still oppose the palm. Due to the asymmetries of the hand, such mixing of oppositions is commonplace and simple (Tubiana 1981). Composite use of oppositions are the key to understanding features of prehensile hand shapes and opposition space.

PARTITIONING OF THE HAND INTO FUNCTIONAL UNITS

A Mechanism for Control: Schemas

Preshaping anticipates the physical constraints that will arise during the prehensile task. Task constraints, which necessarily include object dimensions, determine opposable surfaces which specify *where* the grip can take place. But the issue in preshaping the hand is *how* the grip can take place. Opposition space defines the "how" in terms of the kinematic and dynamic values available at various grasping postures. The system needs to map task constraints (the functional requirements in the task) to opposition space (the functions available at a given hand posture). A controller for preshaping the hand can then send the fingers travelling to an appropriate opposition space.

How may a suitable control system use these constraints to manage functionally effective behavior? Schema theory (Arbib 1981) postulates the existence of coordinated control programs involving both parallel and sequential activation of perceptual and motor structures, or schemas. A schema monitors information relevant to an activity it controls in order to tune its behavior. The overall behavior of the system is the combined behavior of all of its component schemas; thus, control is distributed across interacting parallel units. Schemas might be activated which try to drive action systems into dissimilar activities, in which case the schemas compete for control of the movement.

Preshape Schema as It Relates to Opposition Space

Opposition space is described by its kinematic and dynamic characteristics. For example, a strong force can be applied by palm opposition involving the use of the anterior surfaces of all fingers (as VF2) and the palm (as VF1). Since the object is defined in terms of opposable surfaces, the problem solved by the controlling schemas is to match virtual fingers to opposable surfaces in such a way as to satisfy task constraints. Two opposable surfaces represent a relationship between two virtual fingers (e.g., their size, their distance from each other, etc.). The larger the virtual finger, the more force is passively available from the friction to overcome the torques, as described above.

The *preshape schema* will map task constraints into virtual finger configurations needed to perform the task. In readying the hand for grasping, the *preshape schema* must place the hand into a shape which meets two criteria: (a) the hand shape is large enough to allow a task–specific part of the object to be centered and contained in the opposition space created; and (b) a smooth movement along the opposition axis or axes will then secure the object in a firm grasp.

Task Parameters for Prehensible Movements

To satisfy the task constraints, we suggest that perceptual schemas extract from the visual information relevant values for prehensile parameters about the object within the context of a task. These include:

1. *Relevant object properties.* The appropriate opposable surfaces and opposition axes are task–specific; for example, where and how a pen is picked up for writing is different from where and how it would be picked up for throwing. The relative distance between virtual fingers must be larger than the width of the object along the opposition axis or axes. The size of the virtual finger (e.g., whether two or three real fingers can be mapped into VF2 in the grasp mug task) depends on the size of the opposable surface area as well as its radius of curvature. The relative location of opposition axes on complex objects such as the mug helps define the necessary opposition space with appropriate oppositions, allowing for an error tolerance which can be corrected for during the smooth enclosing movement.

2. *Functional degrees of freedom.* The *preshape schema* will create an opposition space which makes the task relevant *dofs* available, while the others are frozen out (e.g., in the mug task, the functional *dof* needed is rotation, which can be done using wrist rotation, while freezing out all finger *dofs*). The amount of control (i.e., movement resolution) over a *dof*, relative to the anticipated forces within the task context helps define appropriate oppositions as well as how the functional *dofs* are best mapped into actual *dofs* (e.g., finer control of the translational *dofs* is needed in the place cylinder task, more so than in the shake cylinder task). The range of the functional *dofs* also effects the mapping (e.g., the wrist allows an object to rotate further than do the finger pads).

3. *Anticipated forces and torques.* The created opposition space must cancel (or perhaps use) anticipated forces and torques (gravity–induced from the weight of the object, inertial loads that will arise during the task relevant movements, imbalance of forces between the oppo-

sition axis and the object's center of mass, hand–induced moments created by the virtual fingers themselves, etc). For two–phase tasks, such as the shake cylinder task, the *preshape schema* creates an opposition space to account for the initial lifting forces and not the later inertial loads; however, in the grasp mug task the opposition space created accounts for both the initial lifting forces and the later inertial loads (which arise when the mug is rotated toward the mouth).

The units on the dimensions of all the above must be scaled to the dimensions of the hand.

Schemas for Preshaping Virtual Fingers

We suggest that the *preshape schema* will activate, in parallel, opposition schemas which map virtual fingers into real parts of the hand. The three opposition schemas (*pad opposition schema, palm opposition schema, and side opposition schema*) compete and tune their behavior based on values of the task parameters listed above. The behavior of these schemas, all competing for the control of the hand, will produce particular configurations of virtual fingers. *Move schemas* will move the virtual fingers, within a particular time frame and tactile contact expectation, to locations and orientations relative to the object.

We can now reanalyze the two tasks described above in terms of virtual finger configurations available to a schema controller working within opposition space. To complete the analysis, the grasp mug task will also be reviewed once more.

Place Cylinder Task Revisited

In the place cylinder task (Fig. 1a–c), the available opposable surfaces of the cylinder are parallel to gravity. The object is light and long with a moderate radius of curvature, relative to the conformation potential of the hand surfaces. The task of placing it on the circle calls for flexible control of multiple *dofs*, with emphasis on small horizontal translations and rotations about horizontal axes. The center of mass of the object might not remain above the base during the task; however, it is light and so the resulting torques are easy to control. The perceived weight of the object can be overcome by the friction force available parallel to the pads of the fingers and thumb. Therefore, since the many *dofs* available are preferred in this task, and since the normal force required to avoid dropping the object does not exceed that available in the flexible grasp, the constraints converge on a grasp that uses pad opposition. This opposition provides the needed flexibility, in terms of available translational and rotational *dofs* to more precisely fit the cylinder base to the circle, as well as sufficient force to oppose the anticipated light weight of the object. The *pad opposition schema* becomes activated, mapping the thumb into VF1 and the index and long finger into VF2. Opposition space is created by moving the pad of VF1 into opposition with the pad of VF2, with a distance between them larger than the width of the cylinder. An *enclosing schema* will then bring VF1 and VF2 together along the opposition axis.

Shake Cylinder Task Revisited

In the shake cylinder task (Fig. 1d–f), the weight of the object and the reaction forces that will arise during the rapid movement of the object must be cancelled. The length of the object along the direction of the translation axis is longer than the hand width. A more powerful grip is required which will overcome these forces, and which necessarily locks the *dofs* of the hand, leaving only wrist and arm movements. Nevertheless, since these forces will arise later in the task, initially grasping the object in the finger and thumb pads allows the hand better access to the object without requiring adjustment of more proximal elements of the reaching system. It also allows the wrist to stay oriented generally adducted and dorsiflexed, a comfortable and effective position (Kapandji 1982). The result is that opposition space is created for the lifting subtask using the *pad opposition schema*, mapping the thumb into VF1 and the index and long fingers into VF2. The *enclose schema* brings VF1 and VF2 together along the opposition axis. Subsequently, the *shake schema* shifts the hand into palm opposition for the shaking subtask, mapping the palm into VF1 and the four fingers into VF2. The large surface area provided by palm opposition allows the application of both passive (friction) and active (flexor muscles) forces, as well as redundant sensory knowledge about unwanted object movements. The action of the thumb will depend on the competition between the *side opposition schema* and the *palm opposition schema*. The first will try to drive it to the radial side of the object, while the latter will try to drive it into thumb opposition. If the anticipated forces are significant (as they most likely are if the shaking is vigorous), the *palm opposition schema* will move the thumb to a place across the backs of the fingers, thus increasing the surface area of the palm, and creating a stronger grip by helping VF2.

During the transition between subtasks, a gravity–induced torque will rotate the cylinder as the hand turns with it between the thumb and index finger. In order to overcome the torque, the palm is mapped into VF3. Near the end of the transition, the *palm opposition schema* becomes more active, and the new opposition space emerges.

Grasp Mug Task Revisited

In the grasp mug task (Fig. 3a), one functional *dof* is the rotation of the mug when it reaches the mouth. If it is an extremely delicate teacup, the *pad opposition schema* will be active and the rotation will be done at the finger level between VF1 (the thumb) and VF2 (the index finger). If the mug is somewhat heavier and larger, pronation at the wrist would be better, in which case equal activation would occur of the *side opposition schema* (to maintain the correct orientation for and during the rotation) and the *palm opposition schema* (to stabilize the handle against the palm and volar finger surfaces during the rotation). However, a gravity–induced torque will arise directed toward the hand when the mug is lifted, due to the fact that the handle is not above or centered around the center of mass. If any real fingers are available, they are mapped into VF3, and curled in order to cancel the torque. If no real fingers are available, the force being applied on the handle at VF2 is increased. Opposition space is created when the pad of VF1 moves to oppose VF2 (at the side if the *side opposition schema* is active), VF2 curls (driven by the *palm opposition schema*), and VF3 curls to oppose the anticipated torque. The *enclose schema* will then bring VF1 and VF2 together along the opposition axis through the top of the handle, VF2 will enclose around the side of the handle (however, in this case, not moving along the opposition axis, but instead following the contour of the handle), and VF3 will extend opposing the torque.

FUTURE RESEARCH

In the previous section, we have outlined how a schema controller might be mapping functional requirements of tasks onto physical constraints. While still only speculations at this point, we are using these concepts to guide further research in order to verify them. Specifically, we see the basic problems in extracting the relevant features of prehensile behavior as follows:

1. Schema interaction with respect to functional requirements
2. Competition and selection of motor schemas
3. Minimization criteria for opposition space selection
4. Detailed description of opposition in terms of the biomechanics of the hand
5. Center of mass perception and development of grasping strategies to account for its use in hand postures
6. Relative contribution of passive vs active forces

CONCLUSIONS

This paper describes an initial approach to problems in the control of prehensile movements through the use of coordinated control programs involving both parallel and sequential activation of schemas. Preshaping the hand into a posture suitable for interaction with objects forces the controller to have a model of the physical properties of the object and the hand working within a task. We describe flexible functional units called virtual fingers, which are groupings of real fingers or the palm. We define opposition space as the area where three basic oppositions can take place, alone or in parallel, so that opposing forces can be exerted between virtual finger surfaces to effect a stable grasp. We describe opposition schemas activated by the *preshape schema* that implement virtual finger configurations based on such task constraints as object properties, anticipated forces, and functional degrees of freedom. This research will be further developed to model other aspects of human prehensile behavior, and will go on to suggest models for neural mechanisms for control of hand movements.

REFERENCES

Abbs JH, Gracco VL, Cole KJ (1984) Control of multimovement coordination: sensormotor mechanisms in speech motor programming. J Mot Behav 16: 195 – 231

Amontons G (1699) De la résistance causée dans les machines. Memoires de l'Academie Royale des Sciences. pp 206 – 227

Arbib MA (1981) Perceptual structures and distributed motor control. In: Brooks VB (ed) Handbook of physiology – The nervous system II. Motor control. American Physiological Society, Bethesda, pp 1449 – 1480

Arbib MA, Iberall T, Lyons D (1985) Coordinated control programs for movements of the hand. In: Goodwin AW, Darlian – Smith I (eds) Hand function and the neocortex. Springer, Berlin, pp 111 – 129 (Experimental brain research supplement, vol 10)

Bowden FP, Tabor D (1950) Friction and lubrication of solids. Clarendon, Oxford

Chao EY, Opgrande JD, Axmear FE (1976) Three-dimensional force analysis of finger joints in selected isometric hand functions. J Biomech 9: 387-396

Comaish S, Bottoms E (1971) The skin and friction: deviations from Amontons's law, and the effects of hydration and lubrication. Br J Dermatol 84: 37-43

Fearing RS (1983) Touch processing for determining a stable grasp. Department of Electric Engineering and Computer Science, MIT, Cambridge

Glicenstein J, Dardour JC (1981) The pulp: anatomy and physiology. In: Tubiana R (ed) The hand, vol 1. Saunders, Philadelphia, pp 116-120

Hanafusa H, Asada H (1982) Stable prehension by a robot hand with elastic fingers. In: Brady M, Hollerbach JM, Johnson TL, Lozano-Perez T, Mason MT (eds) Robot motion: planning and control. MIT, Cambridge, pp 323-335

Hazelton FT, Smidt GL, Flatt AE, Stephens RI (1975) The influence of wrist position on the force produced by the finger flexors. J Biomech 8: 301-306

Hollerbach JM (1981) An oscillation theory of handwriting. Biol Cybern 39: 139-156

Jeannerod M (1981) Intersegmental coordination during reaching at natural visual objects. In: Long J, Baddeley A (eds) Attention and performance IX. Erlbaum, Hillsdale, pp 153-168

Jeannerod M (1984) The timing of natural prehension movements. J Mot Behav 16: 235-254

Johannson RS, Westling G (1984) Influences of cutaneous sensory input on the motor coordination during precision manipulation. In: von Fuler C, Franzen O, Lindblom U, Ottoson D (eds) Somatosensory mechanisms. MacMillan, London, pp 249-260

Kapandji IA (1982) The physiology of the joints, vol 1. Upper limb, 5th edn. Churchill Livingstone, Edinburgh

Kelso JAS, Tuller B, Fowler CA (1982) The functional specifity of articulatory control and coordination. J Acoust Soc Am 72: S103

Landsmeer JM (1962) Power grip and precision handling. Ann Rheum Dis 22: 164-170

Long C, Conrad PW, Hall EA, Furler SL (1970) Intrinsic-extrinsic muscle control of the hand in power grip and precision handling. J Bone Joint Surg 52A: 853-867

Lyons DM (1985) A simple set of grasps for a dextrous hand. In: IEEE international conference on robotics and automation, march 25-28, St. Louis. pp 588-593

Mardsen CD, Meadows JC, Lange GW, Watson RS (1969) The role of the ballistocardiac impulse in the genesis of physiological tremor. Electroencephalogr Clin Neurophysiol 27: 169–178

Mason MT (1982) Manipulator grasping and pushing operations AI–TR–690. Artificial Intelligence Laboratory, MIT, Cambridge

McCollum G, Horak FB, Nashner LM (1984) Parsimony in neural calculations for postural movements. In: Bloedel J, Dichgans J, Precht W (eds) Cerebellar functions. Springer, Berlin, pp 52–66

McLean WG, Nelson EW (1978) Engineering mechanics, 3rd edn. McGraw–Hill, New York

Moore DF (1972) The friction and lubrication of elastomers. Pergamon, New York

Moore DF (1975) Principles and applications of tribology. Pergamon, New York

Napier JR (1956) The prehensile movements of the human hand. J Bone Joint Surg 38B: 902–913

Nashner LM, McCollum G (1985) The organization of human postural movements: a formal basis and experimental synthesis. Behav Brain Sci 8: 135–172

Quilliam TA (1978) The structure of fingerprint skin. In Gordon G (ed) Active touch. Pergamom, Oxford, pp 1–18

Salisbury JK (1982) Kinematic and force analysis of articulated hands. STAN–CS–82–921. Department of Computer Science, Stanford

Schlesinger G (1919) Der Mechanische Aufbau der Künstlichen Glieder. In: Borchardt M et al (eds) Ersatzglieder und Arbeitshilfen. Springer, Berlin

Thomine JM (1981) The skin of the hand. In: Tubiana R (ed) The hand, vol 1. Saunders, Philadelphia, pp 107–115

Tubiana R (1981) The architecture and function of the hand. In: Tubiana R (ed) The hand, vol 1. Saunders, Philadelphia, pp 19–93

Turvey MT, Shaw RE, Mace W (1978) Issues in the theory of action: degrees of freedom, coordinative structures and coalitions. In: Requin J (ed) Attention and Performance VII. Erlbaum, Hillsdale

Westling G, Johannson RS (1984) Factors influencing the force control during precision grip. Exp Brain Res 53: 277–284

Whiting HTA (1984) Human motor actions: Bernstein reassessed. Elsevier, Amsterdam

Wilkes GL, Brown IA, Wildnauer RH (1973) The biomechanical properties of skin. CRC Crit Rev in Bioengin 1: 453 – 495

Williams M, Lissner HR (1962) Biomechanics of human motion. Saunders, Philadelphia

Wing AM, Fraser C (1983) The contribution of the thumb to reaching movements. Q J Exp Psychol 35A: 297 – 309

Wolter JD, Volz RA, Woo AC (1985) Automatic generation of gripping positions. IEEE Trans Syst Man Cybernetics SMC – 15: 204 – 213

Self-Inhibition and the Disruptive Effects of Internal and External Feedback in Skilled Behavior*

D. G. MacKay

Department of Psychology, University of California, Los Angeles, CA 90024, USA

This paper examines the hypothesis that following activation, the components for producing a skilled behavior undergo a brief period of self-inhibition during which they are difficult to reactivate. Evidence for this self-inhibition hypothesis is reviewed from a wide range of areas: electrophysiology, electromyography, misspellings by dysgraphics, the perception and recall of briefly viewed misspellings by normal subjects, the nature of omission errors in speech production and transcription typing, and between-language regularities in the pattern of phoneme repetition within words.

The paper begins with a theoretical rationale for the concept of self-inhibition, based on a recently proposed theory of the relationship between the perception and production of speech (MacKay 1982, in press). I next outline the evidence for self-inhibition and finally examine some disruptive effects of feedback that can be explained as a side-effect of self-inhibition. I focus in particular on the role of self-inhibition in the disruption of ongoing speech production that results when auditory feedback is amplified and delayed about 200 ms.

SELF-INHIBITION AND THE NODE STRUCTURE THEORY

Nodes are the basic units for perception and production in the node structure theory of MacKay (1982, in press) and some nodes take part in both perception and production, whereas others do not. The unshared components include a system of sensory analysis nodes for representing patterns of sensory experience in perception and a system of muscle movement nodes for representing patterns of muscle movement in production. The shared components are mental nodes which represent phonological units such as segments and syllables, and sentential units such as words and phrases (see MacKay 1982).

Nodes have two states of activity: priming and activation. Activation is required for perception and action and is always sequential in nature. Moreover, activation always requires a special triggering mechanism to determine when and in what order the relevant nodes become activated. The numbers

* This paper was completed while the author was a member of the "Perception and Action" research group at the Center for Interdisciplinary Research (ZiF) at the University of Bielefeld. The author acknowledges the support of ZiF with appreciation. A preliminary version of the paper appeared as Report #32, 1985, Research Group on Perception and Action at the Center for Interdisciplinary Research.

Experimental Brain Research Series 15

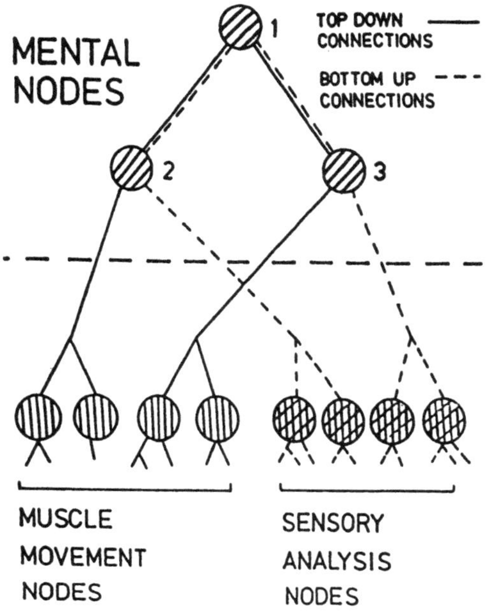

Fig. 1. The relation between mental nodes, sensory analysis, and muscle movement nodes

adjacent to the nodes in Fig. 1 represent this order of activation, but the special activating mechanisms required to achieve this order have been left out. (However, see MacKay 1982.) During its period of self-sustained activation, a node automatically primes all nodes connected to it. Priming refers to a transmission across the connection between nodes. This transmission produces an increased subthreshold activity which readies the connected node for potential activation. Priming summates over time, but cannot by itself cause activation. Activation instead requires application of an activating mechanism which causes the most primed node in its domain to become activated at the time when it is applied. All nodes become activated under this "most-primed-wins principle."

Domains consist of natural categories of nodes (e.g., noun nodes, verb nodes), all of which share the same activation mechanisms. Finally, errors in action and perception occur when an "extraneous" node (i.e., a node from outside a hierarchy such as the one in Fig. 1) receives greatest priming at the time when the activating mechanism is applied. Consider, for example, Freudian slips of the ear such as the misperception of *carcinoma* for *Barcelona* in the case of an individual listening to a sentence about *Barcelona* while simultaneously being concerned or preoccupied with the disease known as *carcinoma*. The misperception occurs because priming for the node representing *carcinoma*, arising from the preoccupation (top-down) and from aspects of the acoustic stimulus (bottom-up) exceeds the priming for the node representing *Barcelona*, arising from the input itself. As a consequence, the extraneous node representing *carcinoma* becomes activated under the most-primed-wins principle rather than the one representing *Barcelona* when the activating mechanism for nouns is applied. The theory predicts a general tendency for errors to involve substitutions from within the same domain. For example, nouns will substitute with nouns and verbs with verbs (see MacKay 1982).

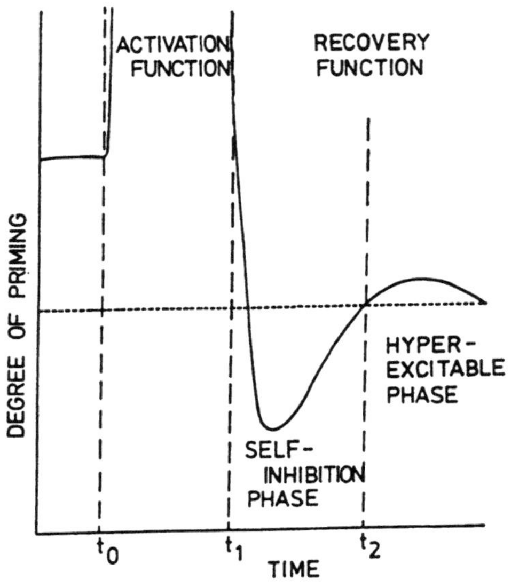

Fig. 2. The recovery cycle for a single node: self–inhibition begins at t_1 following activation at t_0. The period between t_2 and t_3 corresponds to the hyperexcitable phase

A RATIONALE FOR SELF–INHIBITION AND THE RECOVERY CYCLE

Self–inhibition becomes necessary whenever a node receives both top–down and bottom–up connections in an output hierarchy. Because only mental nodes are used both for producing and for perceiving cognitive units such as phrases, words, syllables, and segments, the self–inhibition postulate applies only to mental nodes and not to sensory analysis or muscle movement nodes. The reason can be seen in Fig. 1, which illustrates the detailed relations between sensory analysis nodes, mental nodes, and muscle movement nodes. Bottom–up connections are represented by the broken lines and top–down connections by the solid lines, and, as can be seen, mental nodes receive both top–down and bottom–up connections. These two–way connections lead to the possibility of reverberatory effects at every level in a system. The reasoning can be illustrated for two hierachically connected, but otherwise arbitrary nodes: A (superordinate) and B (subordinate). During production, A becomes activated and primes B via the top–down connection. However, when B subsequently becomes activated, this could lead to reactivation of A because of the bottom–up connection required for perception involving both of these same nodes. The self–inhibition that follows activation therefore ensures that higher level nodes do not become reactivated due to bottom–up priming arising from the activation of subordinate nodes.

I come now to the question of how nodes recover following self–inhibition. My basic assumption is that self–inhibited nodes undergo a normal recovery cycle, including a period of hyperexcitability or postinhibitory rebound during which self–priming first rises above and then returns slowly to resting level. Figure 2 illustrates the recovery cycle for a single node.

Activation begins at time t_0, when the activating mechanism is applied and ends at t_1 with the onset of self–inhibition. During the recovery phase in Fig. 2 priming first falls below resting level and then rebounds (the hyperexcitability phase) before returning to resting level again. The duration of the entire recovery cycle (from self–inhibition until the return to resting level) is assumed to vary with the level of the node in the system. For the lowest level nodes, the entire recovery cycle lasts only a few milliseconds. However, for higher level nodes, the cycle can last up 300 ms with a refractory phase of about 50 ms and a peak in self–priming at about 200 ms.

EVIDENCE FOR SELF–INHIBITION AND THE RECOVERY CYCLE

In this section I discuss various sources of evidence for self–inhibition and recovery with the general characteristics discussed above.

Electrophysiological Evidence

Although the figure of 200 ms for the peak period of hyperexcitability for mental nodes is only approximate, it receives support from the electrophysiological literature. For example, Tunturi (1958) found a postinhibitory rebound that follows activation by as much as 200 ms in central neuronal aggregates.

Electromyographic Evidence

Electromyographic activity without the occurrence of full–fledged muscle movement indicates that muscle movement nodes are being primed in the theory (MacKay 1981). Since spontaneous priming is assumed to peak at about 200 ms following self–inhibition, the theory therefore predicts a peak in electromyographic activity in the appropriate muscles 200 ms after production of a segment. Evidence for such a peak is found in a study by Ohala and Hirano (1967). They had subjects produce the syllable /pa/ while recording electromyographic activity in the obicularis oris musle of the lips and observed two peaks of subthreshold electromyographic activity. The first occurred just prior to the contraction of the lip muscles, and the other, smaller peak occurred about 200 ms later, as if reflecting rebound hyperexcitability.

The Production of Misspellings by Dysgraphics

A dysgraphic is someone who chronically misspells common words with very high probability, due perhaps to cerebral injury, but not to a lack of schooling or a general inability to learn. Lecours (1966) examined the misspellings of one dysgraphic (Lee Harvey Oswald) and discovered a repeated letter effect, which was analyzed further in a study by MacKay (1969a). Oswald often dropped a repeated letter in a word, e.g., misspelling *Elderly* as *Eldery*, but he sometimes added a repeated letter of his own in *Habitituated* and *Decemember*. MacKay (1969a) showed that deletions of repeated letters were significantly more common than additions and that deletions of the second of two repeated letters, e.g., *Eldery* were significantly more common than deletions of the first,

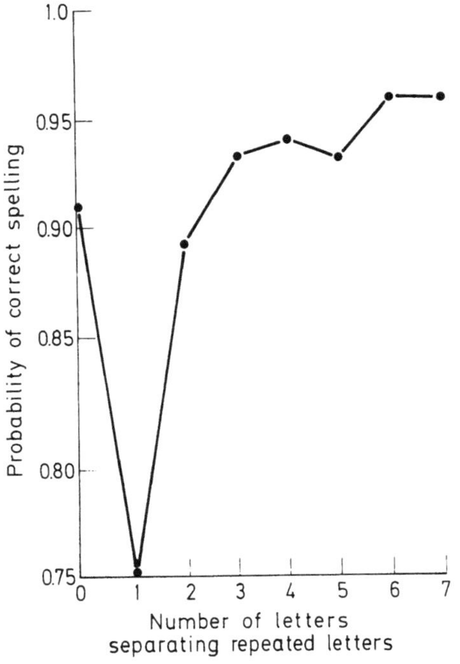

Fig. 3. The probability of correct spelling (vs letter deletion) as a function of the number of letters separating a repeated letter (from MacKay 1969a)

e.g., *Ederly*, as would be expected if repeated letter deletions reflected self-inhibition of nodes which are to be activated in sequence.

However, MacKay's most interesting data bearing on the self-inhibition hypothesis concerned the degree of separation of a repeated letter which became deleted. Oswald frequently misspelled repeated letters which were close together as in *Anlyze*, but he rarely dropped a widely separated repeated letter as in *Misspellng*. The probability of *correctly* spelling a word containing repeated letters in the sample is shown in Fig. 3 as a function of degree of separation of the repeated letters. For example, the repeated letter error in *Coperation* would be assigned Separation 0 since no letters separate the repeated letters in the error, that in *Anlyze* is Separation 1 since one letter separates the repeated letter in the error, and so on up to Separation 7.

As can be seen in Fig. 3, the probability of correct spelling was moderately high for immediately repeated letters (Separation 0), dropped dramatically for repeated letters with Separation 1 and increased to its highest level at Separation 6. Except for the moderately high probability of correctly spelling immediately repeated letters, the function in Fig. 3 can be seen to resemble and be explained by the recovery cycle function in Fig. 2. Moreover, the special status of immediately repeated letters is to be expected: unlike other repeated letters, immediately repeated letters are generally not pronounced in English, do not represent separate phonemes, and are coded via special orthographic rules which do not apply to repetitions with greater separation. All three factors suggest that zero separation letter repetition data should be disregarded or treated separately in this and subsequent figures. This granted, the data suggest that dysgraphic deletions may reflect an abnormally strong and extended recovery cycle following self-inhibition of nodes for writing out the letters in words.

The Perception and Recall of Misspellings

Another source of evidence for the recovery cycle hypothesis comes from a study of the detection of experimentally constructed misspellings. MacKay (1969a) examined how easily normal subjects could perceive and recall misspellings which resembled the ones produced by the dysgraphic discussed above. The results showed that normal individuals experienced greatest difficulty in perceiving and recalling those experimentally constructed misspellings which were similar to the ones that the dysgraphic produced more frequently. The misspellings were planted in sentences which the subjects read at about 77 ms per letter. They then attempted to recall the sentence, writing it out exactly at it was spelled, guessing at misspellings if necessary. Following recall, the sentences were presented again and subjects were shown each spelling error in turn and asked whether they had noticed it when reading the sentence.

The results showed that repeated-letter misspellings such as *Eldery* were harder to perceive than nonrepeated-letter misspellings such as *Eldely*. Similarly, given that the subject claimed to perceive an error when reading the sentence, misspellings that involved repeated letters were more difficult to recall than those that did not. Moreover, the functions relating degree of separation of the repeated letters in both perception and recall resembled both the dysgraphic function (Fig. 3) and the recovery cycle function (Fig. 2). The perception function appears in Fig. 4 (left ordinate), which plots the probability of correctly detecting a repeated-letter misspelling as a function of the degree of separation of the repeated letters. As can be seen there, repeated-letter misspellings (the broken line in Fig. 4) were only harder to detect than nonrepeated misspellings (the solid horizontal line in Fig. 4) when 0 to 3 letters separated the repeated letters. However, when more than three letters separated the repeated letters, detecting a repeated-letter error was as easy or easier than detecting a nonrepeated-letter error.

The recall function also appears in Fig. 4 (right ordinate), which plots the probability of correctly recalling repeated-letter misspellings that have been detected. As can be seen there, repeated-letter misspellings (the broken line in Fig. 4) were only harder to recall than nonrepeated-letter misspellings (the horizontal solid line) when 0 to 2 letters separated the repeated letters. With more than three intervening letters, recalling a repeated-letter error was as easy or easier than recalling a nonrepeated-letter error. This degree of separation effect is readily explained under the recovery cycle hypothesis.

Limitations of MacKay (1969a)

Although the results of MacKay (1969a) support the recovery cycle hypothesis, there are a number of ways that the procedures could be extended and refined using computer-controlled stimulus presentation. Consider, for example, a task where the subject must detect a set of possible target letters in a rapidly presented sequence of displays, as in Shiffrin and Schneider (1977) except for the minor modification that a target letter can be repeated in the sequence. The subject has three tasks. The top priority task is to respond as quickly as possible to indicate detection of the first

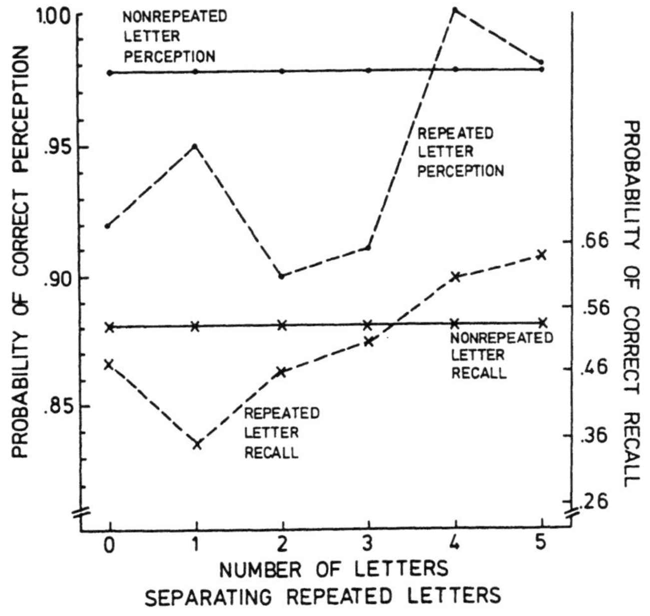

Fig. 4. The probability of detection (left ordinate dots) and recall (right ordinate crosses) for repeated letter mispellings (broken lines) and nonrepeated letter misspellings (solid lines)

target; the other, subordinate tasks are to indicate whether one or two targets were presented and what the target(s) were. In the special case where the same target letter is repeated (in either identical of different location in the display), the node structure theory predicts that occurrence of the first target will reduce detection of the second target if the intervening interval is short (up to 50 ms, say), but will facilitate detection of the second target when the intervening interval approximates some critical value corresponding to the hyperexcitable phase (say 200 ms).

Repeated Phonemes in Fast Speech

As expected under the recovery cycle hypothesis, speakers tend to drop immediately repeated phonemes (Separation 0) when speaking rapidly. However, determining exactly which nodes are responsible for this phenomenon is difficult. For example, when speaking rapidly, speakers of English often drop one of the /k/'s in *take care*, but it is often difficult to tell which /k/ was droped or whether the omission was intentional (Heffner 1964). Moreover, a single segment node cannot be responsible for this effect since these segments are produced by two different segment nodes, /k/ (final consonant group) and /k/ (initial consonant group). The omission can only reflect self–inhibition of nodes below the segment level.

Omission Errors in Speech and Typing

MacKay (1969c) identified a class of speech errors labeled motor masking (Type I) that unlike the rapid speech omissions, are undeniably inadvertent: the speaker unintentionally drops the second of two segments that must be repeated in close succession. An example of a German error from Meringer and Meyer (1895) is *der iese* instead of *der Riese*, where the immediately repeated /r/ has been dropped. Although such errors support the self–inhibition hypothesis (see MacKay 1969c), the use of experimental induction procedures seems advisable for studying these errors, which are relatively rare and require that the observers make subtle perceptual discriminations under less than optimal observational conditions (see Cutler 1982).

However, omission errors in skilled transcription typing exhibit a similar phenomenon which cannot be attributed to observer error. High–speed videotapes indicate that typestrokes proceed in three stages (Grudin 1983): a movement of the finger toward the key, a rapid downstroke to strike the key, and an equally rapid rebound or lift–off from the key. During an omission error the finger either fails to move toward the key or fails to execute the downstroke. Like Type I masking errors, typestroke omissions are usually preceded by an identical letter or finger movement, either in the preceding word or within the word itself. For example, Grudin had skilled typists transcribe a text as rapidly as possible and found that half of the typists had omitted the third /i/ in the word *artificial*. Moreover, in Grudin's overall sample of omission errors, the left–to–right masking (omission) effect was strong enough to override a general tendency for word–initial letters to be correctly typed. Example errors are, *the nitre* for *the entire* and *keep utting* for *keep putting*. Such errors suggest that following activation, the nodes for typing a letter and the movement components for striking the key undergo self–inhibition, which interferes with the subsequent execution of an identical movement due to follow closely in time.

The Structure of Phoneme Repetition

Somewhat indirect, but nevertheless interesting, support for the recovery cycle hypothesis comes from a study of phoneme repetition in the structure of words (MacKay 1970a). The rationale of the study was as follows: If a recovery cycle determines how easy it is to repeat an element in a word, then the structure of repeated elements in the word should reflect that recovery cycle. Immediate repetition of an element should be rare, reflecting the self–inhibitory phase of the recovery cycle. But at some point following self–inhibition, repetition should become more likely than would be expected by chance, reflecting the hyperexcitable phase of the recovery cycle. This of course assumes that speakers normally produce the repeated segments in a word at relatively fixed times on the average and do not alter their rate of speech in the middle of a word, depending say on the nature of its components, but such assumptions seem reasonable in view of phenomena such as constant relative timing (see Shapiro 1977). To summarize, the prediction is that immediate repetition of phonemes in the structure of words should be rare due to the recovery cycle constraint. However, repetition of a phoneme in a word should be highly likely after a degree of separation corresponding to the hyperexcitable phase.

To test this prediction, MacKay (1970a) examined the patterns of phoneme repetition in two very different languages, Serbo–Croatian and Hawaiian. The pattern of phoneme repetition in words of various lengths in the two languages appears in Fig. 5 and is remarkably similar for both

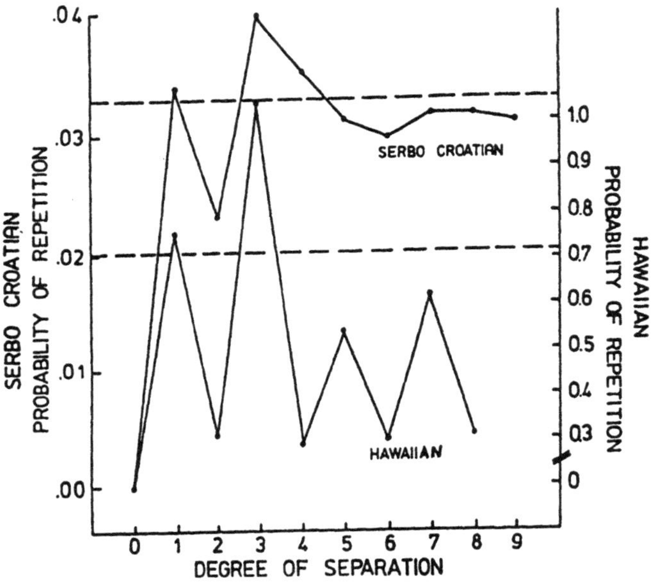

Fig. 5. The probalitity of phoneme repetition in Croatian (left ordinate dots) and Hawaiian (right ordinate crosses) as a function of degree of separation of the repeated phonemes. The horizontal lines represent probabilities under the hypothesis that phoneme repetition is a random event

languages. A very large and random sample of words was examined and the probability of segment repetition at the various separations was corrected on the basis of word length (since long separations for short words are impossible). As can be seen in Fig. 5, immediate repetition of phonemes was highly unlikely and significantly less likely than would be expected if phoneme repetition in a word were a random event (the dashed line in Fig. 5). However, the probability of repetition differed for consonants vs vowels, peaking with one intervening element of vowels in both Serbo–Croatian and Hawaiian, and peaking with about three intervening segments for consonants in both languages.

There were also differences between the two languages. As can be seen in Fig. 5, the repetition pattern was more erratic for Hawaiian than for Serbo–Croatian. The reason is that Hawaiian syllables generally have CV structure, whereas Serbo–Croatian syllables are much more varied and complex in structure. Thus, since vowels and consonants tend to alternate in Hawaiian, segment repetition at even numbered degrees of separation was of necessity infrequent. If syllable structure had been factored out in computing the probabilities of repetition, the Hawaiian and Serbo–Croatian functions would have resembled each other more closely.

As MacKay (1970a) points out, the peaks in Fig. 5 suggest a "law of latent alliteration" and the effect of Hawaiian syllable structure serves to illustrate the statistical nature of that law. Many other factors play a role in determining the phonological structure of words and can thereby override the law of latent alliteration in any one instance. Another important factor, especially for complex or derived words is the nature of the affixes and stems that combine to make up the word. Adding a different prefix could change the structure of repeated phonemes in the word. It is therefore remarkable that a consistent repetition pattern should appear at all and that it should be so similar for Hawaiian and Croatian, languages that differ in the nature of their segments, number of segments (13 vs 35), and average number of segments per syllable and per word. The data therefore suggest, but do not prove, that the pattern of segment repetition in Fig. 5 is representative of all languages. Of course, if general properties of nervous action are responsible for this pattern, then a similar pattern should appear with only minor variations in all languages.

A great deal of further research is required to fully specify how a word acquires its phonological structure and the pattern of segment repetition within it. We know that phoneme repetition is a factor in the phonological changes that words undergo over time, since repeated segments often become dropped as a language evolves, e.g., the Latin word *stipipendium* changed to *stipendium*, dropping the repeated /p/ (see Meringer and Meyer 1895). These observations fit the prediction that phoneme repetition will represent a factor in the evolutionary changes that words undergo in the history of languages, but the extent to which the hypothesized recovery cycle plays a role in the initial invention of words or in the evolutionary changes that occur over time is currently unknown.

Consider now the limits of the revovery cycle hypothesis for theories of speech and other behaviors. One limit is that neither the recovery cycle nor constraints on repetition apply to all units for producing speech. Consider distinctive features, for example. Immediate repetition of distinctive features in adjacent segments is characteristic of many languages, which suggests that distinctive feature nodes either do not undergo self–inhibition, achieve extremely rapid recovery, or undergo activation for longer periods than segment nodes. This difference is consistent with the fact that distinctive feature nodes lack the two–way subordinate connections of other mental nodes: Because distinctive feature nodes receive bottom–up connections from sensory analysis nodes, but send top–down connections not to these same (sensory analysis) nodes but to different (muscle movement) nodes, the main raison d'être of self–inhibition (reverberatory reactivation) does not apply to distinctive features nodes.

Another limit concerns the potential or maximal rate of segment repetition. At best, data on the structure of segment repetition only suggest a repetition pattern that may be *easily* produced at average rates for everyday speech. The data say nothing about the potential or *maximal* rate for moving the speech muscles or for producing syllables. By exerting greater effort, speakers can undoubtedly exceed the natural repetition rates suggested in Fig. 5. A final limit to the recovery cycle hypothesis concerns the nature of the skill being investigated. Similar constraints on element repetition can only be expected for skills that employ shared input–output nodes. Skills that do not engage mental nodes cannot be expected to display the same recovery pattern.

SELF–INHIBITION AND THE DISRUPTIVE EFFECTS OF FEEDBACK

Self–inhibition is also necessary to prevent disruptive effects of external feedback. By way of illustration, consider the node labeled 3 in Fig. 1, a low–level mental node which becomes activated and gives rise to auditory feedback that I will label A. The sensory analysis nodes automatically process A and eventually prime 3, the node responsible for generating the output in the first place. This feedback–induced priming could cause multiple reactivation of 3, unless it arrives during 3's period of self–inhibition. Self–inhibition is therefore needed to prevent another potentially catastrophic effect due to *external* feedback.

Stapedial attenuation plays a similar role in the node structure theory. About 100–165 ms before we begin to speak, the stapedius muscle in our middle ear contracts, thereby attenuating amplitude of eardrum vibration when hearing one's own voice. This attenuation continues throughout the period of overt speech and is conventionally considered necessary to prevent damage to the eardrum that might arise from prolonged screaming. However, if this were its sole purpose, one wonders why attenuation also occurs during everyday speech and whispering at levels which cannot possibly damage the eardrum. Under the node structure theory, the purpose of stapedial attenuation is not just to prevent peripheral damage, but to act as another defence against the disruptive effects of external feedback.

Both of these defences can be overcome by means of a special set of experimental circumstances which are readily explained within the self–inhibition hypothesis. For example, when expert generators of Morse code hear the click of their key amplified and delayed by about 0.2 s, they often make repetition errors, producing (n+1) instead (n) *dits* or *dahs*. Repetition errors also become prominent when normal speakers hear the sound of their own voice amplified over earphones and delayed by 0.2 s. As many investigators have noted, the effects of amplified and delayed auditory feedback are similar in many ways to the errors that occur during stuttering. However, other errors accompany delayed auditory feedback repetitions as well: Speakers involuntarily speak with greater amplitude, prolong and substitute speech sounds, and sometimes produce phonemes that are not part of any language familiar to them (Lee 1950). For adults, the disruptive effects of delayed auditory feedback increase as a function of delay up to about 0.2 s and then decrease with longer delays, but never disappear completely even with delays of 0.8 s (MacKay 1968). The delay producing maximal disruption of speech varies with age: 0.2 s for adults, 0.375 s for children aged 7–9, and approximately 0.85 s for children aged 4–6 (MacKay 1968). However, the peak delay does not vary as a function of language familiarity: When bilinguals produce either their more or their less familiar language under delayed auditory feedback, their peak delay remains the same (MacKay 1970b).

Practice or familiarity influences the *degree* of disruption at any given delay. Bilinguals stutter more when producing their less familiar language under delayed auditory feedback (MacKay 1970b) and practice in producing a sentence reduces the probability of stuttering when subjects subsequently produce the practiced sentence under delayed auditory feedback (MacKay 1970b). Probability of disruption also diminishes with age, since practice, familiarity, or experience in producing speech increases as children grow older (MacKay 1968). Mechanical distortion of the returning auditory feedback likewise reduces the disruptions without influencing the delay producing peak interference (Hull 1952; see also MacKay 1969b). In short, neither practice nor rate of speech nor distortion of the returning feedback influences the delay producing maximal disruption of speech, although all three influence the overall amount of interference. This outcome is exactly what one could expect if an individual–specific physiological factor such as recovery from self–inhibition were determining the delay for maximal interference. That is, under the node structure theory, mental nodes responsible for speech become self–inhibited and then undergo a period of hyperexcitability during which they acquire greater than normal sensitivity. The duration of this recovery cycle differs depending upon the level of the node: a few ms for the lowest level muscle movement and sensory analysis nodes, but about 300 ms for phonological segment nodes. That is, I assume that hyperexcitability peaks at about 200 ms following activation of a phonological segment node and returns to normal spontaneous level about 300 ms following activation. This 200 ms peak would also fit the functions in Fig. 5 for the pattern of phoneme repetition in Hawaiian and Serbo–Croatian.

The maximal influence of delayed auditory feedback with a delay of about 0.2 s therefore seems to reflect an effect of feedback–induced priming arriving with sufficient strength at a critical time in the recovery cycle of just–activated nodes. Amplification of the returning feedback adds further to the priming of the just–activated nodes, and when these combined sources of priming exceed the "top–down" priming for the appropriate nodes in the same domain, these just–activated nodes become reactivated under the most–primed–wins principle, so that the output resembles stuttering (see MacKay and MacDonald 1984).

In summary, the self–inhibition hypothesis not only explains a wide range of relatively directly related phenomena, but also integrates phenomena such as stapedial attenuation and delayed auditory feedback that have hitherto seemed disparate and unrelated.

REFERENCES

Cutler A (1982) The reliability of speech error data. In: Cutler A (ed) Slips of the tongue and language production. Mouton, Amsterdam, pp 7–28

Grudin JT (1983) Error patterns in novice and skilled transcription typing. In: Cooper WE (ed) Cognitive aspects of skilled typewriting. Springer, Berlin Heidelberg New York Tokyo

Heffner FM (1984) General phonetics. Madison, Wis

Hull FM (1952) Experimental investigation of speech disturbance as a function of the frequency distortion of delayed auditory feedback. PhD thesis, University of Illinois

Lecours AR (1966) Serial order in writing: a study of misspelled words in developmental dysgraphia. Neuropsychologia 4: 221

Lee BS (1950) Effects of delayed speech feedback. J Acoust Soc Am 22: 824–26

MacKay DG (1968) Metamorphosis of a critical interval: age–linked changes in the delay in auditory feedback that produces maximal disruption of speech. J Acoust Soc Am 43: 811–821

MacKay DG (1969a) The repeated letter effect in the misspelling of dysgraphics and normals. Perc Psychophys 5: 102–106

MacKay DG (1969b) To speak with an accent: effects of nasal distortion on stuttering under delayed auditory feedback. Perc Psychophys 5: 183–183

MacKay DG (1969c) Forward and backward masking in motor systems. Kybernetik 6: 57–64

MacKay DG (1970a) Phoneme repetition in the structure of languages. Language Speech 13: 199–213

MacKay DG (1970b) How does language familiarity influence stuttering under delayed auditory feedback? Percept Mot Skills 30: 655–669

MacKay DG (1981) The problem of rehearsal or mental practice. J Motor Behav 13: 214–285

MacKay DG (1982) The problems of flexibility, fluency and speed–accuracy trade–off in skilled behavior. Psychol Rev 89: 483–506

MacKay DG (in press) The asymmetrical relationship between speech perception and production. In: Heuer H, Sanders AF (ed) Perspectives on perception and action. Erlbaum, Hillsdale, NJ

MacKay DG, MacDonald M (1984) Stuttering as a sequencing and timing disorder. In: Perkins WH, Curlee R (eds) Nature and treatment of stuttering: new directions. College Hill, San Diego

Meringer R, Meyer K (1895) Versprechen und Verlesen. Göschensche Verlagsbuchhandlung, Stuttgart

O'Hala J, Hirano M (1967) Control mechanisms for the sequencing of neuromuscular events in speech. Paper delivered to the AFCRL meeting, M.I.T., Cambridge, Mass.

Shapiro DC (1977) A preliminary attempt to determine the duration of a motor program. In: Landers DM, Christina RW (eds) Psychology of motor behavior and sport. Human Kinetics, Champaign, Ill

Shiffrin RM, Schneider W (1977) Controlled and automatic human information processing II. Perceptual learning, automatic attending, and general theory. Psychol Rev 84: 127–190

Tunturi AR (1958) Unpublished study referenced in: Field J et al. (eds) The evoked potentials. p 299 (Handbook of physiology, vol 1)

On Knowing How to Do Things

J. Annett
Department of Psychology, Warwick University, Coventry CV4 7AL, England

INTRODUCTION

In the opening symposium of this series on Perception and Action, Wolfgang Prinz referred to the problem of the hyphen in the term 'ideo–motor'. This paper concerns two (and possibly more) hyphens, but particulary those referring to links between perception and action and between verbal and motor processes. I refer to the latter as the action–language bridge (Annett 1983, 1985).

The problem of the 'ideo–motor hyphen' is to elucidate the relationships between the processes involved in passively perceived, imagined, and overt movement. Studies of mental practice (Johnson 1982; Annett 1985) provide evidence of processes which are common to imaginary and overt actions and have shown how secondary tasks may be used to distinguish between processes unique to each and common to both. These psychological studies are complemented by neurophysiological findings (Roland et al. 1980).

The second hyphen is the action–language bridge. There is ample evidence that motor and verbal systems can function independently (Allport et al. 1972; McLeod 1977; Kinsbourne 1983), and can be selectively impaired both in performance and memory (Corkin 1968). The key problem of the action–language bridge is how motor and verbal systems communicate. Tasks which critically involve this connection include following verbal instructions, giving a commentary on events (such as a boxing match), and providing explanations of or instructions on how to perform tasks which do not normally involve speech. Curiously, these everyday tasks have not been systematically explored despite their obvious relevance to skill aquisition and to those forms of apraxia described by Geshwind (1965) as the disconnexion syndrome in which patients are unable to produce actions in response to verbal commands whilst still retaining the ability to produce the same actions when the physical context demands.

This paper describes a series of exploratory studies of how normal adult subjects provide verbal explanations of familiar, but normally non–verbal tasks. They form part of a wider programme of research into the representation of action outlined by Annett (1983). The immediate aim of the investigation is to characterise transactions across the action–language bridge and to try to identify the common code or lingua franca which enables these two dissimilar and apparently independent systems to interact.

Experimental Brain Research Series 15

GENERAL METHOD

The subject is asked, "tell me in as much detail as you can how you...". A number of different tasks were used in pilot studies, but the experiments described here used either "...turn a forward roll or somersault" or "...take two ends of string and tie them together to make a bow".

The subjects' verbal responses to this request are recorded on either audio or video cassette and transcribed. For each transcript the time taken for the explanation and the number of words used is noted. The verbal protocols are then analysed to produce a crude measure of the quality of the explanation. The protocols are divided into sections representing a consensus view of the elementary acts which feature in these descriptions. Each is then given a score of 1 to 3 depending up on the amount of detail provided, 0 if the act does not feature at all. Tying a bow turns out to comprise seven such stages, so a comprehensive explanation would yield a score of 21. The stages are:

1. Grasp the ends of the string
2. Cross them
3. Twist them together
4. Pull the ends to tighten the knot
5. Form a loop or loops from the ends
6. Twist the loops or the loop and the free end together
7. Pull the loops to tighten and knot

In the most recent series of experiments using video recording any gestures by the subjects are noted. Using a special effects mixer an oscilloscope record of the output from the voice channel is superimposed on one corner of the picture and elapsed time in centiseconds is superimposed on another part of the screen to facilitate an analysis of the temporal relations between word and actions.

In addition to these objective data, subjects are questioned about their imagery on completiton of their explanation. This is done without any warning that the experimenter is interested in imagery. The subjects are never instructed to form images prior to the experiment, but simply questioned afterwards on the imagery they experienced. This informal questioning has been refined into a systematic paper and pencil questionnaire.

QUALITATIVE PROTOCOL ANALYSIS

One can ask many kinds of question, but the obviously simple ones are, what units or elements of actions feature in verbal descriptions and can one distinguish between protocols in these terms? Three representative protocols are shown in Fig. 1. One, subject 6, is rather minimal, but subject 8 provides an average 'good' explanation. The explanation of subject 11 is, to say the least, verbose and for this reason, is not given in full. These records are typical with 8 and 11 representing extreme, but nonetheless common, cases and they illustrate the following points.

Subject 6

OK um create a loop holding it between/ oh round my index finger on both hands/ and tie the bow together/ in a double knot/ creates a bow.

Subject 8

Er yes I would pick up the string and slide my hands along one to each end/ er bring the ends up to meet one another/ with my right hand I would put the string over the left hand end/ and then bring it underneath/ and then the piece that I've just brought underneath I would put over the what is now the right hand piece of string/ and fold that over/ and push it through the pull it through the loop/ and pull the whole thing tight.

Subject 11

Um well I would take both pieces of string between er/ the forefinger and thumb of each hand/ er being right handed I would hold the left hand still/ and take the right hand and pass er/ pass the free end of the string in my right hand over the standing portion of the string in my left hand over and under to produce a a a um/ what would it be a 360 degree wrap er that would provide the basis for a bow

um/ having done that um um um/ I would then release the would I um yes I would probably release the um the right hand leaving it free to um/ produce the loop in the um er/ standing end of the string that is in my left hand

in other words um um um I would yeah produce a loop and grip that the loop between the forefinger and thumb of my left hand/ er producing um a loop which projects upwards from the forefinger and thumb/

then again take up the free end of the string between forefinger and thumb of the right hand/

pass it round the base of the loop of thing of the string in the left hand close to the finger and thumb of the left hand to produce a/ a again a 360 degree binding the loop around the base of the er/ of the loop in the left left hand/

having got that far um/ er I would have to produce er have to ensure that there was a length of string between um the binding loop and ....

Fig. 1. Sample Protocols

1. The seven elementary acts used as a basis for scoring are all referred to by S8.
2. It is rare even in a 'poor' protocol for the subject to get the steps in the wrong order. Steps are either not mentioned or referred to obliquely, as in the record for subject 8, but this is probably due to a problem of expression, rather than a failure in generating the procedure, since the subject does manage to get to the end, unlike the rather rare cases who gave up completely.
3. The more detailed explanations illustrated by subject 11 do not generally provide additional steps in the procedure or fine detail of the movements of individual joints and muscles. Three main differences are noticeable. (a) Statements referring to general goals ("create a loop") are further qualified by reference to the part of the body which is involved; (b) the longer explanations include more statements referring to the visuo–spatial characteristics of the task, that is how the knot would appear; (c) the more verbose subjects produce longer descriptions by adding redundant comments and asides. For example S6 refers mainly to generalised goals such as "create a loop" or "tie the bow" and refers only once to a part of the body "round my index finger". S8 refers more often to what the hands do "pick up", "slide along", and to the spatial relationships involved in the task, "put it over what is now right hand piece of string." S11 adds details such as referring to "forefinger and thumb" but also adds comments which are not directly relevant to the description such as "being right–handed" and "what would be a 360 degree wrap or that would provide the basis for a bow." S11 also reverts to statements about general goals (e.g. "produce a loop") rather than describing actions in detail. Thus, even a subject who appears to have no difficulty in keeping the procedure in mind and in producing a flow of descriptive language still refers largely to whole, unanalysed, acts. Whilst the body part used may be referred to, what that body part is actually doing is rarely mentioned.

THE EFFECTS OF SECONDARY TASKS

In providing verbal descriptions of the types illustrated in Fig. 1, it may be presumed that a number of different systems are involved. For example, subjects feel a compulsion to make gestures which suggests that the motor output system is significantly involved even when the task does not strictly require overt action other than talking. Subjects when subsequently asked for their introspections invariably report visual imagery and occasionally halt their descriptions complaining that they are unable to visualise the next step or the outcome of the step just described. Imagery therefore appears to be a mechanism which is crucial to the operation of the action–language bridge. Clearly since the subject is required to talk, some part of the language processing system is required. The question then is whether it is possible to tease out the contributions of these different mechanisms. The method adopted is to require subjects to carry out secondary tasks which can be presumed to make use of these same systems and in this way shed some light on their relative importance for the execution of the task.

Results so far have failed to produce significant decrements in performance with secondary visual tasks such as monitoring a pair of flashing lights or a slowly moving lissajou figure. Both these tasks were shown by Johnson (1982) to produce interference in short–term motor memory. An auditory monitoring task, listening for colour names in a series of common words read at 5s intervals increased the speech rate slightly but insignificantly, but the most surprising negative finding was that tapping alternately between two points about 10 cm apart with the preferred hand did not produce statistically significant decrements. It seems that it is possible to describe an action even while the limb involved is in fact doing something else.

EFFECTS OF 'FACILITATING' CONDITIONS

Subjects often remarked that the task might be easier if they could use real rather than imaginary string and so, following the apparent failure of the secondary task method to produce interference, three different conditions which it was hypothesised would enhance performance were compared with a control. Forty subjects, students at an Open University summer school, were allocated to one of four conditions. One group was given 30 s following the request for an explanation to gather their thoughts before being required to speak. Subjects in a second group were handed a board approximately 30 x 20 x 2 cm to which were attached two pieces of string and they were first asked to tie a bow. When they had done this the board was removed from sight and they were then required to give their explanation. Subjects in a third group were handed the board and asked to tie a bow, simultaneously giving an explanation of what they were doing. A control group gave their explanation under the standard conditions.

Although 30 s warning apparently enabled subjects to produce longer and slightly better explanations and having the string available produced the shortest explanations, none of the three measures was found to distinguish significantly between the experimental and the control groups.

EFFECTS OF IMMOBILISATION

The compulsion to gesture whilst giving explanations suggests that motor activity associated with the task constitutes an integral part of the processes underlying explanation. However the findings that (a) irrelevant activity of the limb presumably involved in the imaginary task did not interfere with the explanation and (b) that using actual string did not appear to significantly enhance performance cast some doubt on the validity of this suggestion. In order to assess the degree to which gestures are critical, 20 additional subjects were recruited and randomly assigned to either the standard condition or an experimental group who were required to sit on their hands whilst explaining how to tie a bow.

Immobilising the hands tends to increase the number of words used and the time taken to provide an explanation and to slightly reduce the quality of the protocols, but none of these differences was statistically significant and we must conclude that overt activity and, hence, involvement of the motor production system is not critical to the performance of the task, despite subjective impressions to the contrary.

The failure to find either facilitatory or inhibitory effects on the production of explanations is probably not due to lack of statistical power in the experiments, but it could be due to the relative insensitivity of the measures chosen. There is always the possibility that more sophisticated protocol analyses might reveal subtle effects.

Action Imagery Questionnaire

1. I visualised myself performing the task clearly/vaguely/not at all.

2. I visualised someone else performing the task C/V/N.

3. I visualised myself from a point outside myself/as it were through my own eyes.

4. I was aware of the colour of the string C/V/N.

5. I was aware of the texture of the string C/V/N.

6. I was aware of an object (e.g. shoe or parcel) to which the string was attached C/V/N.

7. I was aware of the environment (e.g. indoors/outdoors) C/V/N.

8. I could feel the string C/V/N.

9. I could feel myself making the movements C/V/N.

10. The imagery I experienced was predominantly visual/predominantly non-visual feelings/ a mixture of the two.

11. The action was continuous (like a movie)/ a series of stills/ part continuous/part stills.

12. It was easy/difficult/neither easy nor difficult to form images.

13. It was easy/difficult/neither easy nor difficult to 'hold' images once formed.

14. It was easy/difficult/neither to describe the images in words.

15. Talking disrupted the images a lot/a little/ not at all.

Fig. 2. The Action Imagery Questionnaire

IMAGERY

All subjects were questioned after these experiments on the nature of the imagery they experienced in the process of producing explanations. The questionnaire shown in Fig. 2 was developed on the basis of spontaneous comments made by subjects in the pilot stage and refers to the subjective viewpoint, the modality and content of the images, their movement characteristics, ease of image formation and manipulation, and susceptibility to interference. Most questions admit of a three-point rating of image clarity or ease of control. Subjects were invited to add comments and many did so. The principal results may be summarised as follows.

1. No subject who was able to do the task at all denied using imagery. Accounts of imagery were produced by all subjects with virtually no probing.
2. A few subjects (2 out of over 100) failed to complete the task. Both offered inability to retain an image as the principal reason. Long pauses and restarts were frequently attributed to difficulties in attaining, regaining, or manipulating images.
3. Subjects uniformly experience a feeling of voluntary effort, of 'doing the task' which is itself distinguishable from the images, visual, tactile, or kinesthetic, which may have been experienced.
4. The imagery experienced is predominantly visual. Few subjects had any problem specifying the colour of the string, but hardly anyone reported any clear image of its texture.
5. There were differences in the apparent point of view. Almost all subjects reported having imagery as if through their own eyes.
6. About half the subjects reported the images as a series of stills and most of the remainder who reported a continuous movie felt they could control the speed, that is to slow down the action to provide time to produce an explanation.
7. The stills generally referred to or were noted as a completed stage which marked an accomplishment and signalled the need to produce the next step. This place marker function seems subjectively important. On one occasion a subject reported experiencing an image which indicated to him that he had made a mistake and was describing tying the wrong kind of knot. Another subject reported having to stop and start again because his imaginary hand was in such a position as to conceal the imaginary string!
8. Whilst most subjects reported they could, with a feeling of effort, generate and manipulate images at will to help them perform the task, the detailed content of the imagery often came unexpectedly to the subject. For instance, subjects who gave their descriptions shortly after having tied a real bow usually reported imagery of the particular board and string they had just used whilst others reported imagining the laces of a familiar pair of shoes but in neither case was the specific image chosen deliberately.

Relationships between imagery and other features of the performance have yet to be examined in detail and it is not possible to say if the type of imagery affects the quality of explanations. A comparison of the imagery protocols of the 10 subjects who were required to sit on their hands with the 10 controls was made. The distributions of responses between the three categories associated with each question were compared for the two groups using ñ.

Only one difference was found to exceed the 5% confidence level and that is question 8. More subjects in the experimental group reported they were not aware of the feel of the string, ($\chi = 6.92$, $p < 0.05$). The next largest difference, but not significant, ($\chi = 4.1$) is on question 9. Subjects in the control group tended to report that they could clearly feel themselves making hand movements whilst the experimental group were mostly not aware of making hand movements.

With restricted movement there was a shift away from continuous or movie–like imagery to a mixture of stills and continuous movement imagery. Subjects in the experimental group tended to report more difficulty in forming images, but no additional difficulty in holding the image once formed.

You take hold of both ends of the string

Um you take hold of both ends

cross the right one over the left one

cross the right one over the left

and underneath to make a knot

and underneath

make a loop in what is now the right hand string

make a loop with that is now the right hand one

pass the left underneath

cross the left hand one over

and pull the loop through

pull a loop through

tighten by pulling both loops.

tighten the bow by pulling both loops.

Fig. 3. Sample Protocols from Subject 5. Normal typeface – without the string, and italics – with the string

These results are consistent in suggesting that the effect of restriction of movement in this task is to reduce kinesthetic imagery, whilst having no obvious effect on the quality of verbal performance. Thus, despite the fact that the feeling of movement is commonly reported it is probably not critical to the production of a verbal explanation. The fact that it tends to be reduced when movement is restricted suggests that it may arise from small movements and in that sense the reported kinaesthetic sensations may be real rather than imaginary .

THE ROLE OF GESTURES

These results on the value of making gestures tend to contradict the intuitions of the subjects and they are also of interest to the classical theory that imaginary actions necessarily involve making

Forearms resting on table, hands bent, back fingers open; shaking both hands synchronously with stressed words

Hands held upwards and outwards

Hands approach each other, right forefinger pointing

Hands brought together, string held by fingertips

Both forefingers make circling movements, right hand beats to emphasise "knot"

Twisting string with left hand

Hands apart, right hand circling movement

Hands collaborate to make loop

Left finger and thumb move slightly

Left hand moves string over

Slight pulling movement

Loop made with finger and thumb of left hand

Outward pulling movements, synchronised with stressed words

Shaking both hands for emphasis

Fig. 4. Gestures made by Subject 5. Normal type shows gestures without string and italics indicates movements made whilst demonstrating with string

small movements. The next part of the investigation looked at the gestures subjects actually make and their relationship to the words they produce. The method was simply to examine video recordings of subjects giving their explanations. Recordings have been made of about 50 subjects so far, each one in three different conditions, the standard condition in which the task is simply to explain how to tie a bow, giving an explanation whilst actually tying the knot, and thirdly with the subject tying a bow but without giving a simultaneous explanation.

The records of 12 subjects have so far been analysed. The descriptions of bow tying given by a typical subject first under the control condition and then whilst actually tying a bow are shown in Fig. 3. The similarities are striking. Figure 4 compares the movements made whilst giving these explanations and Fig. 5 the movements made whilst actually tying a bow, but without any accompanying verbal commentary.

Starting with fingers slightly curled and pointing down both strings quickly grasped and immediately brought together	(1 s)
String held at finger tips and quickly crossed with small finger movements	(2 s)
Hands close together form loop with left fingers then another with right fingers	(4 s)
Pull loops through with quick finger movement	(1 s)
Twisting movement with right hand	(2 s)
Hands moving apart with short jerk at end	(1 s)

Fig. 5. Subject 5's actual movements in tying a bow, without explanation

The analysis is necessarily subjective, but there are clearly some striking differences between the three sets of movements which contrast with the considerable degree of similarity between verbal protocols collected in the two principal conditions. Consider the movements associated with the first stage, picking up the string. The actual grasping movement is rapid and bringing the strings together is a direct continuation of grasping. When giving a description without the string the arms stay resting on the table and hands remain open in a pre-grasping posture. The movement is primarily from the wrist and consists of shaking synchronously with the stressed words.

When actually tying the bow and giving a simultaneous explanation, the hands are held in a quite different position, up and away from the table with palms facing up rather than down. One characteristic difference between actual tying movements and gestures made in the absence of real string is that gestures often appear to symbolise the purpose or intention of the act rather than simply replicating a real performance. For example, the subject uses the forefinger to symbolise the direction of movement with the forefingers making circling movements to symbolise "knot." The movements made when actually manipulating the string are quite different.

A second way in which gestures differ from actual movement is that they are not infrequently synchronised with key words. These movements are distinctive insofar as they are not just morphologically different from the relevant real movement, but that their function seems to be one of emphasis rather than meaning.

The results in Table 1, based on 12 subjects who each completed the three tasks, demonstrate some of the principal differences between explaining, demonstrating with verbal commentary, and normal

Table 1. Comparison of real bow tying (without explanation) and explanations with string (demo) and without string (image)

	Real	Demo	Image	P
No. of words	-	72.92	111.17	< 0.05
Time(s)	9.05	30.23	52.42	< 0.01
Protocol score	-	17.70	18.10	NS

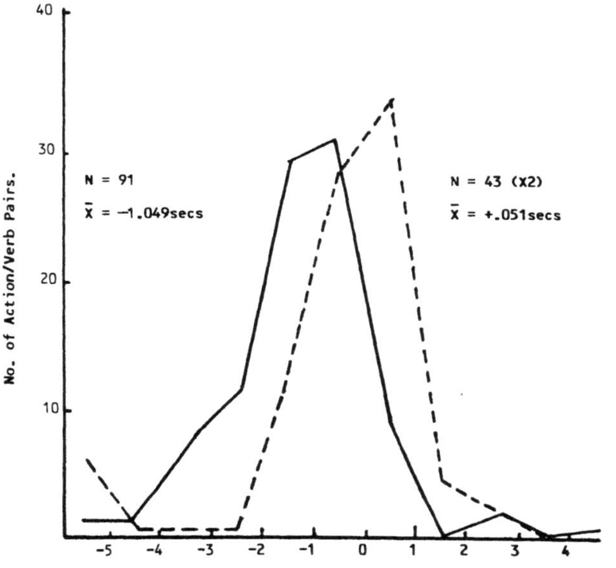

Fig. 6. Distribution of time of onset (s) of action described by a verb at onset time 0. Solid line, whilst giving commentary of actual task; dashed line, gestures whilst giving explanation

performance of the task. The 'demonstration' task takes over three times as long as the actual bow tying and if it has to be done in imagination the process takes six times as long. The latter difference is interesting in view of the similarity between the verbal protocols in the two cases. One possibility is that this represents the additional time required to create and 'read' images. However, we see that this group of subjects uses 50% more words and perhaps it just takes longer to say more words. Looking at the sample protocols for S5 we see that whilst the ideas expressed are very similar whether real string is used or not, the use of words is more economical in the former case, presumably because the speaker assumes the listener shares the same visual data.

A further difference emerges from the detailed timing of the video records. The distributions in Fig. 6 show the temporal relationships between the time of onset of a sample of 134 verbs and the commencement of the actions to which they relate taken from 12 subjects. Time zero is the time at which the verb begins as indicated by the oscilloscope trace of the voice signal superimposed on the video recording. The solid line represents the onset time of the action to which the verb refers when the subject is actually performing the task, i.e. giving a demonstration, and the mean is at just about -1s, that is the action precedes the verb. The dotted line represents the onset of gestures which appear to relate to a given verb and it will be seen that the mean is at about $+0.5$ s. Not everyone makes gestures and only the actual movement data of those subjects who did gesture have been used. Nevertheless, it would appear that in giving a demonstration, actions typically precede the words used to describe them. When the action is largely imaginary, however, those parts of it which break surface are more nearly synchronous as if generated by some common internal process.

DISCUSSION

Although there are wide variations in the quality and quantity of verbal explanations of familiar skills, all subjects appear to be referring to an internal model composed of a series of basic units or primitive acts such as grasping, twisting, looping, and so on. These primitive acts are essentially defined not so much by a specific set of movements as by the attainment of a set of criteria, for example the strings being crossed. In these serial tasks each criterion state is a prerequisite for the subsequent stage of the task. When there is no explicit reference to one of these acts the rest of the explanation proceeds as if it has been included and it seems probable, therefore, that limitations on these explanations are more to do with verbal expression than with any failure of the underlying model. Elaboration of the verbal protocols does not take the form of describing ever finer details of these primitive acts, but rather may refer to the manner in which the act is accomplished (e.g. which fingers are involved), or a description of the effect of the act in terms of how the scene has been changed, or in some cases a rationale for the act or the particular way in which it was performed. These primitive acts, which are not described in fine detail would appear to be the currency both of the internal representation or model of the performance and of the words used to describe it.

In the absence of overt performance, the translation from action to language codes is mediated by imagery. Since the quality of verbal protocols is not markedly affected by overt activity, these results suggest that limitations on explanations are imposed by the capacity of the verbal system rather than any capacity for generating and controlling actions and images. Imagery is essential to giving an explanation and, at least in the task used in this study, it is primarily visual, although hardly affected by mild visual distractors. A competing motor activity does not interfere with the ability to produce a verbal explanation of a series of motor actions, nor does the inhibition of involuntary gestures which frequently accompany the explanations.

These findings together with the result that explanations accompanying overt performance are no better is evidence against the theory that imaginary actions which occur in mental practice are due simply to a sub-threshold activation of the motor output system. The imagery itself is related to the stages or primitive acts and seems to play an essential role in confirming that stages have been completed, and the essential criteria attained, in the correct order. Of course the nature of bow tying is such that a series of criterion conditions (the ends twisted or a loop made) have to be achieved and the imagery is used to ensure that this is so. In this respect images perform the same function as perceptions in the real task. Although activation of the motor output system is not necessary (in this task at least) for the production of visual images, it may play a role in the generation of kinaesthetic images since subjects with immobilised hands are less likely to report kinaesthetic imagery. William James (1890) reports that auditory imagery for labial and dental speech sounds is made less distinct by holding the mouth open whilst thinking words containing these sounds. If this were the case, it would be tempting to conclude that kinaesthetic imagery is the product of sub–threshold activation of the musculature.

The apparent dissociation of visual imagery from the motor output system is further illustrated by qualitative differences between movements made when actually performing a task and the involuntary gestures accompanying explanations. Gestures which are closely similar to the actual movements are relatively rare and quite often a movement is made by a different hand or in a different direction or of grossly different extent. Gestures seem often to fulfil a symbolic function, that is they represent the intention of the act rather than its precise form. A further interesting feature of gestures lies in their temporal relationships with words. A distinction has been made by McNeil and Levy (1982) between 'iconic' and 'beat' gestures, the latter being characterised by synchrony with the relevant words. Both types occur as an accompaniment to explanations and their relative timing may have a bearing on the relationships between the mechanisms which produce actions, images and words.

Although it is not possible to time occurrence of images, it may be suggested that in the explanation task images must precede words since the production of appropriate words appears to be dependant upon the availability of images. These results point to the hypothesis that an action must be generated, but not necessarily overtly expressed, as a prerequiste of image generated substitute for external feedback. The small sample of data in Fig. 6 is consistent with this view in the case of 'demonstration' gestures which are initiated on average about 1 s before the verbs to which they refer. However, where the explanation is based on imagery alone, the actions are initiated on average half a second after the relevant word. Of course, both overt actions and gestures ought to be regarded as surface manifestations of some underlying representation of the act. Words, images and gestures may all be generated by an underlying 'proto–action' system rather than by each other. However, some gestures of the kind described by McNeil and Levy as 'beat' gestures may be generated by the speech production system insofar as they appear to relate more to verbal emphases than to meaning as such. If we can distinguish between these beat gestures and the iconic type which represent some aspect of the meaning of the action, it could be predicted that the former are generated synchronously with the stressed word whilst the latter are initiated shortly before the word emerges. The grounds for this prediction are that the role of iconic gestures is to help produce images which are in turn necessary for the production of words.

This paper represents an early progress report on some exploratory investigations. The results so far encourage the belief that a careful combination of introspective and objective techniques may clarify the complex relationships between verbal and non–verbal knowledge as exhibited by the process of giving explanations. Further, and possibly more revealing, ways of analysing the verbal protocols may be necessary and the temporal relationships between words and different kinds of gestures would seem to merit more detailed study. Looking further ahead it may be possible to cross–check explanatory models by using the other experimental paradigms which involve translation across the action–language bridge.

References

Allport DA, Antonis B, Reynolds P (1972) On the division of attention: a disproof of the single-channel hypothesis. Q J Exp Psychol 24: 225-235

Annett J (1983) Motor learning: a cognitive psychological viewpoint. In Rieder H, Bos K, Mechling H, Reischle K (eds) Motor learning and movement behavior. Karl Hofmann, Schorndorf

Annett J (1985) Motor learning: a review. In Heuer H, Kleinbeck U, Schmidt K (eds) Motor behavior: programming, control and acquisition. Heidelberg: Springer, Berlin Heidelberg New York Tokyo

Corkin S (1968) Acquisition of a motor skill after bilateral medial temporal lobe excision. Neuropsychologia 6: 255-265

Geschwind N (1965) Disconnexion syndromes in animals and man. Brain 88: 585-644

Johnson P (1982) The functional equivalence of imagery and movement. Q Exp Psychol 34A: 349-365

James W (1890) Principles of psychology. Holt, New York

Kinsbourne M (1981) Single channel theory. In Holding D (ed) Human skills. Wiley, Chichester

McLeod P (1977) A dual task response modality effect: support for multi-processor models of attention. Q J Exp. Psychol 29: 651-667

McNeil D, Levy E (1982) Conceptual regulation in language activity and gesture. In Jarvella RJ, Klein W (eds) Speech, place and action. Wiley, New York

Roland, PE, Lassen NA, Skinhoj E (1980) Supplementary motor area and other cortical areas in organisation of voluntary movements in man. J Neurophysiol 43: 118-136

Do Units of Motor Action Really Exist?*

P. Viviani

Instituto di Fisiologia dei Centri – Nervosi CNR, Via Mario Bianco 9, 20131 Milano, Italy

INTRODUCTION

Human movements are characterized jointly by the trajectories of the limbs being displaced and by the time course of their positions. In the particular case of drawing movements to be here considered, the parametric laws of motion $x = x(t)$, $y = y(t)$ of the pen's tip afford a complete description of the motor act. At least in principle, the trajectory described by the pen is instead independent of the specific kinematics of the movement, in the precise mathematical sense that one planar trajectory $y = f(x)$ can be generated by an infinite number of different laws of motion.

To some extent, this formal independence of geometrical and kinematical aspects of planar movements seems to be reflected at the level of the processes that organize and execute drawing movements. Several hints of this relative independence can be cited. For instance, it has long been known (Merton 1972) that, when a movement is scaled up or down in size, the geometrical relations among its different parts are preserved, even in the case that the scale change requires the mobilization of different limbs. This suggests that the form of the trajectory is centrally represented in strictly geometrical terms, independent of the specific muscular commands issued in any one instantiation of the movement. Moreover, it is also known that a trajectory whose form and size are specified can be executed at different velocities, the inference being that the overall tempo of the movement can be modulated independently of the parameters which specify the trajectory.

Despite the aforementioned phenomena – which suggest a relative independence of the geometrical and kinematical levels of representations – some evidence is accumulating of deep-seated relations between these two levels. In particular, a number of recent papers (Viviani and Terzuolo 1982; Viviani and McCollum 1983; Lacquaniti et al. 1983, 1984; Viviani and Cenzato 1985) have suggested the existence of at least two types of relations: a local interaction between the differential properties of the trajectories and the instantaneous velocity, and a global interaction which involves instead the size, the linear extent, and the overall tempo of the movement.

This paper will only be concerned with the latter type of relations. My aim is to bring together and assess both previous work on this topic and some ongoing research.

* This research was partly supported by CNR research funds.

SOME QUESTIONS AND A FEW ANSWERS

Previous studies (Lacquaniti et al. 1983, 1984; Viviani and Cenzato 1985) have shown that, in the course of spontaneous drawing movements, the instantaneous angular velocity is approximately proportional to the two-thirds power of the instantaneous curvature C (t):

$$A(t) = K \cdot C(t)^{2/3} \tag{1}$$

For obvious reasons, this relation was dubbed the "two-thirds power law." Except in the proximity of inflection points, this empirical equation can also be written in terms of tangential velocity V(t) and radius of curvature R(t) = 1/C(t):

$$\log V(t) = \log K + 1/3 \log R(t)$$

The parameter K in this expression is approximately constant over successive segments of the trajectory and its average takes generally different values over each segment. A rather strong validation and qualification of these statements is provided by considering random continuous movements produced extemporaneously within a circular frame (Fig. 1A). Part B of this figure shows several records of the gain time course calculated directly from the digitized samples of the movement coordinates. The plots in the left column are from one subject and correspond to three different sizes of the encompassing frame. The plots in the right column are from three different subjects and only one frame size. Zero crossings from positive to negative values of the gain correspond to transitions from clockwise to counterclockwise rotations via an inflexion point ($R = \infty$). As indicated above, the parameter K is approximately constant between any two successive zero crossings. Moreover, in many instances its absolute average value changes significantly after a sign reversal. This type of results have led us to propose that Eqs. (1) and (2) not only describe satisfactorily an empirical finding, but also represent the structural relationship which exists between geometrical and kinematical aspects of human drawing movements. In such procedural sense, Eq. (2) states that the instantaneous velocity depends jointly on the differential properties of the trajectory being executed and on a velocity gain factor K which is specified piecewise in the course of the movement.

If one accepts such a factorization of the velocity, a number of questions arise quite naturally concerning the mechanisms responsible for the specification of the velocity gain factor and the implications that these mechanisms have for the timing control of the movement:

1. What relation, if any, can be established between the overall average of the gain factor and the metric properties of the trajectory?

2. Are local changes in the absolute value of K related to some identifiable parameter of the movement?

3. Do these changes provide the rather loose notion of segment evoked above with some operational meaning?

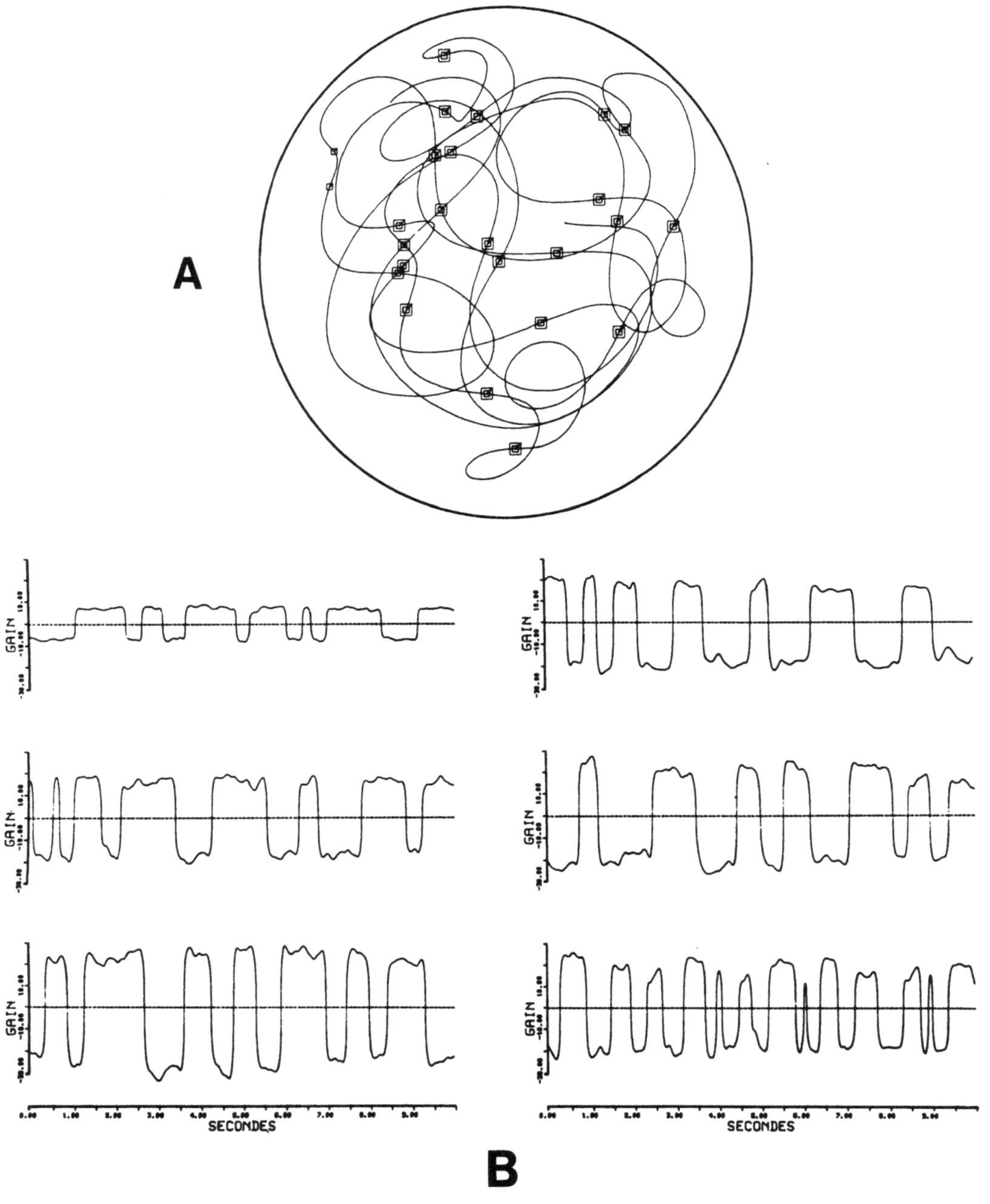

Fig. 1. Analysis of a random movement. In A a representative example of the trajectory produced extemporaneously for 10 s within the indicated frame (radius: 7.79 cm). Points of inflection are identified. In B six examples of the time course of the instantaneous velocity gain factor K(t). In the left column results in one subject for three different frame sizes. In the right column results in three different subjects and one frame size. Transitions from positive (counterclockwise rotations) to negative (clockwise rotations) values of the gain occur at every inflection point. The gain between any two successive inflections is almost constant. In some cases, the average absolute gain changes significantly upon crossing an inflection. These changes are positively correlated with the length of the corresponding stretch of trajectory (see text)

4. If so, may we conceive of the movement as a sequence of segments planned and executed somewhat independently of each other?

5. Ultimately, what is the status of the velocity gain factors and of the segments within the context of the motor plan?

In what follows I will try to gather those pieces of evidence which may lead to some plausible line of reasoning for each one of these issues.

Question 1. The Isochrony Issue: A Global View

With respect to the temporal planning of drawing movements, it has been proposed (Lacquaniti et al. 1984) that a distinction should be made between two general classes of movements. The first class includes all the movements for which it is likely that the whole trajectory is planned before beginning the execution (for instance, small chunks of connected handwriting or simple geometrical figures). The second class includes movements whose trajectory is much more complex and for which complete preplanning becomes impossible. In the first case, well-established evidence (see Viviani and Terzuolo 1983, for a review) shows that the total execution time T is fairly insensitive to changes in the overall size of the trajectory (isochrony). For instance, in a recent experiment (Viviani and Cenzato 1985) in which subjects were asked to draw folded-up elliptic figure eights, a more than tenfold change in pattern size only resulted in a 50% lengthening of execution time.

In trying to assess the role that the gain factor K plays in this compensatory mechanism, ellipses provide a convenient starting point inasmuch as the gain factor for this type of patterns remains practically constant throughout their execution (Lacquaniti et al. 1983). In fact, one can demonstrate (Viviani and Cenzato 1985) that, if K is constant, then:

$$\log T = F(e) + 2/3 \log P - \log K \tag{3}$$

where P is the perimeter of the ellipse and F(e) is a mildly varying function of the eccentricity e. If we assume T = const., it necessarily follows that the gain factor is a power function of the perimeter P:

$$\log K = \log K_0 + 2/3 \log P$$

where K_0 is almost independent of the eccentricity of the pattern. Fig. 2 reports data from an experiment in which the size and the eccentricity of elliptic patterns were systematically varied (Lacquaniti et al. 1984). As predicted, the perimeter P of the pattern affects the gain factor according to a power law, while the eccentricity has no appreciable influence. Thus, in the particular case of ellipses (K = const.), isochrony is mediated by a tuning mechanism which regulates the values of K as a function of the perimeter.

When considering more complex patterns, the velocity gain factor may vary during the movement. However, inasmuch as we make the assumption that a global plan is available for the entire movement, we expect that the average gain factor bears with the total trajectory length the same

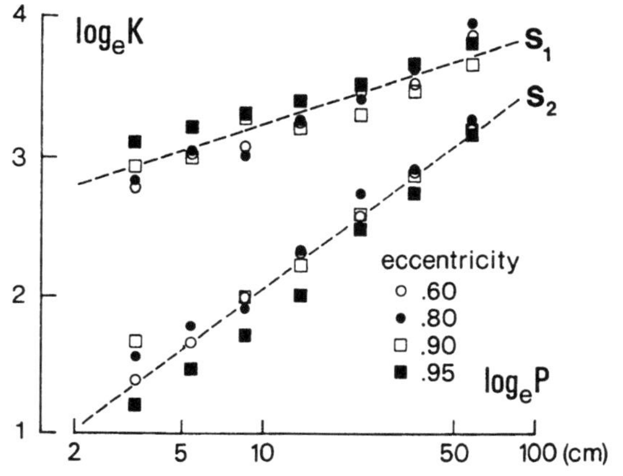

Fig. 2. Analysis of elliptic movements. Results in two subjects (S_1 and S_2) from an experiment in which ellipses with different sizes and eccentricities are executed at a freely selected tempo. Data points describe the relation between the average gain over several movement cycles and the perimeter of the ellipse. As shown by the regression lines, this relation is accurately described by a power law whose parameters are independent of the eccentricity (from Lacquaniti et al. 1984)

type of relationship that is valid for ellipses. The analysis of the double ellipse executed repeatedly (Viviani and Cenzato 1985) confirms this expectation. In fact, a power law of the type:

$$\log K = a + b \log P$$

describes quite accurately the relation between the gain factor K and the perimeter P, both for the entire pattern and for the two ellipses that compose the entire pattern.

We come now to discuss in greater detail the second type of patterns mentioned above for which complete preplanning is unlikely, either because of their excessive complexity or because of the modality of execution. An extreme instance of such patterns is provided by scribbles such as the one already shown in Fig. 1A which represents a 10-s sample of an extemporaneous movement that the subject was producing continuously within the indicated frame. Under such conditions, the very notion of preplanning is actually meaningless. In particular, we can no longer hypothesize that the total length of the trajectory directly provides a control parameter for the velocity gain factor. The remarkable fact, however, is that – even in the absence of any regularity and constraint, besides that due to the frame – the average gain K_T is finely tuned to the length of the trajectory. This is demonstrated by an experiment in which we induced systematic variations of both the trajectory length L_T and the average gain factor K_T, by varying the radius r of the frame. For each frame size, the trial-by-trial and across-subjects variations of L_T and K_T are highly correlated via the same type of relation obtained for double ellipses:

$$\log K_T = a + b \log L_T$$

Moreover, in the entire range of L_T values covered by the experiment (50 – 400 cm) the exponent b of the power law is almost invariant ($b = 0.945 \pm 0.040$). Notice that this result cannot be taken to suggest a causal dependency of the average velocity gain factor from the linear extent of the trajectory, as we did for simple closed patterns, because the movement was being produced continuously and without a definite scheme. A more sensible interpretation might be that the gain factor in these framed random movements depends on the size of the frame, and that the length of the trajectory produced in a given time interval is in fact a dependent variable. Actually the overall velocity gain factor can be expressed directly and accurately as a function of the frame radius by the logarithmic regression:

$$\log K_T = 0.701 + 0.634 \log r$$

An interesting implication of this interpretation will be discussed in a later section.

To summarize the data presented so far, it appears that indeed the velocity gain factor depends on the geometrical properties of the movement. However, the nature of the movement and the conditions of execution largely determine which specific geometrical aspect is taken into account. In a sense this is quite unsatisfactory, for one would like to be able to single out one general rule that applies to all cases. Whether this can be achieved, is left for future research.

Question 2. The Segmentation Issue

In many instances, summarizing the behavior of the gain factor K by its overall average value is a crude oversimplification of reality. If we draw a spiral, for instance, K decreases systematically as the movement winds up toward the center (Lacquaniti et al. 1984). Moreover, as we will show shortly, within-movement modulations of K may be contingent upon the modalities of execution. It is therefore necessary both to describe these effects and to try to identify their determinants. A possible clue comes again from the timing analysis of relatively simple closed patterns for which gain factor changes are both systematic and obviously related to some figural regularities of the trajectory. A typical example is shown in Fig. 3 which describes the execution of a figure-eight pattern. Since the angular velocity of the movement $d\alpha(t)/dt$ remains almost constant throughout the movement (sign reversals are immaterial), the two-thirds power law implies that the gain factor is an increasing function of the perimeters of the two circular subpatterns that compose the figure. In this and similar examples the natural partition of the trajectory into figural subpatterns maps into a clear modulation of the gain factor. Thus, also in view of the tentative answer given to Question 1, it seems worthwhile to explore the hypothesis that in all cases local modulations of the gain factor are related to some geometrical aspect of the movement. Once again, scribbles afford an ideal - albeit extreme - context to tackle the question.

Some of our earlier studies on handwriting, as well as the analysis of K-plots such as those in Fig. 1B, indicate that changes in the velocity gain parameter frequently occur at points of inflection of the trajectory. While a change in the direction of rotation is neither a necessary nor a sufficient condition for a significant change of K, the association between these two events is significant enough to justify - as a first order approximation to the underlying segmentation rule - the assumption that all stretches of trajectory between successive points of inflection represent distinct

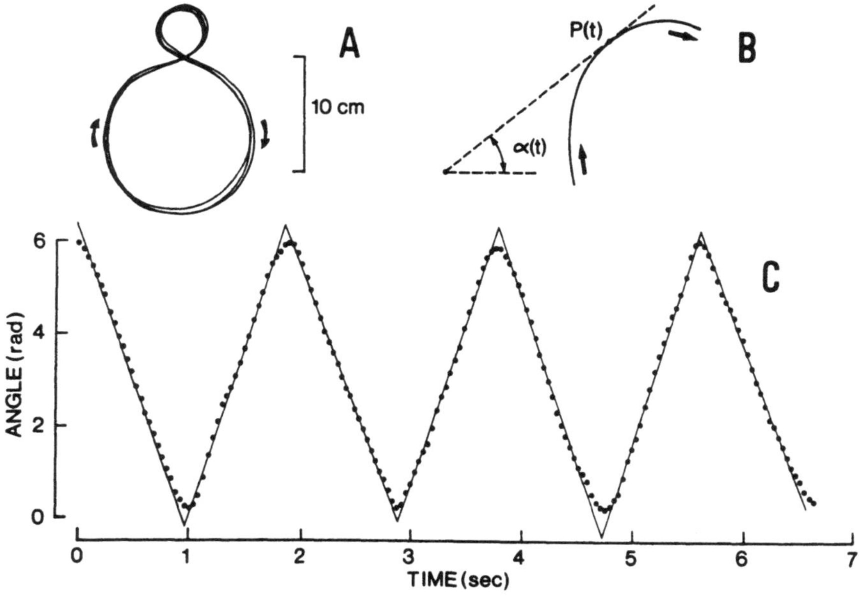

Fig. 3. Illustration of the isochrony principle. In A several cycles of execution of the figure – eight pattern. In C the time course of the angle that the tangent to the trajectory makes with an arbitrary reference (B). The absolute values of the angular velocity (derivative of the angle) in the two loops of the pattern are almost identical. Thus, despite the large size difference, the execution time of the loops is not much different. The example shown is typical of relatively fast execution rates (from Lacquaniti et al. 1983)

units of motor action. The results of the scribbling experiment already mentioned above lend some credit to this tentative segmentation rule. Let L_s and K_s denote the trajectory length between inflections (segments) and the within-segment average gain factor, respectively. Moreover, let L_{st} and K_{st} be the means of these quantities over all segments of one trial (these means depend on the subject and the frame). A simple way to support the stated assumption is to show that indeed K_s and L_s covary across segments. In order to prevent an artefactual correlation from being injected into the results, we actually performed a regression analysis on the normalized quantities K_s/K_{st} and L_s/L_{st}. The analysis shows that for all frame sizes and all subjects the power function:

$$\log (K_s/K_{st}) = 0.209 \log (L_s/L_{st}) \tag{4}$$

provides a satisfactory description of the data. Therefore, it is demonstrated that a direct relationship exists between the geometrical and kinematical parameters of each segment. A more detailed description of the K–L relation may be obtained by assuming that the within-trial averages K_{st} and L_{st} can be replaced by the corresponding general averages over all trials and all subjects. Substituting in Eq. (4) the empirical expressions that can be derived for these general averages, one gets:

$$\log K_s = 0.702 + 0.307 \log r + 0.209 \log L_s \tag{5}$$

Thus, while at the global level (cf Question 1) only the frame size could be reasonably credited with a direct influence on the overall trial gain, gain control within segments appears to depend on both frame size and segment length. Notice that at the segmental level it is again reasonable to assume that the length of trajectory to be executed is represented centrally.

Two last points on the scribbling experiment. First, multiple correlation analysis can be used to show that also the within-segment curvature C_s depends on both frame size and segment length:

$$\log C_s = 0.567 - 0.565 \log r - 0.387 \log L_s \tag{6}$$

Second, it can also be shown that the within-trial average segment length increases almost proportionally with the frame size. However, a more than tenfold increase of L_{st} is matched by a corresponding increase of the average velocity. As a consequence, the average duration of a segment remains almost invariant across all frame sizes. An interpretation of these two points is postponed to the next section.

Question 3. Isochrony Revisited and the Representation Problem

In simple regular movements the modulations of the velocity gain factor across clearly distinct figural subpatterns induce a local form of isochrony which manifests itself in the relative invariance of the execution time for each subpattern. This is illustrated in the example of Fig. 3 by the approximate constancy of the angular velocity, and will again be demonstrated in the double ellipse to be reconsidered shortly (see Fig. 4A). From all such simple instances, one is tempted to infer the following general picture: An abstract plan for the movement exists centrally which is characterized by a sequence of figurally distinct units. Upon execution, the mechanisms responsible for specifying the kinematics of the movement append to each unit a specific value of the velocity gain factor that is related to the length of the corresponding trajectory segment. The relation is such that, in conjunction with the two-thirds power law, it results in a velocity profile that makes the execution of each unit almost isochronous (Lacquaniti et al. 1983). At the same time, the set of all gain factors is also such that the execution time for the entire pattern is approximately invariant across size changes.

As argued above, this schematic picture cannot accommodate the case of complex and unpredictable movements for which it is unlikely or even impossible that a global plan is available for the entire movement. Even though the total length of the trajectory is not available and cannot, therefore, provide the basis for isochrony as defined in closed and/or simpler patterns, we still observe, however, the existence of a compensatory mechanism for the average velocity, which is related to the area encompassed by the movement. Moreover, since local modulations of the gain factor are positively correlated with the length of the corresponding segments (cf Eq. 5), a form of local isochrony is also at work in these complex movements. In fact, we can capitalize on the results of the previous section to demonstrate how the two-thirds power law provides the link between this local form of isochrony and global isochrony. Let again C_s represent the average segment curvature and let us rewrite Eq. (2) for within-segment averages in the form:

$$\log V_s = \log K_s - 1/3 \log C_s$$

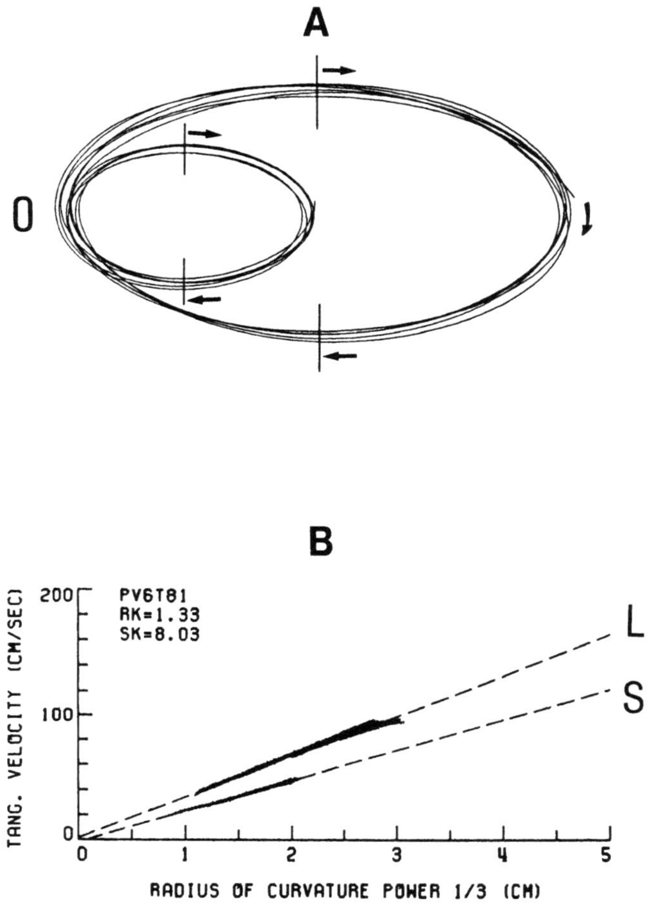

Fig. 4. The coupling experiment. In A several cycles of execution of the double folded ellipse. In B the instantaneous relation between the tangential velocity V(t) and the power 1/3 of the radius of curvature. The two bundles of points describe this relation in the two half–trajectories opposite to the junction point 0. The excellent fitting of the regression lines to the data points demonstrates the validity of the two–thirds power law (Eq. 2 in the text) throughout the movement. The slope of the regression lines provides a least squares estimate of the gain factor in each ellipse. The example shown is typical of fast execution rates (about one complete cycle per second). The significant difference between the velocity gains indicates that the two ellipses are executed almost independently as distinct units of motor action. At slower rates this difference disappears (cf Fig. 5) (modified from Viviani and Cenzato 1985)

Upon replacing the expressions for K_s and C_s supplied by Eqs. (5) and (6) respectively, one gets:

$$\log V_s = 0.512 + 0.496 \log r + 0.338 \log L_s \tag{7}$$

This equation states that the average velocity in each segment is determined jointly by a global scale factor (r) and a local scale factor (L_S). Since the two-thirds power law is instrumental in specifying the relative weight of these two factors, the close agreement of Eq. (7) with the multivariate fitting of V that can be calculated directly from the results confirms the validity of the law.

In view of these results, the tentative picture outlined above should be generalized by admitting that, while local compensatory mechanisms are common to all movements, global ones derive their control parameters from several different geometrical aspects of the movement according to selection rules which are context-dependent.

The difficult question that remains largely open in this general scheme is that of the relationship between the concept of figural unit which is defined on the basis of geometrical criteria, and the concept of segment which is based on kinematic criteria (cf point (a) in Question 2). In patterns like those shown in Figs. 3 and 4A, the two criteria overlap nicely (see, however, the qualifications discussed in the next section), and the hierarchical preeminence of the first over the second one makes good sense: together, these concepts provide two complementary specifications for the notion of unit of motor action (Viviani and Cenzato 1985). In general, however, figural units and segments need not coincide. We already warned, for instance, that segmenting a scribble on the sole basis of the sense of rotation (i.e., assuming that segment endpoints coincide with points of inflections) is only a first-order approximation. A closer analysis of the results shows in fact several departures from this hypothesis. In particular, fast waving movements, in which the sense of rotation alternates regularly several times, are sometimes executed with an almost constant gain factor. Although these wiggles may be thought of as being composed of several distinct arc segments, their kinematic analysis suggests instead that they represent but one single unit of motor action. On the other hand, relatively abrupt transitions from one value of the gain factor to another, which are taken to indicate boundaries between distinct units, may occur away from points of inflection. In all these cases kinematic criteria seem more reliable than figural ones in providing clues to the segmentation problem.

At a very general and speculative level, reconciling these apparently conflicting criteria can be attained if one admits that relevant definitions of the figural units are not necessarily couched only in terms of the most familiar geometrical entities, but must also include higher-order principles of organization. Thus, for instance, the discrepancy mentioned above while considering the case of fast wiggles would disappear if one assumes that, at the representational level, a wiggle cannot be decomposed into simpler units, being already a self-contained figural unit. Indeed psychologists have proposed and discussed many such principles of perceptual organization which either presuppose or imply the existence of higher-order figural units (Zusne 1970; Reed 1973). It should be clear, however, that the import of these speculations vis-à-vis the problem of perceptuomotor organization still awaits experimental scrutiny.

To conclude on this point, I would like to mention the definite possibility that, at such general level, trying to establish a hierarchical relation between the perceptual representation of a pattern and its motokinematic representation may turn out to be like trying to solve the chicken dilemma, the two representations being just but two different aspects of the same thing.

Question 4. The Independence Issue

Suppose we accept the notion that, as soon as a movement becomes sufficiently complex, the necessity arises to decompose its execution into simpler chunks of motor action. The evidence presented so far suggests that the velocity gain factor - which is considered one qualifying attribute of the chunk - is related to both local and global metric properties of the trajectory. The question can then be posed as to whether each chunk is planned independently of all the others. Since the hypothesis of independence is clearly untenable as far as the geometry of the trajectory is concerned, we will only consider the question from the point of view of kinematics. If indeed the kinematics of each chunk were to be controlled independently, the global relationships illustrated in the first section should derive entirely from the local ones, and - at least in principle - they should be easily predictable. The analysis of circular figure-eight patterns (Viviani and McCollum 1983) has strongly suggested that this simple null hypothesis is false and has led us to investigate the nature of the existing interactions. A first attempt to tackle this problem has been made in the specific case of the folded double ellipse (Fig. 4A) (Viviani and Cenzato 1985). To begin with, we show that the double ellipse is well adapted to the analysis envisaged, in the sense that its execution is in fact decomposable into identifiable motor chunks. This is demonstrated in the scatter diagram of Fig. 4B in which the instantaneous values V(t) of the tangential velocity are plotted against the corresponding values of R(t)**1/3. The example is typical of what is observed at fast execution rates. The two bundles of points appearing in this plot are relative to the execution of the two half ellipses opposite to the junction point 0. According to Eq. (2) the slope of the bundles provides the numerical value of the velocity gain factor K in the inner and outer ellipse respectively. Thus, the two ellipses forming the complete pattern are executed with two sharply different gain factors. Within our conceptual framework they can then be considered as two distinct units of motor action. Notice that in this case the transition between the two units occur at the junction point 0 which is not an inflection point.

To pursue further the issue of this section, let us suppose that each unit of motor action (i.e., each ellipse) is planned independently with the same principles described for isolated ellipses (cf Fig. 2). In this case the two velocity gain factors K_1 and K_2 should only depend on the corresponding perimeters P_1 and P_2. Then, simple manipulations of Eq. (3) permit one to show that the ratio of the mean velocities V_2 and V_1 in the two ellipses should be close to P_2/P_1 for all rates of execution. This prediction does not pass the experimental test: as shown in Fig. 5, the ratio V_2/V_1 varies systematically as a function of the overall period T and only approaches the hypothetical value 2 for very fast movements (T = 1.0 s). At all other rates of execution we can no longer accept the simple expression $K = K_0 P^b$ as a valid predictor of the gain factor. We must instead envisage the possibility that the kinematic planning of each segment also takes into account the metric properties of the other segment.

A number of plausible interaction mechanisms can be invoked to explain this mutual dependency. In the simple case of the double ellipse, which is only composed of two segments, we explored the hypothesis that the gain factor in each segment depends on both the perimeter of the ellipse being executed and the perimeter of the total pattern. The relative weight of these two components is supposed to be determined trough a coupling function E whose value ranges between 0 and 1, depending on the overall rhythm of execution. When the rhythm is high, the coupling is weak and the two segments are executed almost independently as in the example of Fig. 4B. As the rhythm slows down, the coupling increases progressively toward its maximum value of 1 and both ellipses are executed with the same velocity gain which only depends on the total perimeter $P_1 + P_2$ of the pattern. In this limiting case, the subdivision of the movement into identifiable chunks of motor action vanishes altogether and the pattern is traced out as a unit. Notice that, under conditions of high coupling, the average velocity across successive units is almost the same. Thus, also the ratio

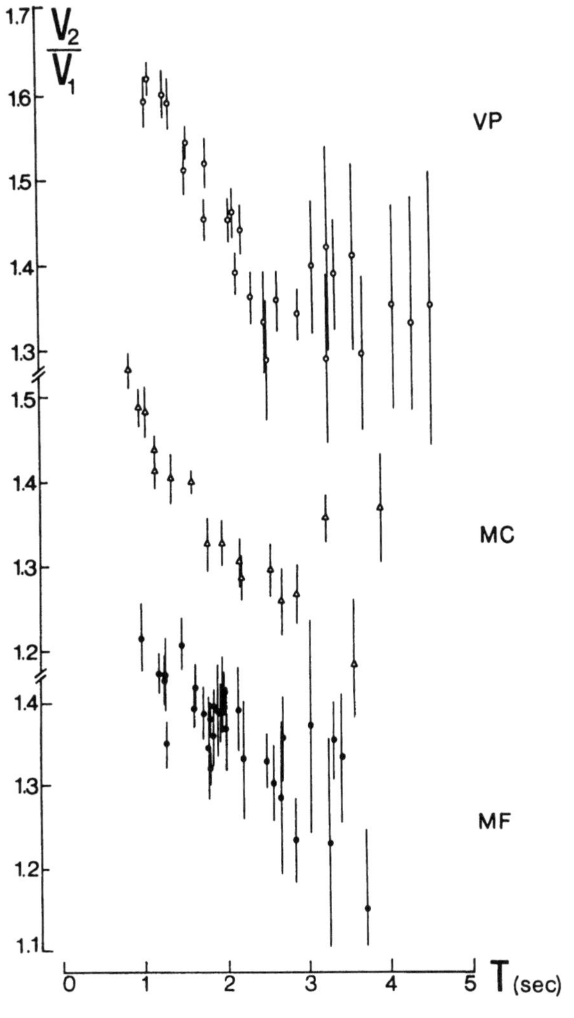

Fig. 5. The coupling experiment. Results in three subjects from an experiment in which the double folded ellipse was executed at different rates. Data points describe the relation between the ratio of the average velocities in the two ellipses and the duration T of one complete cycle. Vertical bars encompass 2 standard deviations. The ratio V_2/V_1 decreases continuously from a value which indicates weak coupling between the execution of the ellipses, to a lower value which indicates strong coupling (see text). In the presence of weak coupling (cf Fig. 3) the gain factors are positively correlated to the perimeters of the two parts of the pattern. Thus, the corresponding execution times are not much different, as in the figure–eight example of Fig. 4. When the coupling is strong the gain factors are almost equal and only depend on the overall perimeter of the double ellipse

between the duration of the units is invariant and equal to the ratio of the respective lengths. This relative invariance of the temporal structure of complex movements has actually been observed in many different motor performances: handwriting (Viviani and Terzuolo 1980, 1983), typing (Terzuolo and Viviani 1979, 1980), dance (Cadopi and Viviani, in preparation), gait (Shapiro et al. 1981), and wrist movements (Carter and Shapiro 1984).

The quantitative model outlined above for unit interaction has been tested on the basis of the available data (for details see the quoted paper) and proved quite successful in predicting a number of experimental results. However, it should be obvious that a much more articulated version of the coupling hypothesis will be needed to cope with more complex patterns. In particular, the desired formulation should be able to encompass also those cases in which one cannot suppose that the total perimeter of the pattern is centrally represented (see above).

Question 5. Posing as Conclusion, the Teleonomic Issue

In a nutshell, the take-home message from the preceding sections can be summarized in two points:

1. Drawing movements can be decomposed into units of motor action which are distinguished on the basis of a single parameter (the velocity gain factor) relating form and velocity of the movement.

2. The velocity gain factors are so specified that both the overall average velocity throughout the movements, and the average velocity within each segment increase with the size of the trajectory. This covariation results into that particular form of timing control that was dubbed isochrony.

For each of these two points, it is tempting to raise what might be called the teleonomic issue: If indeed things are as suggested, can we adumbrate why this is so? How different would our conception of motor control would be if things *were not* as suggested?

Much of what I will have to contribute on this issue revolves around the single pivotal concept of hierarchical procedure which, in the present context, refers to any sequence or array of processes whereby a perceptuomotor representation couched in terms of one set of primitive concepts is transformed into another representation with a different set of primitives. The primitives in the two representations must be related either by logical inclusion, or by some form of causative dependency. Such notion of hierarchical procedure has been invoked in various theories of visual perception (Broadbent 1977; Rumelhart 1970; Bouwhuis and Bouma 1979). Moreover it proved to be a fitting frame of reference for motor learning theories. In particular, the learning process leading to professional morse code tapping (Bryan and Harter 1899), typing (Book 1908), and piano playing (Shaffer 1981) has been described as the progressive coalescence of relatively low–ranking units of motor action into hierarchically higher units.

The main argument of this discussion is that all possible construals of the notion of hierarchical procedure share a common feature: They all imply that both the primitives which are acted upon by the procedure, and the result of the procedure constitute a finite, discrete set of logically identifiable entities. In fact, I am here arguing that the very notion of hierarchy is meaningless unless some categorization principle is available for the variables relevant to the process to be described. Thus, one can surmise that the observed decomposition of movements into discriminable units is related to the hierarchical nature of the representation and control system inasmuch as it provides the basic categorization principle. To be sure, this hypothesis does not rule out the possibility of construing some notion of unit of motor action without committing oneself to a specific hierarchical scheme. For instance, a general conception that can be traced to the Moscow school (e.g., Bernstein, Feldman) conceptualizes the problem of motor control in terms of dynamic principles and constraints akin to those that regulate all physical dissipative systems (Turvey et al. 1978; Kugler et al. 1980; Bizzi and Abend 1983). Although this conception is not explicitly hierarchical, it does assume that coordinated muscular synergies represent the basic unit of motor action in the production of many complex motor behaviors (cf. Kelso and Tuller 1984). Again, my basic contention is only that the existence of such units is a necessary prerequisite for a system to possess a hierarchical architecture. Since, however, this is not a sufficiency condition, whatever evidence I have marshalled in favor of the segmentation hypothesis can only be suggestive of a hierarchical structure but cannot prove it.

Coming now to the phenomenon described as isochrony, an old but recurring theme in motor studies is the intermittency or timing hypothesis (Young and Stark 1963; Young 1981; Shaffer 1976) according to which perceptuomotor coordination is paced by an internal clock, and movement plans can only be updated and executed in synchrony with the beating of this clock. While this hypothesis is certainly not valid in some cases, its basic tenet is in keeping with the observation that segments tend to be executed in a constant time.

At least two alternative lines of reasoning can be pursued from this surmisal. If one assumes that (a) units of motor action come into being (either in the course of phylogenesis or in the learning process) as a prerequisite for a hierarchical architecture of the motor planning processes (cf above), and (b) that a pacemaker is actually at work during the execution of movements, it could be argued that the purpose of the compensatory velocity modulations that give rise to isochrony is to allow each unit to be completed within one clock cycle. The other possibility is that changes in the velocity gain factor – which are used as a criterion for identifying the units of motor action – are simply the consequence of the intermittency introduced by the pacing mechanism.

The second hypothesis is more parsimonious. It is hard put, however, in explaining why a correlation exists between the velocity gain factors within segments and the linear extent of the corresponding stretch of trajectory (cf Question 3). The first line of reasoning seems instead entirely congruent with the available evidence.

Harking back to the original, teleonomic issue and summing up the arguments presented so far, it appears that the two questions raised at the beginning of this section can be condensed into a single one: Why would it be desirable for the motor planning mechanisms to possess a multilevel system of representation? Clearly, trying to provide an answer to this question would require the concepts and tools of robotics, control theory, and cybernetics, all of which lay well outside the scope of this presentation. The only remaining merit of this discussion may be that it has pointed out the level at which it would be appropriate to assess the import of the experimental findings.

REFERENCES

Bizzi E, Abend W (1983) Posture control and trajectory formation in single-and multi-joint arm movements. In: Desmedt JE (ed) Motor control mechanisms in health and disease. Raven, New York, pp 31-45

Book WF (1908) The psychology of skill. Montana, Missoula

Bouwhuis D, Bouma H (1979) Visual word recognition of three-letter words as derived from the recognition of the constituent letters. Percept Psychoph 25: 12-22

Broadbent DE (1977) The hidden preattentive processes. Am Psychol 32: 109-118

Bryan WL, Harter N (1899) Studies in the telegraphic language: the acquisition of a hierarchy of habits. Psychol Rev 6: 345-375

Carter MC, Shapiro DC (1984) Invariant properties of sequential movements. J Neurophysiol 52: 787-796

Kelso JAS, Tuller B (1984) A dynamical basis for action systems. In: Gazzaniga MS (ed) Handbook of cognitive neuroscience. Plenum, New York, pp 321-356

Kugler PN, Kelso JAS, Turvey MT (1980) On the concept of coordinative structures as dissipative structures: I. Theoretical lines of convergence. In: Stelmach GE, Requin J (eds) Tutorials in motor behavior. North-Holland, Amsterdam, pp 1-47

Lacquaniti F, Terzuolo C, Viviani P (1983) The law relating kinematic and figural aspects of drawing movements. Acta Psychol 54: 115-130

Lacquaniti F, Terzuolo C, Viviani P (1984) Global metric properties and preparatory processes in drawing movements. In: Kornblum S, Requin J (eds) Preparatory states and processess. Erlbaum, Hillsdale, pp 357-370

Merton PA (1972) How we control the contraction of our muscles. Sci Am 226: 30-37

Reed SK (1973) Psychological processes in pattern recognition. Academic, New York

Rumelhart DE (1970) A multicomponent theory of confusion among briefly exposed visual displays. J Math Psychol 7: 191-218

Shaffer LH (1976) Intention and performance. Psychol Rev 83: 375-393

Shaffer LH (1981) Performances of Chopin, Bach, and Bartok: studies in motor programming. Cognitive Psychol 13: 326-376

Shapiro DC, Zernicke RF, Gregor RJ, Diestel JD (1981) Evidence for generalized motor programs using gait pattern analysis. J Motor Behav 13: 33-47

Terzuolo C, Viviani P (1979) The central representation of learned motor patterns. In: Talbot R, Humphrey DR (eds) Posture and movement. Raven, New York, pp 113–121

Terzuolo C, Viviani P (1980) Determinants and characteristics of motor patterns used for typing. Neuroscience 5: 1085–1103

Turvey MT, Shaw R, Mace W (1978) Issues in the theory of action: Degrees of freedom, coordinative structures and coalitions. In: Requin J (ed) Attention and performance VII. Erlbaum, Hillsdale

Viviani P, Cenzato M (1985) Segmentation and coupling in complex movements. J Exp Psychol: Hum Percept Perform 11: 828–845

Viviani P, McCollum G (1983) The relation between linear extent and velocity in drawing movements. Neuroscience 10: 211–218

Viviani P, Terzuolo C (1980) Space-time invariance in learned motor skills. In: Stelmach GE, Requin J (eds) Tutorials in motor behavior. North-Holland, Amsterdam, pp 525–533

Viviani P, Terzuolo C (1982) Trajectory determines movement dynamics. Neuroscience 7: 431–437

Viviani P, Terzuolo C (1983) The organization of movement in handwriting and typing. In: Butterworth B (ed) Language production, II - Development, writing and other language processes. Academic, New York, pp 103–146

Young LR (1981) The sampled data model and foveal dead zone for saccades. In: Zuber BL (ed) Models of oculomotor behavior and control. CRC Press, Boca Raton, pp 43–74

Young LR, Stark L (1963) Variable feedback experiments testing a sampled data model for eye tracking movements. IEEE Trans Hum Factors Electron 4: 38

Zusne L (1970) Visual perception of form. Academic, New York

Gearing Action to the Environment*

D. N. Lee and D. S. Young
Department of Psychology, University of Edinburgh, Edinburgh EH8 9JZ, Scotland

To survive, an organism must gear its actions to the environment. This requires perceptual information about the position, orientation and movement of the body and body parts in relation to environmental surfaces and objects. The information required is not simply exterospecific, about relative sizes, shapes etc., nor is it simply propriospecific, about the configuration of the body itself. What is needed is information about the extero–proprio relationship, i.e., expropriospecific information (Lee 1978). Here we review the exproprioceptive function of vision in stabilizing the head and eyes, maintaining balance, steering, controlling timing in locomotion and controlling timing in intercepting or avoiding moving objects.

STABILIZING HEAD AND BODY

The pied kingfisher in Fig. 1 is demonstrating elementary gearing – anchoring its head to the environment. The bird is hovering above the water ready to dive on a fish. Though its wings and body are moving rhythmically up and down, it keeps the head pointing down by neck adjustments. In film analysis over seven wing beats we found that the standard deviation of head orientation was a mere 0.5° compared with 4.6° for the body. The head is anchored presumably to facilitate sight of the prey with the temporal fovea (Walls 1967), since the bird can turn its eyes very little. Expropriospecific information for controlling orientation and position of the head is available in the optic flow field at the bird's eye: distinct types of optic flow fields result from rotation (Fig. 1, left) and translation (Fig. 1, right) of the head (Lee and Young 1985). Vestibular information about angular accelerations of the head and the direction of gravity might also be used in controlling head orientation.

Pigeons and other head–bobbing species such as doves, hens, magpies and starlings, anchor their heads to the environment while walking. Dunlap and Mowrer (1930) showed by film analysis that when the head appears to be bobbing backward it is in fact stabilized on the environment, with the body moving forwards beneath it. When the head can retract no further it is snapped forward to the next anchor position. As with the kingfisher, the purpose is probably to keep the fovea directed to whatever is of interest in order to optimise information pick–up.

* The work reported here was supported by the Medical Research Council, Science Research Council, General Accident and Scottish Health Education Group.

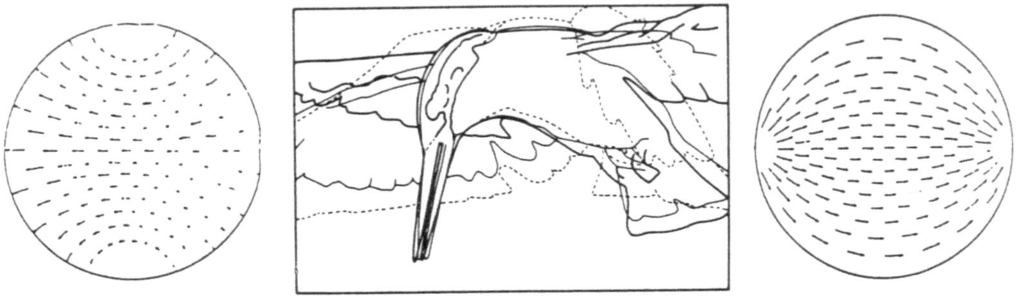

Fig. 1. Pied kingfisher demonstrating elementary gearing

Humans, with their more mobile eyes, generally stabilize gaze when walking, running, turning and so on by rotating both head and eyes. The precision of control is evident in the ease with which one can read from a noticeboard while twisting the trunk to and fro. As the body rotates to the left, the head turns to the right relative to the shoulders and the eyes to the right relative to the head, so that fixation is maintained. Not much seems to be known about this precise coordinative control of head and eyes, which is essential for visual guidance in many activities. Studies of constrained behaviours such as maintaining fixation with eye movements alone when the body is turned and the head immobilized (e.g., Herman et al. 1982; Regal et al. 1983) indicate that both visual and vestibular expropriospecific information are implicated in controlling the head and eyes. However, vision appears the ultimate arbiter in that it can modify the meaning for action of vestibular information (Melvill Jones 1977).

A spinning ballerina behaves rather like a head-bobbing pigeon. She fixates a point until her head and eyes can turn no further, then rapidly rotates her head to catch the point again as she spins round. This she does for balance control (perhaps it helps the pigeon's balance too). Fixation removes the high-speed rotary component of the optic flow field at the eye, thus making any small translatory movement of the head or rotary movement about a different axis, which would indicate a shift from equilibrium, more detectable visually. In addition, the nystagmic movements of the head when spinning mean that the vestibular system is not subject to unusual stimulation and so post-rotary nystagmic eye movements and dizziness are obviated.

The importance of vision in balance control (and in head bobbing) is illustrated in Fig. 2. A human subject stands on a fixed floor while the walls and ceiling of a surrounding "room" are e.g. moved forward. This provides visual information that the head is moving backward and so the subject compensates by swaying forwards. Toddlers can be literally bowled over by the visual information (Lee and Aronson 1974). Adults, however, normally just sway more, apparently because they get better expropriospecific information through their feet than do toddlers. If that information is rendered less precise by, e.g. balancing crossways on a beam, when the feet can tip, then adults too can be bowled over (Lee and Lishman 1975). The eye is very sensitive to head movement, as the graphs in Fig. 2 show: while standing on foam pads, an adult's body sway was visually driven by anterior-posterior sinusoidal oscillation of the room of amplitude 3mm (Lee and Lishman 1975). Larger oscillations cause a standing dove to bob its head, but if the bird is yoked to a room (rather smaller than that illustrated!), so that when it walks the room moves with it, then the bird does not bob its head (Friedman 1975). The head is anchored to the visual world rather than the inertial world.

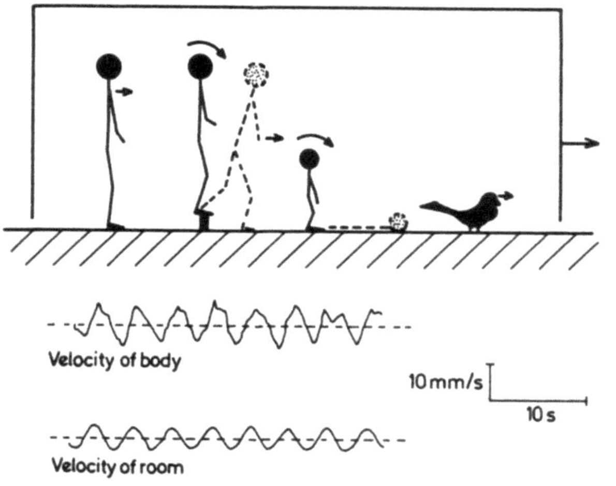

Fig. 2. Vision in balance control

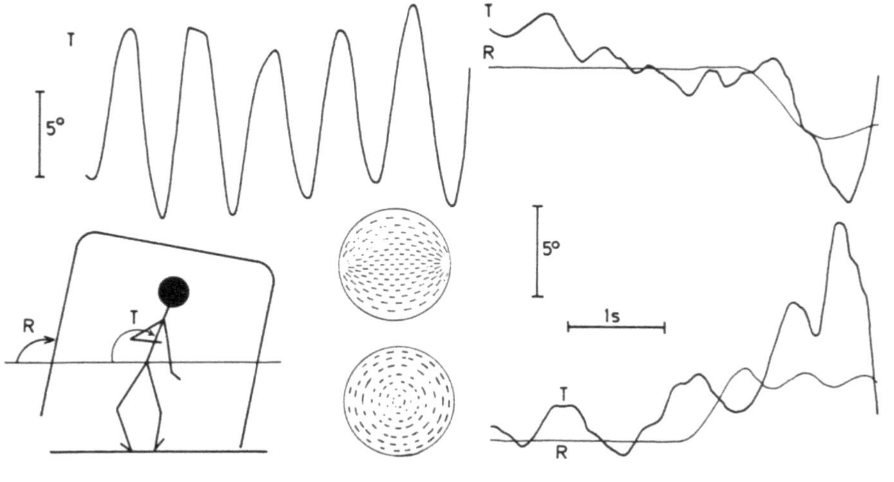

Fig. 3. Experiments with a movable room supported over a treadmill

When a person runs, the trunk tilts from side to side. Vision helps keep the tilt within stable bounds so that balance and course are maintained. This is shown by the experiments illustrated in Fig. 3, carried out with Michael Anderson (paper in preparation). The movable room was supported over a treadmill. As the subject ran on the treadmill the room was unexpectedly tilted through 4° about an horizontal axis roughly through the ankles. A Selspot system monitored the inclination of the room and of the subject's trunk, defined by the line joining nape of neck to coccyx. The curve on the left of Fig. 3 shows the oscillations of the trunk as the subject ran with the room stationary. These oscillations were subtracted from the trunk movement which occurred in the neighbourhood of a room rotation to reveal how the rotation affected the pattern of trunk

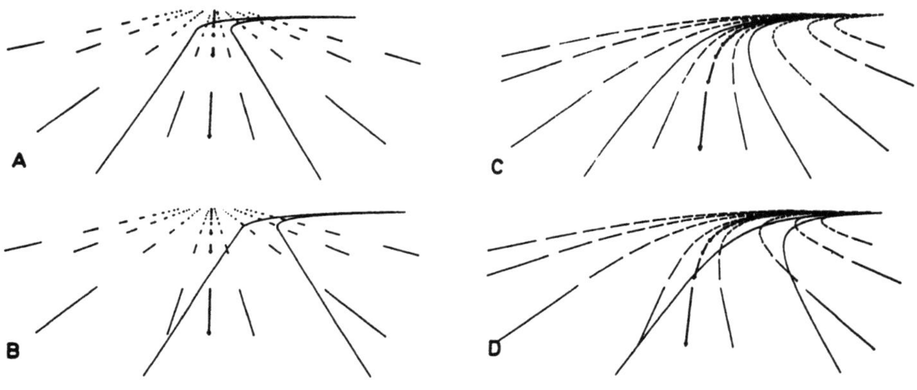

Fig. 4. The optic field of an automobile driver

movement (right–hand graphs). The subject tilts with the room, apparently compensating for the visually simulated body tilt in the opposite direction.

STEERING

Locomotion entails getting around obstacles and this requires prospective control. When driving, it is no use waiting until you hit the kerb before adjusting the steering. You need to see ahead of time whether you are going to hit the kerb if you maintain the current steering angle. Thus, McLean and Hoffmann (1973) found that drivers based steering adjustments primarily on heading angle – indicating potential future course – rather than on lateral position on the road.

Potential future course is optically specified (Lee and Lishman 1977). In Fig. 4, the optic flow field of a driver is depicted as the projection onto a plane perpendicular to the instantaneous direction of travel, like the film plane of a camera that is pointing forward. The solid lines in Fig. 4 represent the edges of a road; the broken lines represent the optic flow lines.

The central heavily drawn line is the "locomotor flow line": it specifies the potential future course of the vehicle if the current steering angle were to be maintained. In (A) the driver is on course straight for a bend. In (B) the steering is straight, but off course. In (C) the driver is on course steering around a bend of uniform curvature and in (D) is off course on the same bend.

BODY–SCALED INFORMATION

The optic flow field affords a wealth of information about the layout of the environment and about the organism's movement. Figure 5a illustrates the basic optical geometry. As in Fig. 4, we consider a schematic eye with a flat retina orthogonal to the instantaneous direction of movement. This we do for ease of description and without loss of generality since the flow pattern on the

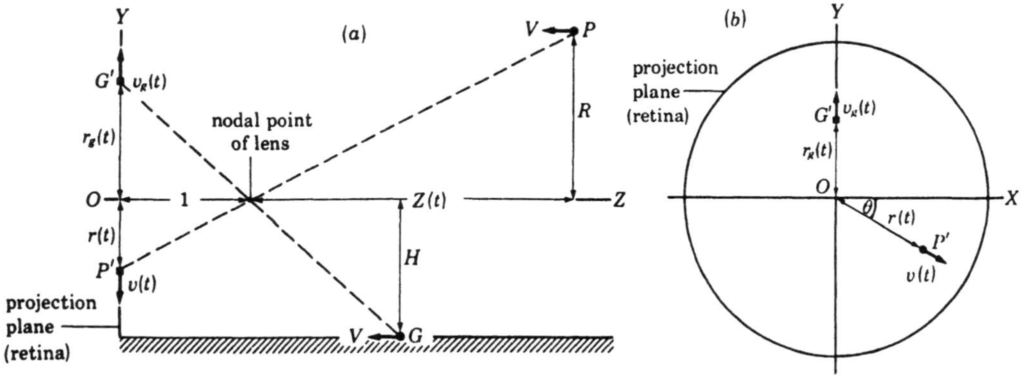

Fig. 5. Geometry of the optic flow field

schematic retina can be deduced from the flow pattern on a real retina of any shape, which may also be rotating (Lee and Young 1985). In Fig. 5a, the schematic eye is considered to be stationary and the environment moving towards it with velocity V in the direction Z to O. P and G denote texture elements on surfaces in the environment, G being on the ground. Light reflected from P and G passes through the nodal point of the lens, giving rise to the moving optic texture elements P′ and G′ on the "retina". Figure 5b gives a view of the retina from behind. The element P′, distant r(t) from O, moves outwards with optic velocity v(t). In general, the densely textured environment yields a densely textured optic flow field where all optic texture elements move outwards along radial flow lines emanating from O.

The position of P relative to the eye at time t is defined by the distance coordinates Z(t) and R together with the angle between the OZP and OZX planes. The angle is optically specified by the angular coordinate θ of the optic element P′. As regards the distance coordinates from similar triangles

$$Z(t)/R = 1/r(t) \tag{1}$$

differentiating Eq. (1) with respect to time

$$V/R = v(t)/r(t)^2 \tag{2}$$

and dividing Eq. (1) by Eq. (2)

$$Z(t)/V = r(t)/v(t) \tag{3}$$

Thus, from Eqs. (2) and (3), V needs to be known before Z(t) and R are specified. When locomoting over level ground, V (hence Z(t) and R) is given in terms of the body unit H, the height of the eye above the ground; applying Eq. (2) to element G

$$V/H = v_g(t)/r_g(t)^2 \tag{4}$$

Hence, one way that the distance coordinates (R, Z(t)) of point P might be perceived is by picking up optic flow information from P and scaling information from the optic flow of G. There are other possible ways, but in general perceiving distance always requires picking up scaling information.

Fig. 6. Ski–jumper approaching the lip of the jump

By contrast, Eq. (3) shows that expropriospecific temporal information is available at every point in the optic flow field without the need of additional scaling information. The quantity Z(t)/V is the time to contact of element P with the plane through the nodal point of the lens and parallel to the retina, given that the approach velocity V remains constant. We have named this time interval the tau–margin, the optic variable, r(t)/v(t), which gives the value of the tau–margin, we have named tau. The tau–margin can be perceived without perceiving distance or speed (Schiff and Detwiler 1979), which indicates that the optic variable tau is sensed directly. The tau–margin is important information for controlling action, whether or not approach velocity is constant, as we will now illustrate.

TIMING LOCOMOTOR ACTS

Figure 6 shows the action a ski–jumper has to perform when approaching the lip of the jump at about 90 km/h (56 mph). The rapid forceful straightening up from the streamlined crouch has to be precisely timed for a proper take–off. Film analysis of 14 ski–jumpers on the 70 m jump showed that they started straightening on average 194 ms before their feet passed over the lip and the standard deviation of this interval across the jumpers was only 10 ms (Lee et al. 1982). As the skier nears the lip of the jump, the in–run levels out and so the velocity of travel is about constant. The optic variable tau for the lip therefore provides precise information for timing the act.

Long jumpers face a similar timing problem, but the action that needs to be geared to tau is more complex. When sprinting at about 10 ms^{-1} the athlete has to strike a 0.2 m wide take–off board in the right posture for a powerful jump. Film analysis showed that the last few strides to the board are visually regulated. Furthermore, the data were consistent with the hypothesis that control consists in the optic variable tau for the take–off board modulating the vertical impulse applied to the ground during the support phase (Lee et al. 1982). We will examine this hypothesis more closely in terms of a further experiment that we ran in order to obtain more detailed data about stride regulation.

As illustrated in Fig. 7, stride length S is made up of trail distance T (the horizontal distance between a reference point on the trunk, e.g. at the coccyx, and the toe of the supporting foot at the

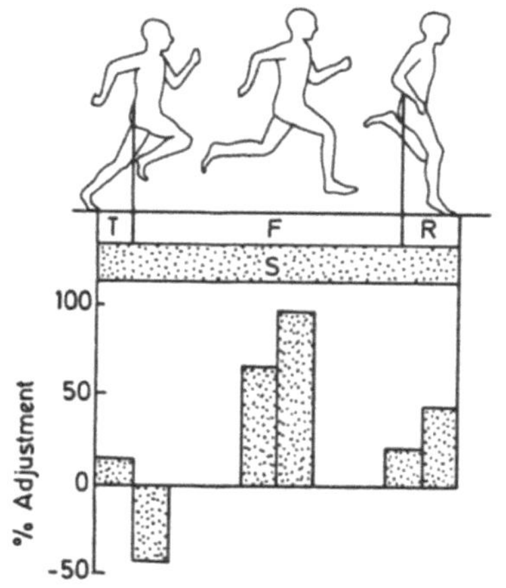

Fig. 7. Stride length

toe–off), flight distance F (the horizontal distance travelled by the coccyx when the body is airborne), and the reach distance R (the horizontal distance between the coccyx and the toe of the supporting foot at heel–strike).

An estimate of the percentage adjustment to S that is attributable to a change in R, say, is the linear regression coefficient of R on S. These estimates could not be obtained in the long–jump study, because of limitations of the filming, though approximate measures indicated that flight distance F was primarily adjusted. In an experiment with Warren (1985) we investigated the question in more detail. Subjects ran at 4 ms^{-1} over "stepping stones" on a treadmill. The stones were marks on a tape driven by the treadmill belt; they were unevenly spaced in groups of five with about 9 m between groups (Fig. 8). The three segments of a stride were measured using a Selspot system, with infra–red emitting diodes placed on the subject's toe, heel and coccyx. Estimates of the percentage adjustment to stride length made on each segment are shown in Fig. 7. (The results are for two subjects on the approach strides to the stepping stones; the strides on the stones gave similar results.) Flight distance was primarily adjusted, as for the long jumpers. Further analysis showed that 95% of the change in flight distance was brought about by changing flight time; flight velocity was kept about constant.

In sum, the results of the long jump and the stepping stones experiments indicate that striking a specific point on the ground entails adjusting the durations of the approach strides so that they just span the time remaining to reach the point. Since tau for the point specifies the time to reach it (at

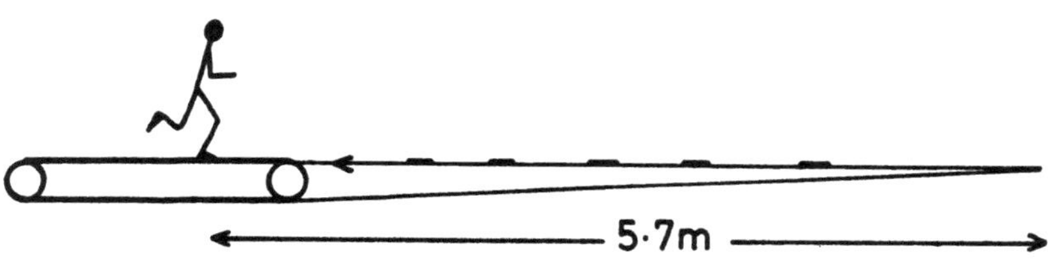

Fig. 8. Subject running over "stepping stones" on a treadmill

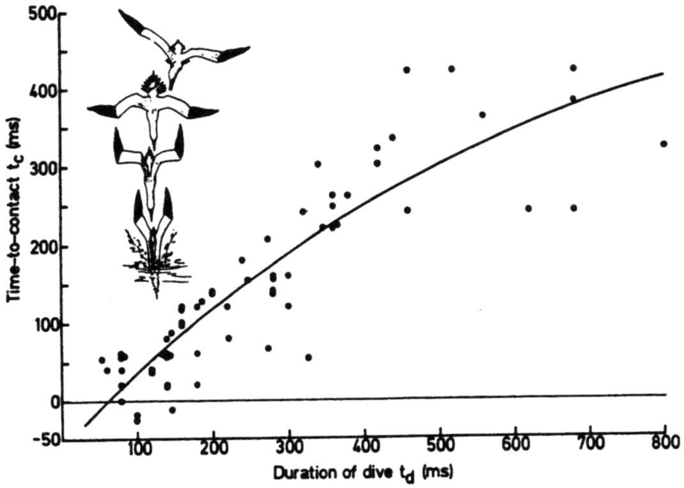

Fig. 9. Locomotion of the diving gannet

constant speed) and stride duration is proportional to the vertical impulse applied to the ground (Lee et al. 1982), the results are consistent with the hypothesis that tau modulates vertical impulse.

The diving gannet (Fig. 9) appears to have a more difficult locomotor timing problem. As it plummets toward the sea, wings half open for steerage, it needs to see when to streamline so that it enters the water without injury. Since it is accelerating under gravity, the tau–margin (which assumes constant speed) overestimates its time to contact with the water. Does the bird, therefore, have to perform a complex visual computation, taking into account acceleration? Or might is simply start streamlining when the tau–margin reaches a certain trigger value? If the bird were to follow the latter, easier and more robust "tau strategy", then the longer the dive the sooner before contact would it streamline. Briefly, this is because a higher velocity is reached in a longer dive and so, as the bird approaches the water, gravity does not produce as large a proportionate increase in velocity as it does with a shorter dive. Consequently, the time to contact is closer to the tau–margin. Thus, for example, in a 3 s dive, when the tau–margin reaches a trigger value of 1 s, the time to contact will be 0.84 s, whereas in a 1 s dive, when the tau–margin reaches 1 s, the time to contact will be 0.59 s.

Data from film analysis of about 50 dives were consistent with the hypothesis that the birds follow the tau strategy (Lee and Reddish 1981). The curve in Fig. 9 is the best–fitting curve derived from the theory. The parameter values of the curve indicate that, on average, the birds started streamlining 60 ms after the tau–margin had reached a value of 820 ms.

Landing on a surface requires precise visual timing to cushion the impact. Dietz and Noth (1978) found that human subjects falling on their hands, against a platform tilted at different angles, activated their arm extensors to act as shock absorbers about 70–150 ms before contact. As with the gannets, muscle activity started sooner before contact the greater the tilt and so the longer the fall, consistent with the hypothesis that the subjects' timing was geared to the tau–margin. Monkeys leaping head down to a platform likewise activate their arm extensors – about 50 ms

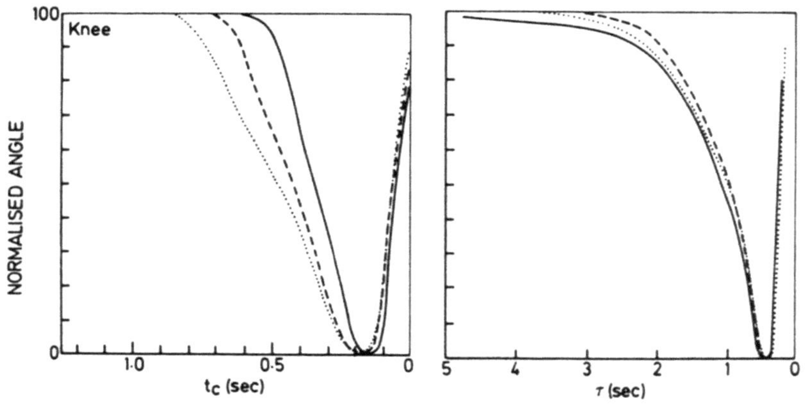

Fig. 10. Knee angle changing over time for different drop heights (left) and against the tau–margin at the different times before contact (right)

from contact for drops of 60–165 cm (Dyhre–Poulsen et al. 1980). Flies, on the other hand, slow down when landing. They start decelerating, with about the same 60 ms visuomotor delay as gannets, after the tau–margin for the surface reaches a value of 80 ms or so (Wagner 1982).

INTERCEPTING MOVING OBJECTS

Intercepting moving objects is a similar problem to timing locomotor acts. As young as 4 months, babies start developing catching skill and by 9 months can achieve 50 ms precision (von Hofsten 1980, 1983). When an object is approaching at a constant speed, the tau–margin gives exact information for timing the grasp or strike. To test whether people gear interceptive actions to the tau–margin when an object is accelerating toward them, we dropped a soccer ball down to a subject standing directly below, who had to leap and punch it back up (Lee et al. 1983). Figure 10 (left) shows how the angle at a subject's knee changed over time for different drop heights (3 m, solid lines; 5 m, broken; 7.2 m, dotted). Mean normalised angles (100, maximum extension; 0, maximum flexion) are plotted against the time to contact (t_c) of the fist with the ball. The patterns for knee and elbow were similar. Both were as predicted if the subject were using the tau–strategy: flexion occurs sooner before contact the higher the drop (as with the gannets), but as the moment of contact is approached the curves become increasingly similar as the tau–margin gives a closer and closer estimate of the time to contact.

As a stronger test, Fig. 10 (right) plots the knee angle against the value of the tau–margin at the different times before contact. If the subject's actions are geared to the tau–margin, then the curves for the different drop heights should not be significantly different. This is so: the difference between the curves is no more than expected from the within–condition variance. The same was true for the elbow.

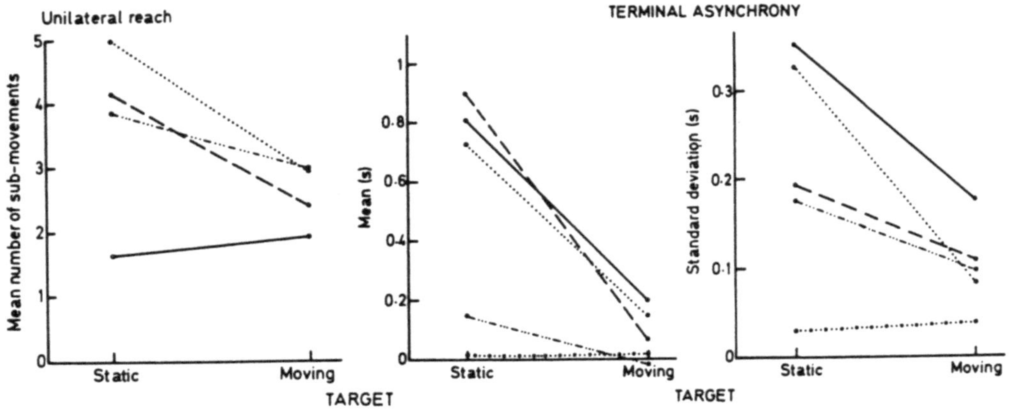

Fig. 11. Four left–hemisphere stroke patients reaching for a static or moving target

The experiment shows how the temporal sequencing of muscular activity can be guided *extrinsically* by visual information: the changing value of the tau–margin specifies a time course for the motor system to follow. At Edinburgh, Lough (1985) has recently applied this idea in a study of hemiparesis. After partial recovery from a stroke, when reaching to stationary objects movements of the affected arm are frequently jerky and slow, and uncoordinated with the other arm, as if the agonist and antagonist muscles are at loggerheads (Miller and Hammond 1982). This could be due, at least in part, to a disruption of intrinsic information from joint and muscle receptors for regulating agonist/antagonist timing. If so, then providing the patient with *extrinsic* timing information should improve smoothness and coordination of movement.

Four left–hemisphere stroke patients with no comprehensional or perceptual difficulties participated in the study. A soccer ball was rolled down an inclined track toward them and they had to reach out to either catch it with both hands or to hit it with one hand. As a control, they performed the same acts with the ball stationary. Arm movements were analysed using a Selspot system, with infra–red diodes attached to the wrists.

As predicted, performance was better when the ball was moving and thereby providing extrinsic timing information. First, movement of the affected arm was generally smoother when reaching for the moving ball. Figure 11, left–hand panel, shows that three of the patients had fewer jerky submovements (i.e. acceleration/deceleration cycles) in their affected arm when reaching to the moving target as opposed to the static target. Second, bilateral coordination was better with the moving target for each of the four patients. This can be seen in Fig. 11, centre and right–hand panels. (In each panel, the upper four graph lines are for individual patients and the lowest line is for a normal control. "Terminal asynchrony" is the difference between the time of arrival of the unaffected hand at the ball and the time of arrival of the other hand.) The terminal asynchrony was not only shorter, which could have simply been due to the fact that the affected arm speeded up; it was also less variable, which indicates that the arms were being jointly timed by the visual information from the moving ball. The experiment suggests that recovery from stroke may be aided by exercises, such as catching, where limb movements are visually driven. A learning study needs to be run to test this hypothesis.

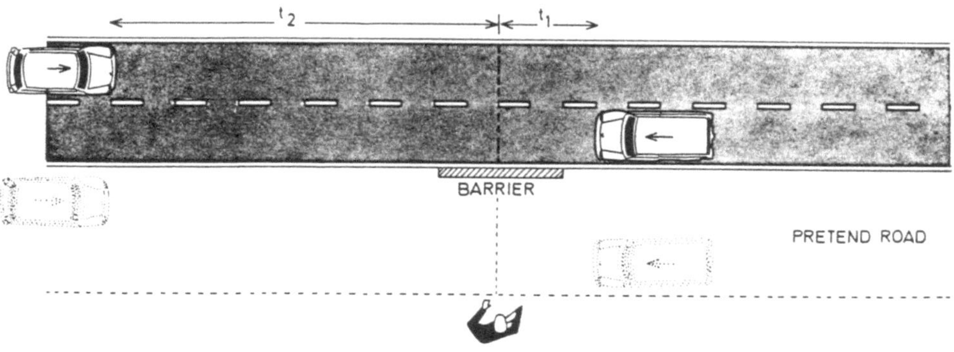

Fig. 12. "Pretend road" crossing task

PEDESTRIANS AND DRIVERS

Crossing the road safely very much requires gearing actions to the environment. For example, crossing in the gap between two vehicles properly entails seeing the difference between the time to arrival of the first vehicle (tta_1) and that of the second (tta_2) – estimated by the values of tau for the vehicles – and relating this to walking speed (s) and the distance across (d). The gap is too small if $tta_2 - tta_1 < d/s$. Thus, the ability to judge gaps – or whether it is safe to cross in front of an approaching vehicle, i.e. whether $tta > d/s$ – requires establishing visuomotor calibration through practice. A skilled adult is often not aware of this until walking speed has to be reassessed – as when temporarily lame, carrying a heavy bag or wearing a tight skirt.

Children have to learn the skill from scratch. They need a safe way of practising and their performance has to be assessed. Crossing the road under adult guidance, which is the only way children currently get supervised practice, does not allow the child to properly explore and develop her capabilities. Because of the danger, she has to be constrained from making mistakes and so she does not have the opportunity to learn from her mistakes, unlike when learning most other skills.

In an attempt to provide safe practice for children, we have devised a "pretend road" crossing task that is performed at the roadside and requires no special equipment (Lee et al. 1984). The "pretend road" is simply a stretch of pavement alongside a road. An adult chooses a suitable site which allows adequate view down the road and erects a short barrier alongside the kerb. The situation is illustrated in Fig. 12. The child stands where indicated and pretends that for each vehicle seen coming along the real road there is one in a corresponding position on the pretend road (the pretend road vehicles are indicated by dotted outlines in Fig. 12). The child's task is to cross the pretend road as far as the barrier without being knocked down by any of the pretend vehicles. This means, in actuality, that the child must not be in the process of crossing the pretend road when a vehicle passes the barrier. The child can thus easily see how well she is performing, where she is making mistakes and where she needs to improve. The adult supervisor can likewise assess the child's ability and judge when it is safe to introduce the child to crossing real roads.

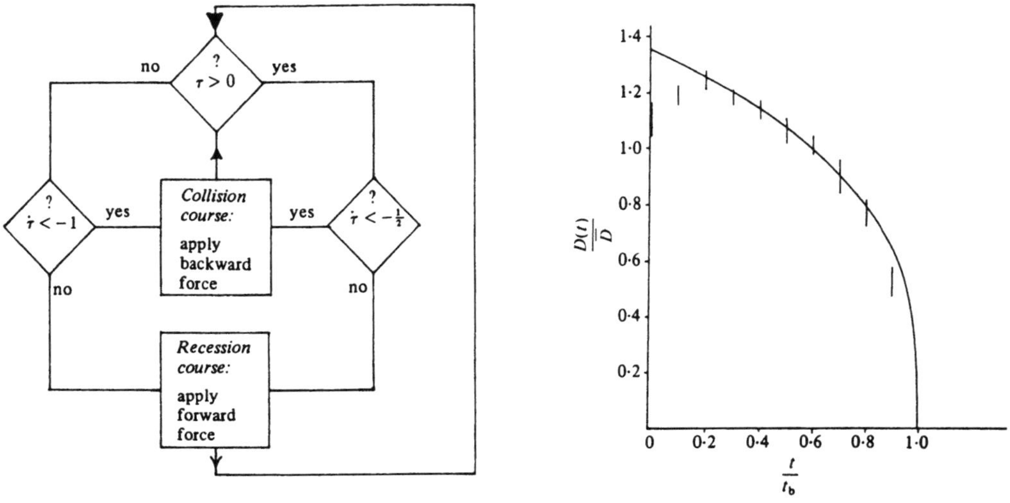

Fig. 13. Use of tau in controlling braking for an obstacle

Our experiments (Lee et al. 1984; Young and Lee 1986) have shown that 5-year-olds can readily understand the pretend road crossing task and perform sensibly on it. Furthermore, after but a few short training sessions, the children improved markedly in their ability to time their movement relative to vehicles to make efficient use of gaps in the traffic. This suggests an improved ability to detect tau and couple the information to action. In general, the 5-years-olds developed a degree of competence normally shown by older children, whose experience on roads puts them at lower risk. This suggests that pretend road crossing, which allows children to safely explore and develop their capabilities more fully than when actually crossing roads, would be a valuable addition to road safety training, whether supervised by parents or professional instructors.

Drivers too appear to use tau in controlling their braking for an obstacle. Figure 13 (left) shows how tau (τ) and its rate of change over time ($\dot{\tau}$) provide sufficient information to control deceleration to avoid collision (for details, see Lee 1976). In particular, in stopping for a stationary obstacle, the value of $\dot{\tau}$ specifies whether or not braking is hard enough: safety requires $\dot{\tau} > -1/2$. Thus, a safe strategy would be to regulate braking force so that $\dot{\tau}$ remained at a safe margin value greater than $-1/2$. The deceleration curve in Fig. 13 (right) was generated by this strategy, with $\dot{\tau} = -0.425$ (where D(t) is deceleration at time t, D is mean deceleration, t_b is stopping time). It closely fits the data points for a driver stopping at speeds up to 100 km/h (62 mph), except for the initial rise in deceleration which is not taken into account in the theory (Lee 1976).

CONCLUDING REMARKS

Motor control and perception are commonly treated as separate fields of study with their own dedicated laboratories. But if actions are of necessity geared through information to the environment, then any theory of action must incorporate perception. A theory which does not can only be about an animal in a vacuum. As Bernstein (1967, p. 107) put it: "The central effectors achieve co-ordination of movements only by plastically reacting to the totality of the signals from the afferent field".

Action both requires expropriospecific information and brings it to light. As an animal moves, the optic array forms a continuous stream of up–to–date information about the animal's dynamic relation to its surroundings. In this paper we have shown some of the action–geared riches of optical information. The more that is discovered the closer can we hope to get to understanding how actions are controlled. For, as Gibson (1979, p. 225) wrote: "Control lies in the animal–environment system", not simply the brain. The optic flow field expresses that system.

REFERENCES

Bernstein N (1967) The co–ordination and regulation of movements. Pergamon, Oxford

Dietz V, Noth J (1978) Pre–innervation and stretch responses of triceps bracchii in man falling with and without visual control. Brain Res 142: 576–579

Dunlap K, Mowrer OH (1930) Head movements and eye functions of birds. J Comp Psychol 11: 99–113

Dyhre–Poulsen P, Laursen AM, Jahnsen H, Djorup A (1980) Programmed and reflex muscular activity in monkeys landing from a leap. In: Desmedt JE (ed) Spinal and supraspinal mechanisms of voluntary motor control and locomotion. Prog Clin Neurophysiol, vol 8. Karger, Basel, pp 323–329

Friedman MB (1975) Visual control of head movements during avian locomotion. Nature 255: 67–69

Gibson JJ (1979) The ecological approach to visual perception. Houghton Mifflin, Boston

Herman R, Maulucci R, Stuyck J (1982) Development and plasticity of visual and vestibular generated eye movements. Exp Brain Res 47: 69–78

Lee DN (1976) A theory of visual control of braking based on information about time to contact. Perception 5: 437–459

Lee DN (1978) The functions of vision. In: Pick HL Jr, Saltzman E (eds) Modes of perceiving and processing information. Erlbaum, Hillsdale

Lee DN (1980) The optic flow field: the foundation of vision. Philos Trans R Soc Lond [Biol] 290: 169–179

Lee DN, Aronson E (1974) Visual proprioceptive control of standing in human infants. Percept Psychoph 19: 529–532

Lee DN, Lishman JR (1975) Visual proprioceptive control of stance. J Human Movement Stud 1: 87–95

Lee DN, Lishman R (1977) Visual control of locomotion. Scand J Psychol 18: 224–230

Lee DN, Reddish PE (1981) Plummeting gannets: a paradigm of ecological optics. Nature 293: 293 – 294

Lee DN, Young DS (1985) Visual timing of interceptive action. In: Ingle D, Jeannerod M, Lee DN (eds) Brain mechanisms and spatial vision. Nijhoff, Dordrecht

Lee DN, Lishman JR, Thomson JA (1982) Regulation of gait in long jumping. J Exp Psychol: Hum Perc Perform 8: 448 – 459

Lee DN, Young DS, Reddish PE, Lough S, Clayton TMH (1983) Visual timing in hitting an accelerating ball. Q J Exp Psychol 35A: 333 – 346

Lee DN, Young DS, McLaughlin CM (1984) A roadside simulation of road crossing for children. Ergonomics 27: 1271 – 1281

Lough S (1985) Visuo – motor control following stroke: a motor skills perspective. Ph.D. thesis, Edinburgh University, Edinburgh

McLean JR, Hoffmann ER (1973) The effects of restricted preview on driver steering control and performance. Human Factors 15: 421 – 430

Melvill Jones G (1977) Plasticity in the adult vestibulo – ocular reflex arc. Phil Trans R Soc Lond [Biol] 278: 319 – 334

Miller S, Hammond GR (1982) Neural control of arm movements in patients following stroke. In: Van Hof MW (ed) Recovery after damage to central nervous system. Elsevier, New York

Regal DM, Ashmead DH, Salapatek P (1983) The coordination of eye and head movements during early infancy: a selective review. Behav Brain Res 10: 125 – 132

Schiff W, Detwiler ML (1979) Information used in judging impending collision. Perception 8: 647 – 658

von Hofsten C (1980) Predictive reaching for moving objects by human infants. J Exp Child Psychol 30: 369 – 382

von Hofsten C (1983) Catching skills in infancy. J Exp Psychol: Hum Perc Perform 9: 75 – 85

Wagner H (1982) Flow – field variables trigger landing in flies. Nature 297: 147 – 148

Walls GL (1967) The vertebrate eye and its adaptive radiation. Hafner, New York

Warren WH, Young DS, Lee DN (1985) Visual control of step length during running over irregular terrain. J Exp Psychol: Hum Perc Perform (to be published)

Young DS, Lee DN (1985) Training children in road crossing skills using a roadside simulation. Accident Analysis and Prevention. (to be published)

Rapid Limb Movements and Their Modulation with Task Demands

D. C. Shapiro

Motor Control Laboratory, Department of Kinesiology, University of California,
Los Angeles, CA 90024, USA

Several types of movements are required to carry out our normal daily activities and the present chapter will focus on three such movements: unidirectional, sequential, and bimanual. The purpose is to provide some approaches to the study of human motor control and raise some concerns about designing behavioral experiments. In addition, two experiments for each type of movement will be discussed.

EXPERIMENTAL APPROACH

The experiments to be described examine movements performed with both spatial and temporal requirements imposed by the experimenter. This approach, termed here the "two-dimensional" paradigm can be contrasted with the "one-dimensional" paradigm in which either time or space is specified. The latter method allows the subject to determine the unspecified dimension. For example, recent investigations have manipulated only the spatial requirement (e.g., Ghez 1979; Viviani and Terzuolo 1980). In this case subjects are required to move to specific target locations, however they are not restricted in the times they select to arrive at the targets. This approach is interesting if one wants to examine strategies adopted for the dimension unspecified by the experimenter. One must be careful, however, not to overgeneralize these results to explain the control of movements with both spatial and temporal goals (Shapiro and Walter 1985). Moreover, it is critical to document both the spatial and temporal characteristics of the responses, to ascertain whether the unspecified dimension remains constant or varies with the specified dimension. Also it is important to note whether all subjects use the same strategy. Individual differences may be evidenced in motor control experiments and these effects should be identified before collapsing the data. Once the movement dimensions are determined, then inferences concerning motor control can be suggested.

Two behavioral manipulations within the two-dimensional paradigm will be discussed. The first type includes a systematic variation of one dimension while holding the other constant. For example, subjects may be instructed to move to the same target in different movement times, thus keeping distance constant. The second manipulation involves a perturbation to the movement, such as the unexpected addition of an inertial load. The goal of these studies is to investigate constancy or invariance in the system; that is, things that do not change under different experimental conditions. Invariances are interesting because they suggest how the nervous system simplifies the control problem and may also imply the structure of motor responses represented in memory. More

Experimental Brain Research Series 15

specifically, invariances may suggest how the nervous system achieves changes in speed, adapts to environmental factors, and coordinates interlimb tasks.

Invariance is investigated through a multilevels of analysis approach. Three levels are examined: the kinematic level (displacement and its derivatives), the neuromuscular level (EMG activity), and the kinetic level (net forces generating the movement). Whenever possible, all three levels are investigated in an attempt to determine what remains constant. Attention is directed to systematic changes in movement patterns at the different levels, to make inferences concerning motor control.

UNIDIRECTIONAL POSITIONING MOVEMENTS

Although several studies have examined unidirectional positioning movements (e.g., Brown and Cooke 1981; Hallet and Marsden 1979; Lestienne 1979), a one-dimensional approach was utilized and the strategy adopted by the subject to produce the other dimension (movement time) was not always reported. This approach has led to conflicting interpretations of the control of unidirectional positioning movements. For example, Freund and Büdingen (1978) and Hallet and Khoshbin (1980) asked subjects to move different distances as quickly as possible. The results indicated that time to peak force and EMG duration were held constant while amplitudes were modulated to achieve the various task demands. Thus, one can conclude from these studies that the amplitude of the controlling signal is modulated to produce movements of varying distances. Subjects appeared to keep movement time constant while varying distance, although, this was not numerically reported. In a study by Wadman et al. (1979), subjects were also required to achieve different distances; however, in this case, EMG duration varied with changes in movement distance. Without providing actual movement times, Wadman et al. (1979) reported that subjects increased movement time with increases in distance. Therefore, different strategies were adopted to achieve the same specified dimension (distance), producing different modulations of the control signal. The former studies suggest that in order to achieve movements of different distances, the amplitude of the control signal is modulated while the latter study suggests burst duration is modulated. Perhaps, the inconsistency may be explained in that the Wadman et al. (1979) study utilized a two-joint movement and the remainder of the studies examined one-joint movements. It is apparent, however, that both dimensions must be studied systematically to understand the control of undirectional positioning movements.

A two-dimensional approach was utilized to alleviate limitations of previous studies. Movement time and distance were independently manipulated in two experiments (Shapiro and Walter 1985). The task required subjects to perform horizontal flexion movements of a lever to targets. EMG activity was recorded from two muscles responsible for the action, the biceps (agonist) and the lateral triceps (antagonist); however, only the agonist activity was quantified. The forces applied to the lever were expressed as output from a strain gauge attached to the lever. In the first experiment, subjects performed movement amplitudes of 20, 30, 40, and 50 degrees while instructed to keep movement time constant at 200 ms. Only the trials where the subjects were within a specific range of the goal movement time and displacement were further analyzed.

When movement time was held constant and displacement increased, EMG durations remained constant while EMG amplitudes increased. Decelerative force durations and time to peak dF/dt remained constant with increasing displacements. As displacement increased, peak force and peak dF/dt also increased. These findings are consistent with pulse step (Ghez 1979), speed control (Freund and Büdingen 1978), and impulse timing models (Schmidt 1976, 1982; Wallace 1981) of motor control. Pulse step and speed control predict that changes in displacement or isometric force are produced by an initial control pulse which is modulated in amplitude, but not in duration. Pulses produce a constant time to peak force, and the pulse step model (Ghez 1979) suggests that time to peak dF/dt remains constant as well. Therefore, proponents of this view would assume that burst durations are invariant. The results also support an impulse timing view (Schmidt 1976, 1982; Wallace 1981); a more formalized version of the generalized motor program concept (Schmidt 1975). The impulse timing view suggests that the durations of neural and kinetic events vary proportionally with overall movement time. Since movement time is constant in this experiment, EMG burst durations and force durations should remain constant.

Accelerative force duration and time to peak force decreased with increases in movement displacement, which is not predicted by any of the models discussed. The incompatibility between the increase in the accelerative force duration without an increase in agonist EMG duration is perhaps due to the cocontraction of the initial agonist and antagonist burst exhibited while performing this type of task. The resultant force exerted is a function of the differences in magnitude of activity of the opposing muscles during contraction. None of the aforementioned models take into account cocontraction of the agonist and antagonist burst, which is frequently found in the performance of rapid movements (e.g., Lestienne 1979). Schmidt et al. (1985) examined reversal movements to reduce the amount of cocontraction; however, it is not clear if reversal movements are controlled in the same manner as unidirectional movements.

The second experiment provided a test of how positioning movements of different durations are controlled, since movement time varied while distance remained constant. Subjects were required to vary the times they moved to a single 30 degree target and the data were later grouped into blocks of trials containing five movement time ranges with means of 150, 180, 210, 240, and 270 ms. In this experiment both EMG durations and amplitudes varied with movement time. Durations increased with increases in movement time while EMG amplitudes decreased. Wallace and Wright (1982) found similar results. Time to peak force and time to peak dF/dt also increased with increases in movement time, while peak force decreased. Strictly speaking, the pulse step and speed control models do not have predictions for movements that vary along the time dimension and are therefore limited to explaining the control of movements produced in a constant duration. These data tend to support impulse timing (Schmidt 1976, 1982; Wallace 1981), although one aspect of this model was not supported. Impulse timing predicts a proportional relationship between movement time and EMG burst duration. Although a linear relationship was found, this relationship was not proportional, as seen in Fig. 1.

It appears from the earlier studies, that when movement time is constant, time to peak force and EMG duration are also invariant while the amplitudes of the signals are modulated. Many of these studies required movements of maximum velocity (e.g., Freund and Büdingen 1978; Hallett and Koshbin 1980) or allowed subjects to select their own response durations (e.g., Brown and Cooke 1981; Ghez 1979) while moving different distances, and the strategy employed typically led to constant movement times. The present study used the two-dimensional approach and forced subjects to manipulate movement time.

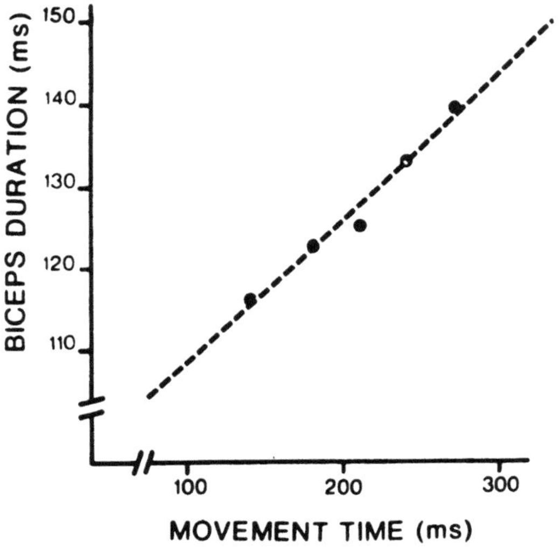

Fig. 1. Average biceps (agonist) durations plotted with movement time. There is a linear increase in biceps duration with increases in movement time, however, this increase is not proportional (from Shapiro and Walter 1985)

When changes in movement time are required, however, both amplitudes and durations of EMG vary. EMG durations vary linearly, though not proportionally, to movement time. Somehow, time must be represented as a component of the motor command for theses unidirectional movements. These experiments also demonstrate that it is critical to report both spatial and temporal variables when using the one-dimensional approach because one could easily be misled in developing general models of motor control.

SEQUENTIAL MOVEMENTS

The next set of experiments extended the uniarticular evidence by examining movements consisting of more than one element in a sequence. Indirect evidence from a previous study (Armstrong 1970) demonstrated that when subjects performed a movement sequence either faster or slower in time, they tended to "compress" or "stretch" the movement sequence in time, while keeping the "relative timing" pattern constant. The notion of relative timing predicts that the time between any two landmarks in a movement sequence, such as the beginning of a segment and its end, is proportional to the total movement time of the sequence regardless of the absolute movement time. This prediction was suggested by the generalized motor program viewpoint (Schmidt 1975) and later by the impulse timing model (Schmidt 1976, 1982; Wallace 1981), as discussed in the previous section.

To investigate the control of sequential movements, a novel movement sequence was practiced which required subjects to perform lever rotations by pronating and supinating their forearm (Shapiro 1979). Using the two-dimensional paradigm, subjects performed nine reversal movements to targets and were given specific times for performing each movement segment. Movement time

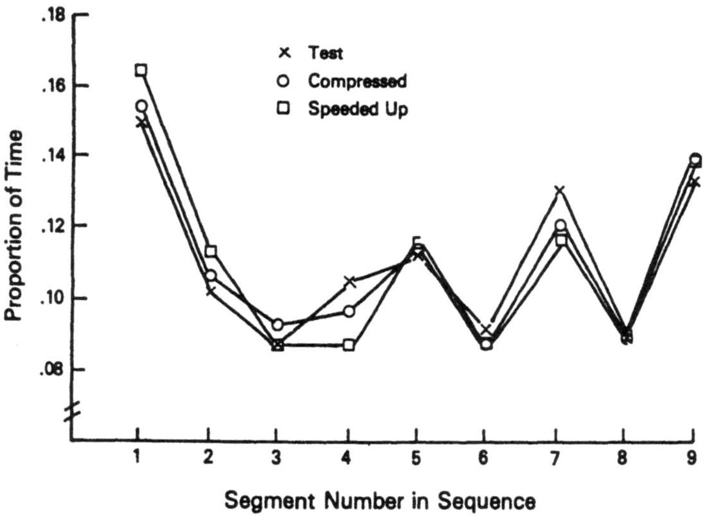

Fig. 2. Proportions of the total movement time required to traverse each segment of the movement sequence. The proportions are similar across the experimental, compressed, and sped-up trials (from Shapiro 1979)

and amplitude feedback were provided after each training trial. After the training session, subjects were required to perform the "experimental" trials, which were the trials in which they were attempting to repeat the spatiotemporal pattern they had practiced. On a few occasions, they performed "compressed" trials, where they were asked to complete the movement sequence in a faster total time while keeping the relative timing of the sequence invariant. Subjects were also asked to speed up the movement and ignore any timing they had learned and perform the task as fast as possible. These latter trials were identified as the "sped-up" trials. If relative timing is an invariant feature of a movement sequence, then the proportion of time to arrive at each target should be the same across all three speed manipulations regardless of whether a one- or a two-dimensional approach is utilized.

The results of this study demonstrated that the proportion of time to arrive at each target was similar across the speed manipulations (Fig. 2). Subjects were able to perform the entire movement sequence faster; however, they tended to keep the proportion of time to arrive at each target the same, even when requested to ignore timing and move through the sequence as quickly as possible. Therefore, relative timing appeared to be maintained. In a previous study (Shapiro 1976), a one-dimensional approach was utilized at the outset, whereby subjects were allowed to select their own timing pattern for a specified movement sequence. When subsequently asked to ignore the timing they had learned, relative timing was still maintained. Therefore, regardless of whether the experimenter imposes a particular timing pattern to learn or the subject selects their own timing pattern, relative timing is maintained.

The next question examined was whether relative timing is also maintained at the neuromuscular level. In an experiment using a similar apparatus and approach (Carter and Shapiro 1984), subjects learned a spatiotemporal pattern consisting of four reversals. EMG activity from the biceps and pronator-teres muscles were measured during the experimental trials and during the trials where the subjects were asked to speed up the movement and ignore the timing they had learned (sped-up). The proportion of time each muscle was active during the movement was similar between the experimental and sped-up trials, demonstrating the maintenance of relative timing at the neuromuscular level. This level of analysis more clearly represents the commands from the central nervous system. A similar finding appeared in a task requiring one reversal (Schmidt et al. 1985). As movement time decreased to a minimum, agonist burst durations decreased linearly although not proportionally with movement time.

After demonstrating the invariance of relative timing in novel movements, we explored locomotor patterns (Shapiro et al. 1981). Locomotion is a highly practiced skill and somewhat innate. The interlimb kinematics of the lower extremity of humans walking and running on a treadmill were recorded while the speed of the treadmill was gradually increased from 3 to 12 km/h. The running cycle was divided into the four phases of the Philippson step cycle (1905) and the percentage of the total cycle time to perform each phase of the step cycle across speeds was calculated. Within a gait (walking and running), relative timing was generally invariant as speed increased. Thus, relative timing is also observed in a highly practiced movement, when the step cycle was divided into four components. This finding is not always apparent in the animal literature on gait (e.g., Goslow et al. 1973); however, it is difficult to determine the relative timing from the data presented. The maintenance of relative timing has been found in other sequential movements, such as typing (Terzuolo and Viviani 1979, 1980; but, see Gentner 1982 for a different interpretation), handwriting (Viviani and Terzuolo 1980), and speech (Tuller et al. 1982) as well.

It appears that both novel and highly skilled sequential movements display the invariant feature of relative timing regardless of whether a one-dimensional or a two-dimensional approach is utilized. Relative timing is also evident at the neuromuscular level. These data tend to support the notion of a generalized motor program (Schmidt 1975) - where overall duration is a variant feature or "parameter" of a motor program and relative timing is an invariant one - as well as the impulse-timing model (Schmidt 1976, 1982; Wallace 1981). Time is again demonstrated to be a critical component of the movement characteristics, suggesting it is encoded in the structure of the motor command.

BIMANUAL MOVEMENTS

The final section is concerned with the production of two simultaneous positioning movements. Previous evidence concerned with interlimb coordination (Cohen 1970; Klapp 1979; Yamanishi et al. 1980) demonstrates that two movements are performed with little interference when they are harmonically related. Kelso et al. (1979) have also shown that simultaneous movements that would not normally be performed in the same response times individually, tend to become temporally related when dual task performance is required (see Marteniuk and MacKenzie 1980 for another interpretation of the latter study).

The studies to be presented used the simplest case of the two-dimensional approach with both movements generated with the same spatial and temporal requirements. Thus, the coordination of timing in these experiments is not a difficult control problem. In separate experiments, a single limb was either completely prevented from moving or was perturbed slightly during a bimanual task.

The disturbed limb was examined to determine the extent to which the movement is carried out as planned, in order to make inferences concerning the control of positioning movements. The contralateral limb not interfered with was also examined to determine whether any changes occurred as a result of the perturbation. This provides a test of the coupling or behavioral linkage between the limbs.

In the first experiment (Shapiro and Walter 1982), subjects performed rapid horizontal 30 degree flexions to a target in a specified time (200 ms). After training, a mechanical block was randomly introduced to the left limb on 10% of the experimental trials (200 trials). This block prevented subjects from moving the left lever from the start position. EMG activity was recorded from the agonist (biceps) and antagonist (lateral triceps) on all trials for both arms. When the block was introduced, the contralateral limb demonstrated an increased movement time and an overshoot of the target. Examination of the EMG data from the moving limb, however, did not demonstrate any significant changes between the trials where no block was introduced as compared with the block trials. Perhaps other muscles were contributing to the change or surface EMG was not sensitive enough to pick up changes in motor unit firing. There is a possibility that the change in the observed movement was due to some type of associative reaction (Brunnstrom 1970) from the blocked limb. This is a reaction used in physical therapy sessions when a patient has only one lower limb affected. The therapist pushes against the intact limb and a movement is experienced in the affected limb. Thus, the total blockage of one response during a bimanual task affects the contralateral limb. Perhaps this coupling between the limbs is a basic constraint of the motor system.

It is also interesting to examine the blocked limb. Wadman et al. (1979) examined subjects performing maximum velocity positioning movements and randomly introduced a mechanical block. The results demonstrated that during the trials when the block occurred, the first 100 ms of EMG activity was identical to an unblocked movement. They concluded that the first 100 ms of a motor program cannot be influenced by peripheral feedback. The present data also demonstrated that the first 100 ms of EMG activity did not change; however, careful examination of this invariance demonstrates that it occurs before the movement would normally have started. Therefore, proprioceptive feedback should not be different during this time period. About 20–30 ms after movement would have normally occurred, a rapid reflex appeared. Thus, subjects can utilize proprioceptive feedback in less than 100 ms as revealed through this multilevels of analysis approach.

In the final experiment an inertial load was unexpectedly added to the end of a single lever while utilizing a paradigm similar to the aforementioned experiment (Shapiro and Young 1984). The load changed the moment of inertia of the lever by about 35%. In terms of the limb not loaded (right limb), subjects performed the movement with no significant difference in movement time; however, there was a significant overshoot. Although the overshoot was only 1 degree, every subject demonstrated this overshoot and it was statistically significant. Therefore, loading a limb affects the contralateral limb slightly.

In terms of the limb unexpectedly loaded (left), there was both a statistically significant increase in movement time and an overshoot. Therefore, the load greatly affected the limb by slowing it down and causing an overshoot. Although the overshoot was small, this finding does not support the mass-spring model (Polit and Bizzi 1978; Schmidt and McGown 1980). An interesting finding was evident when examining the right and left limbs as they moved during the loaded condition. To reiterate, both limbs were affected by the block in terms of displacement. There was, however, no significant difference in displacements between the limbs. Thus, it appears that both limbs are linked together in some manner which may attempt to retain the symmetry between the limbs.

The EMG activity was examined to observe whether any of the changes evident in the kinematics were reflected in the EMG patterns. There did not appear to be a significant change in the EMG activity in either limb for the agonist or antagonist. The only apparent difference was that the trials that were loaded demonstrated a greater electromechanical delay, as one would expect (Norman and Komi 1979). It appeared that the EMG activity did not change very much. In this experiment we were also able to examine the horizontal force exerted on the lever by attaching a strain gauge at the end of the lever. In the right limb, there was not much difference between the trials where no load was presented as compared with the trials where the contralateral limb was loaded. In the left limb (loaded limb), however, there was a drastic and immediate change in the force curves. An increase in peak force as well as an increase in force duration resulted. One explanation of this findings is based on the well-known force-velocity relationship (Hill 1970). Since the limb was moving slower with the added load and the same level of activation was being produced as evident in the EMG patterns, then given the same level of activation with less velocity, more force should have been exerted on the lever. The effect was a mechanical compensation of the load resulting from muscle dynamics.

The results of the studies on interlimb coordination suggest a coupling or some type of constraint between the limbs during a positioning task with spatiotemporal requirements. It appears that when a single limb is disturbed during a bimanual task, the effect is not only evident in the perturbed limb, but is also manifested in the contralateral limb. During the movement, information about the disturbance is transmitted to the unperturbed limb, modifying the intended movement. This suggests that the limbs are acting as a coordinative structure (Kelso et al. 1984; Turvey 1977).

CONCLUSION

The previous experiments demonstrate the need to experimentally control or document the spatio-temporal characteristics of the movements investigated either with a one- or two- dimensional paradigm. Also, whenever possible, a multilevels of analysis approach facilitates the search for the structure of the motor command through determining the invariances or contraints of the system. Several of the investigations reported indicate that time is a component of motor programs for unidirectional and sequential movements. This finding has been evident at each of the three levels of analysis discussed earlier. Relative timing has been demonstrated to be an invariant feature of a motor program for sequential tasks. The bimanual studies suggest that there is a fundamental linkage between two movements performed simultaneously. This linkage may be thought of as a constraint within the system. By looking for invariances or constraints in different types of movement behaviors, we gain a better understanding about the control and coordination of human movement.

Acknowledgements. The author would like to acknowledge Charles Walter and Richard Schmidt for their valuable input on an earlier draft and the UCLA Academic Senate for supporting this research.

REFERENCES

Armstrong TR (1970) Training for the production of memorized movement patterns. Technical report no 26, Human Performance Center, University of Michigan, Ann Arbor

Brown SCE, Cooke JD (1981) Amplitude-and instruction-dependent modulations of movement-related electromyogram activity in humans. J Physiol 316: 97-107

Brunnstrom S (1970) Movement therapy in hemiplegia: a neurophysiological approach. Harper and Row, New York

Carter MC, Shapiro DC (1984) Control of sequential movements: evidence for generalized motor programs. J Neurophysiol 52: 787-796

Cohen L (1970) Interaction betwen limbs duration bimanual control. Brain 93: 259-272

Freund HJ, Büdingen HJ (1978) The relationship between speed and amplitude of the fastest voluntary contractions of human arm muscles. Exp Brain Res 31: 1-12

Gentner DR (1982) Evidence against a central control model of timing in typing. J Exp Psychol: Hum Percept Perform 8: 793-810

Ghez C (1979) Contributions of central programs to rapid limb movement in the cat. In: Asanuma H, Wilson VJ (eds) Integration in the nervous system. Isaku-Shoin, Tokyo, pp 305–319

Goslow GE, Reinking RM, Stuart DG (1973) The cat step cycle: hindlimb joint angles and muscle lengths during unrestrained locomotion. J Morphol 141: 1-41

Hallet M, Koshbin S (1980) A physiological mechanism of bradykinesia. Brain 103: 301-314

Hallet M, Marsden CD (1979) Ballistic flexion movements of the thumb. J Physiol 294: 33–50

Hill AV (1970) First and last experiments in muscle mechanics. Cambridge University Press, London

Kelso JAS, Southard DL, Goodman D (1979) On the coordination of two-handed movements. J Exp Psychol: Hum Percept Perform 5: 229-238

Kelso JAS, Tuller B, Vatikiotis-Bateson E, Fowler C (1984) Functionally specific articulatory cooperation following jaw perturbations during speech: evidence for coordinative structures. J Exp Psychol: Hum Percept Perform 10: 812-832

Klapp ST (1979) Doing two things at once: the role of temporal compatibility. Memory Cognition 7: 375-381

Lestienne F (1979) Effects of inertial load and velocity on the braking process of voluntary limb movements. Exp Brain Res 35: 407-418

Marteniuk RG, MacKenzie CL (1980) A preliminary theory of two-hand co-ordinated control. In: Stelmach GE, Requin J (eds) Tutorials in motor behavior. North-Holland, Amsterdam, pp 185–197

Norman RW, Komi PV (1979) Electromechanical delay in skeletal muscle under normal movement conditions. Acta Physiol Scand 106: 241-248

Philippson M (1905) L'autononie et la centralisation dans le systeme nerveux des animaux. Trav Lab Physiol Inst (Bruxelles) 7: 1-208

Polit A, Bizzi E (1978) Processes controlling arm movements in monkeys. Science 201: 1235–1237

Schmidt RA (1975) A schema theory of discrete motor learning. Psychol Rev 82: 225-260

Schmidt RA (1976) Control processes in motor skills. In Keogh J, Hutton RS (eds) Exercise and sport science reviews. Journal Publishing Affiliates, Santa Barbara, pp 229-261

Schmidt RA (1982) Motor control and learning: a behavioral emphasis. Human Kinetics, Champaign, IL.

Schmidt RA, McGown CM (1980) Terminal accuracy of unexpectedly loaded rapid movements: evidence for a mass-spring mechanism in programming. J Motor Behav 12: 149–161

Schmidt RA, Sherwood DE, Walter CB (1985) Rapid movements with reversals in direction: I. The control of movement time. (submitted)

Shapiro DC (1976) A preliminary attempt to determine the duration of a motor program. In Landers DM, Christina RW (eds) Psychology of sport and motor behavior I. Human Kinetics, Champaign, pp 239-252

Shapiro DC (1979) The learning of generalized motor programs. Unpublished Doctoral Dissertation, USC

Shapiro DC, Walter CB (1982) Control of rapid bimanual aiming movements: the effect of a mechanical block. Soc Neurosci [Abstract] 8: 733

Shapiro DC, Walter CB (1985) An examination of rapid positioning movements with spatiotemporal constraints. (submitted)

Shapiro DC, Young DE (1984) Bimanual positioning movements: the effect of an unexpected load on dual task performance. Soc Neurosoci [Abstract] 10: 340

Shapiro DC, Zernicke RF, Gregor RJ, Diestel JD (1981) Evidence for generalized motor programs using gait-pattern analysis. J Motor Behav 13: 33-47

Terzuolo CA, Viviani P (1979) About the central representation of learned motor patterns. In: Talbott R, Humphrey DR (eds) Posture and movement. Raven, New York, pp 113-121

Terzuolo CA, Viviani P (1980) Determinants and characteristics of motor patterns used for typing. Neuroscience 5: 1085-1103

Tuller B, Kelso JAS, Harris KS (1982) Interarticular phasing as an index of temporal regularity in speech. J Exp Psychol: Hum Percept Perform 8: 460-472

Turvey MT (1977) Preliminaries to a theory of action with reference to vision. In Shaw R, Bransford J (eds) Perceiving, acting, and knowing. Erlbaum, Hillsdale NJ, pp 211 – 265

Viviani P, Terzuolo C (1980) Space-time invariance in learned motor skills. In Stelmach GE, Requin J (eds) Tutorials in motor behavior. North-Holland, Amsterdam, pp 525-533

Wadman WJ, Denier van der Gon JJ, Gueze RH, Mol CR (1979) Control of fast goal- directed arm movements. J Hum Mov Studies 6: 19-37

Wallace SA (1981) An impulse-timing theory for reciprocal control of muscular activity in rapid discrete movements. J Motor Behav 13: 144-160

Wallace SA, Wright L (1982) Distance and movement time effects on the timing of agonist and antagonist muscles: a test of the impulse-timing theory. J Motor Behav 14: 341-352

Yamanishi J, Kawato M, Suzuki R (1980) Two coupled oscillators as a model for the coordinated finger tapping by both hands. Biol Cybern 37: 219-225

On Information Processing and Performing a Movement Sequence*

J. T. Massey, A. B. Schwartz, and A. P. Georgopoulos

Philip Bard Laboratories of Neurophysiology, Department of Neuroscience,
The Johns Hopkins University, Baltimore, MD 21205, USA

INTRODUCTION

This paper deals with two main questions. First, is it possible to appreciably increase the speed of the movement without degrading its accuracy? Or, according to the information-theoretical formulation of Fitts (1954), is it possible to increase the information transmission capacity of the motor system? In a way, of course, this characterizes skilled motor performance. For example, a good tennis player is frequently both fast and accurate, especially so if one considers that the accuracy required under these circumstances extends to several aspects of movement, i.e., orientation of the hand, force of hitting the ball, precise timing, etc. How is that increase in information processing being accomplished? We suggest that a key factor responsible for the improved performance in skills that involve sequence of movements may be a simultaneity of motor processing. It seems that, under certain conditions, generating a movement *while* another movement is evolving or being generated "primes" the perceptual-motor system so that, instead of being constrained it actually processes information more efficiently and emits movements that are faster than, and as accurate as, movements produced in isolation. We examined some of these questions in the experiments that we describe below.

The second question deals with the explanation of the sequential movement pattern that we observed in this experiment. In particular, we tested the idea that the two-movement trajectory observed in our experiments could have been the outcome of a superposition of two movement plans. (The idea was pointed to one of us (A.P.G.) by P. Morasso during this meeting.) Assuming that each movement plan is specified in kinematic terms (Morasso 1981), we identified two plans plausibly dictated by the stimulus configuration. Representative estimates of velocity traces were obtained from within the experiment, and were superposed according to the experimental condition and algebraically summed. The resulting profile was then compared to that observed. We found that the observed profiles deviated systematically from those predicted.

These results led us to a close examination of the dynamic aspects of the movements in the task. A detailed analysis of the force patterns applied by the subjects during these movements revealed that both monkey and human subjects followed a more or less stereotyped strategy in performing the task. This strategy simplified decision making and minimized calculations in the task, although at

* This research was supported by US Public Health Service grants NS17413, NS07226, and MH18030, which we gratefully acknowledge.

the expense of physical work, and ensured that subjects would achieve the behavioral goal of capturing the target as soon as possible.

METHODS

The behavioral apparatus consisted of a 25 cm x 25 cm planar working surface (tilted 15° from the horizontal in a direction toward the subject) and a light weight, low friction articulated manipulandum. The device has been described in detail previously (Georgopoulos et al. 1981). The subject grasped the distal end of the manipulandum and captured within a 10 mm diameter transparent plexiglass circle attached to it, a lighted red light emitting diode (LED) on the working surface. Three male rhesus monkeys (3.5 - 5 kg) were trained for approx. 1 month in a reaction time, LED capturing task (see Georgopoulos et al. 1981) before performing in the task of the present study. Three men performed 20 trials in that same task before performing the task described below.

The present task was as follows. A trial started after a 2 s intertrial interval (lights off) by turning on a LED at the center of the working surface. The subject captured it, and waited there for a 1 to 3 s variable period of time (center hold time), after which that LED was turned off as another one was turned on at either a 12 or a 6 o'clock position, each 8 cm away from the center. The LED (first target) stayed on at that position for a time (interstimulus interval, ISI) and then was turned off while another one (second target) was turned on at the symmetrically opposite position, at 6 or 12 o'clock, respectively. The monkey was rewarded by a drop of juice when he captured this last LED. The instruction to the human subjects was "to capture the LED that is on as soon as possible". If the subject moved out of the center hold area (a circle 10 mm in diameter) within the first 100 ms after the onset of the first target, the trial was labeled "early reaction time error"; if he did not move after 700 ms had elapsed, it was a "late reaction time error"; if the hand did not arrive at the target area (a 25-mm diameter circle centered on the target) within 1500 ms after he started moving, it was a "movement time error"; finally, if the subject did not hold within the final target area for at least 250 ms, the trial was labeled "target hold error".

ISI was 50, 100, 150, 200, 250, 300, and 400 ms in different trials. The probabilities of occurrence of task events were balanced, so that the subject could not predict the position (12 or 6 o'clock) of the first target LED (p = 1/2), whether that target would shift position or stay on as in a control trial (p = 1/2), or, in the case of target shift, the time of the shift (p = 1/7). Ten to fifteen successful target shift trials were performed by each subject. Control (nontarget shift) trials were seven times as numerous because of the balanced probabilities in the task described above.

The experiment was controlled using a PDP11 laboratory minicomputer. The position of the distal end of the manipulandum was sampled every 10 ms. Average position values were calculated for each type of trial. Velocity and acceleration traces were derived by successive numerical differentiation of the average position values. The total force (F_{tot}) acting on the distal end of the manipulandum was the sum of two components: the force exerted by the subject (F_s) and the force due to gravity (F_g):

$$F_{tot} = F_s + F_g \qquad (1)$$

The time course of F_{tot} was calculated using the mass and acceleration of the center of mass of the articulated manipulandum, and F_g was computed by taking into account the 15° tilt of the working surface. Finally, F_s was obtained by subtracting F_g from F_{tot}. The term "force" used below refers to F_s.

RESULTS

Some of the results of the experiments with the monkeys were described before (Georgopoulos et al. 1981). The results of the experiments with the human subjects are based on previously unpublished data (A.P. Georgopoulos, J.F. Kalaska, and J.T. Massey, unpublished observations).

General Results

The results described below come from trials in which all subjects performed the task for the first time. All subjects performed the task in a similar way, from the beginning. During a target shift trial, the hand moved initially toward the first target for a period of time depending on the time for which that target stayed on; then, the subject changed direction and moved at a high speed toward the second target, which he captured successfully (see Fig. 2 in Georgopoulos et al. 1981). No errors were committed by the monkeys. The human subjects made errors in 5% of the trials: one-half of these errors was made during control, nontarget shift trials, and were target hold errors. This suggests that the subjects moved as fast as possible because in these trials the subjects went through the target window without stopping; in fact, the peak velocity of the movement in those target hold error trials was about 30% higher than that attained in the trials in which the subjects stopped within the target window. The results below are based on successful trials alone and refer to kinematic and dynamic variables along the 6 to 12 o'clock (Y) axis, on which movements were made. Motion along the X-axis either did not exist or was very small compared with the motion on the Y-axis.

Kinematic and Force Profiles

Figure 1 shows the average position trajectory of movements made by a monkey during target shift trials (ISI = 150 ms, first target at 12 o'clock). Zero time is the onset of the first target. The force profile corresponding to Fig. 1 is shown in Fig. 2. The kinematic and force profiles were initially identical to the control ones, from which they later deviated. A control force profile is illustrated in Fig. 2 by an interrupted line. We analyze in detail below the force profiles because they are sensitive descriptors of the dynamic commands underlying performance in the task.

First Part of the Force Profile

We distinguish two functional parts in the force vs time profile in a target shift trial. The first is from the onset of force application in the direction of the first target (instant b in Fig. 2) until the deviation of the force pattern from the control one (instant c). At trial onset (interval a–b), a steady

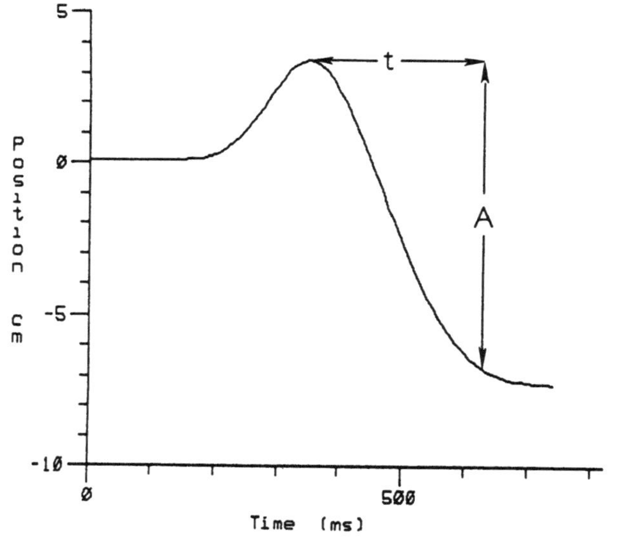

Fig. 1. Position profile fitted on averaged (n = 15 trials) data points. Target shift occurred at ISI = 150 ms. First target location was at 12 o'clock, second at 6 o'clock. Data from one subject. Amplitude (A) of the second movement is from reversal of position to entrance to target window. Movement time (t) is the corresponding time interval

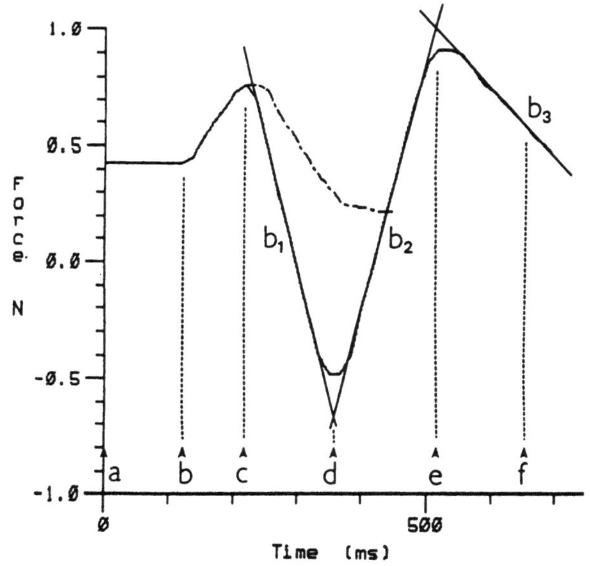

Fig. 2. Force profile fitted on averaged (n = 15 trials) data points, corresponding to those of Fig. 1. (See text for explanation of symbols.)

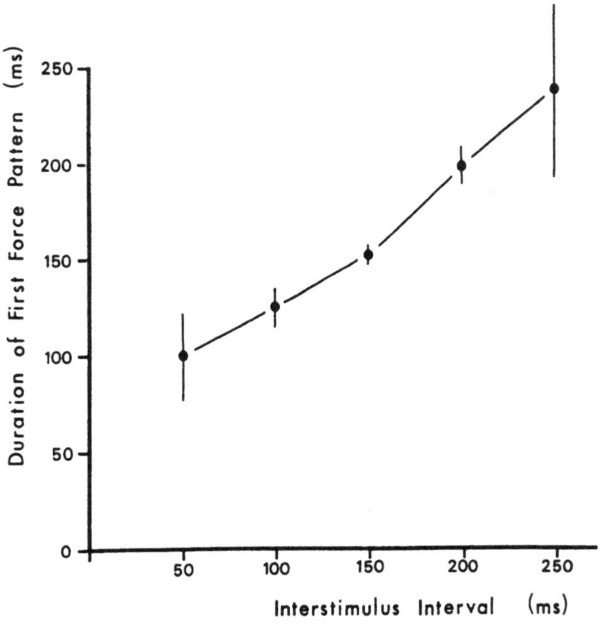

Fig. 3. Duration of the initial force pattern as a function of ISI. Data are mean ± SD from three human subjects

force was exerted by the subject in order to keep the manipulandum steady at the center LED position. At time b a new force was applied in the direction of the first target (at 12 o'clock). The rate of force application in this interval was approximately constant and almost identical to that of the control case. This was true for all the subjects. These results indicate that the subject's strategy was to start the movement as if no change of target would occur. This is understandable for the probability that the target would change was 0.5, so that the subject could not predict whether a particular trial was a control or a target shift trial. The initial, control-like force pattern was applied for a time (interval b–c) that increased monotonically with the ISI, that is, the time for which the first target stayed on. This is shown in Fig. 3.

Second Part of the Force Profile

This part corresponds to the time interval c–f of Fig. 2, beginning at the deviation from the control pattern (c) and continuing until the manipulandum arrives and stops within the target area (f). This profile comprises, in turn three parts, each approximating a straight line. The first slope b_1 (Fig. 2) corresponds to the time interval c–d (T_1), extending from the onset of deviation (c) to the slope reversal (d). The second slope b_2 occurs in the time interval (d–e) (T_2), from the slope reversal (d) to its next reversal (e). Finally, the third slope b_3 occurs in the time interval e–f, during which the force returns to the static level associated with holding the manipulandum at the target position. This last slope was difficult to determine because of marked variability in the individual values. In contrast, slopes b_1 and b_2 were easy to measure and shared several common characteristics. First, they were large and of similar absolute magnitude. Slope b_1 was -9.2 ± 2.03 Newtons/s (mean ± SD, n = 3 human subjects), and b_2 was 10.5 ± 2.1 Newtons/s. The difference $|b_1| - |b_2|$ was not statistically significant (paired t−test). Second, the slopes b_1 and b_2 had almost the same value among target shift trials with different ISIs, i.e., they did not vary with the time for which the first target stayed on. Finally, they were applied for approximately the same amount of time. Thus, b_1 was applied for a time T_1 = 151 ± 32 ms (mean ± SD, n = 3 human subjects x 6 types of target shift trials = 18) and b_2 for a time T_2 = 136 ± 52 ms. The two distributions were not statistically different (Mann−Whitney test). However, the variance of T_2 between trials with different ISIs was three to six times greater than that of T_1, even in the same subject ($p < 0.05$, F test). It is interesting to note that b_1 was different from, and larger than the equivalent (decelerating) slope in control trials. These results suggest that the subjects adopted a new and stereotyped strategy in this part of the task; they applied large forces to brake the first movement, if needed, and to implement the second movement as fast as possible. Since the slopes b_1 and b_2, and the corresponding intervals c–d and d–e did not vary appreciably in trials with different ISIs, it can be argued that the distance covered during the interval c–e was approximately the same. In fact, variation of the slope b_3 and its interval e–f were responsible for the increased distance traveled in trials with increasing ISI.

Information Processing of the Second Movement

The large forces applied to implement the second movement resulted in an appreciable increase of the speed of that movement, in spite of the fact that accuracy requirements remained the same. This was achieved by all subjects, monkey and human alike. We calculated the rate of information transmission under these conditions according to Fitts' equation (1954) as follows:

$$Ip = -1/t \log_2(W/2A) \tag{2}$$

where Ip is the information transmitted in bits/s, t is the duration of the second movement time in seconds, W is the diameter of the target window, and A is the amplitude of the second movement. The term $-\log_2(W/2A)$ is the index of difficulty. We measured t and A from the reversal of position until the entrance to the target area (see Fig. 1); W was 2.5 cm. Representative results obtained from one monkey for different target shift trials and control trials corresponding to the direction of the second movement are shown in Fig. 4. It can be seen that the rate of information transmission was greatly increased during the target shift trial, as compared with the control values (C). Moreover, the time course of this increase in the rate of information processing was such that a peak was observed at ISI = 150 ms. In contrast, the index of performance was the same in the control case and when the target shifted after an ISI of 400 ms. In the latter case, the first movement was completed and the second was made all the way from the first to the second target location after the hand had already rested at the first target location for a short time. Thus performance returned to control levels. Similar findings were obtained from the remaining subjects. These results indicate that when a movement is generated while another one is being processed, information is processed more efficiently. This has obvious implications as an explanatory principle of the performances in skills involving sequences of movements.

A correlate of this increased efficiency in information processing might lie in the observation by Soechting and Lacquaniti (1983) that under similar conditions of rapid response to a target change

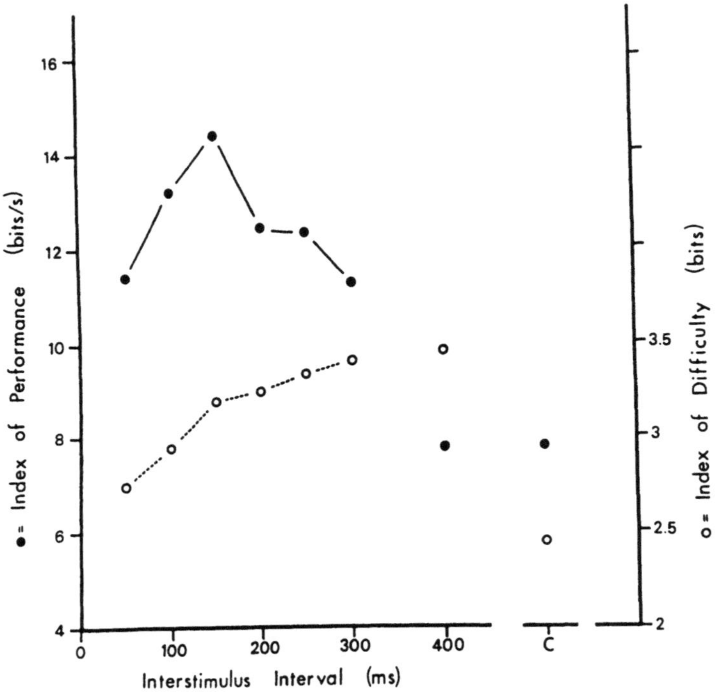

Fig. 4. Index of performance (filled circles) and index of difficulty (open circles) as a function of ISI. Data from one subject. First target at 6 o'clock

there is a reduction in the degrees of freedom of the movement; this is achieved by imposing constraints on the kinematic variables (angular deceleration at the shoulder and elbow joints), and by generating more stereotyped patterns of activity in the muscles acting on these two joints.

Psychological Refractory Period

The time at which the force profile started to deviate from the control profile can be taken as the beginning of the second movement, in response to the shift of target location. Therefore, the time from the onset of the target shift to the beginning of the force deviation can serve as an indication of the time it took the system to process the signal of the second stimulus and to start generating the second movement. This time can then be compared with the corresponding time in control trials, i.e., from the onset of the stimulus until the instant when the force begins to build up. The results are shown in Fig. 5. It can be seen that the times in the target shift trials resemble those of control trials. There was an increase of 46 ms in the case of ISI = 50 ms. Although this increase was statistically significant (t-test), it was much smaller in magnitude than that predicted by the equation (Davis 1956):

$$\mathrm{PRP} = \mathrm{RT}_1 - \mathrm{ISI} \tag{3}$$

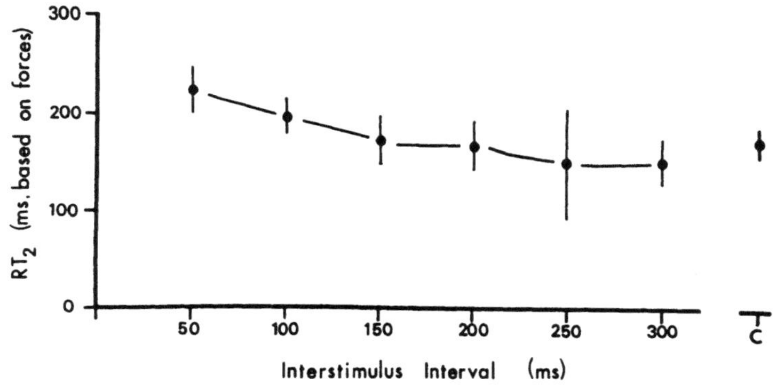

Fig. 5. "Second reaction times" (see text) based on the deviation of the force pattern from the control one, as a function of ISI. Data are mean ± SD from three human subjects

Based on our force measurements, this predicted psychological refractory period (PRP) was 124 ms. These results are very similar to those described previously regarding the reaction time (Georgopoulos et al. 1981), and to others regarding the torque and electromyogram (EMG) onset times at the shoulder and elbow joints during responses to change of target location in a pointing task (Soechting and Lacquaniti 1983).

A Note on Superposition

The kinematic outcomes of the present experiments were within the framework of a recent hypothesis concerning the formation of the trajectories of hand movements (Morasso 1981). According to that hypothesis, movements are programmed as dome- shaped velocity profiles. A single such profile is considered as representative of a single motor program; by extension, velocity profiles with more than one peak can be regarded as the output of linearly superposed individual profiles (see also Morasso and Mussa Ivaldi 1982). We analyzed the results of the present experiments in light of this idea as follows. We assumed that the velocity profile we observed during target–shift trials is the outcome of two superposed profiles. The first would correspond to the control profile, i.e., the one the subject would make toward the first target in the absence of a target shift. Superposed upon that is a second velocity profile that corresponds to the movement all the way from the location of the first to that of the second target. These profiles are dictated by the respective target location. The second velocity profile was derived from the data; it was the second profile from trials with ISI = 400 ms, in which the first movement was completed up to the location of the first target, and the second movement was made all the way from the location of the first to that of the second target. The second profile was superposed upon the first after a time interval equal to the time from the onset of target to the change of force in the corresponding control trials. The two profiles were added algebraically. An obvious prediction of the superposition is that the peak of the second velocity profile cannot be more than that of the second constituent velocity profile, since the first and second profiles that are added are of opposite signs (corresponding to movements in opposite directions). This prediction was not observed in the actual data. Indeed, peak second velocities in the target shift trials often exceeded those of the movements made all the way from the

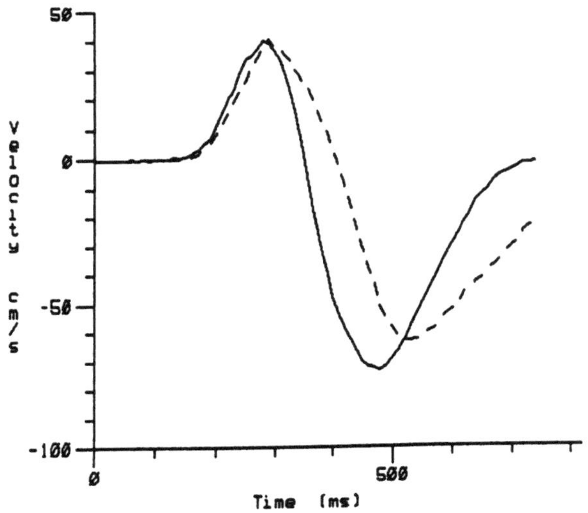

Fig. 6. Observed (solid line) and predicted by the superposition hypothesis (dashed line) velocity profiles. Data from one subject (ISI = 150 ms). (See text for explanation.)

first to the second target. This is illustrated in Fig. 6. In addition, the slopes of the experimentally observed second velocity profile often differed appreciably from, and were steeper than those predicted by the hypothesis. Therefore, we conclude that the results of the present experiments cannot be explained by the superposition hypothesis.

DISCUSSION AND CONCLUSIONS

The subjects were asked to move to a target that changed location during the reaction or movement time. They moved initially toward the first target location, but the movement was truncated and was followed by another toward the second location. This second movement was completed successfully within the accuracy constraints of the task. This task presents a challenge to the motor system, for the generation of the appropriate sequence of forces to successfully perform the task might require, for example, foreknowledge of the dynamics of the initial movement, the relation between force and displacement under the varying dynamic conditions, and of the accuracy constraints. It is remarkable that all the subjects, monkeys and humans alike, employed a strategy from the first trial that dealt with these complexities simply, uniformly, efficiently and effectively. The strategy was *simple*: "move to the target that is on until it shifts; then, move to the new location as soon as possible". How was such a two-part strategy implemented? Since the time of target shift was uncertain, the dynamic characteristics of the movement that was interrupted (e.g., velocity and acceleration at the time of target shift) were also uncertain. Again, a simple solution was found: "apply a large force in the new direction, as soon as possible after the target shift; stop the initial movement and start the new one". This force sequence that was emitted after the target shift appears to be a "default" sequence, consisting of two steep slopes of force change that are opposite in direction, have the same absolute magnitude, and are applied sequentially for about the same interval of time. Moreover, this sequence was *uniformly* applied without distinction in all cases, in seemingly the same form during trials with different times of target shift. Was this strategy efficient? Maybe not, if physical work is considered, for these large forces would probably be excessive in cases of short target shift times. Yet, it was quite *efficient in terms of information processing*, for it unburdened the system from calculating or predicting the upcoming dynamic parameters and displacements, and, positively, resulted in an appreciable increase of information transmission, as judged by Fitts' index of performance. Finally, the strategy was behaviorally *effective*. Performance was practically errorless from the very beginning, and yet faster. The strategy seems to have been chosen on behavioral rather than mechanical grounds. It apparently took into account the probabilities in the experimental design: control and target shift trials occured on a 50%:50% basis, hence the initial movement toward the first target. After the target shift, the goal was straightforward: "correct the error (i.e., stop the initial movement, for there is no reward in going to the first target location), and go to the new target as soon as possible (to get the reward)". Under these conditions, one could suppose that the switching of motor commands would be impeded due to limitations of the nature of the PRP. On the contrary, the switching of the commands was not only possible but quick and successful, resulting in accurate performance. It is noteworthy that the switching of patterns of cell discharge was also unimpeded in the motor cortex of monkeys behaving under the same conditions (Georgopoulos et al. 1983). What allowed this remarkable performance? We argued before (Georgopoulos et al. 1981) that the high stimulus-response compatibility in the spatial domain and the fact that only one anatomical unit (the same hand) was the performer lie in the root of these results. It is expected that tasks and devices which take into account these principles would result in improved performance even in more complicated sequential movement tasks.

REFERENCES

Davis R (1956) The limits of the psychological refractory period. Q J Exp Psychol 8: 24-38

Fitts P (1954) The information capacity of the human motor system in controlling the amplitude of the movement. J Exp Psychol 47: 381-391

Georgopoulos AP, Kalaska JF, Massey JT (1981) Spatial trajectories and reaction times of aimed movements: effects of practice, uncertainty, and change in target location. J Neurophysiol 46: 725-743

Georgopoulos AP, Kalaska JF, Caminiti R, Massey JT (1983) Interruption of motor cortical discharge subserving aimed arm movements. Exp Brain Res 49: 327-340

Morasso P (1981) Spatial control of arm movements. Exp Brain Res 42: 223-227

Morasso P, Mussa Ivaldi FA (1982) Trajectory formation and handwriting: A computational model. Biol Cybern 45: 131-142

Soechting JF, Lacquaniti F (1983) Modification of trajectory of a pointing movement in response to a change in target location. J Neurophysiol 49: 548-564

Requirements for a Generalized Model of the Control of Walking Behaviour in Hexapods at Any Orientation to the Gravity Vector

D. Graham
17a, Hastings Road, Bexhill-on-Sea, Sussex, England

In the investigation of the manner in which muscles are controlled to achieve the movements of limbs, the complexity of vertebrate articulation and sense organs -which alter their properties as a direct result of changes in muscle length- make precise analysis of the significant parameters difficult (Stein 1982).

Insect movements often show similar precison and complexity to those found in vertebrates and possess two important advantages. In many insects the limb articulations are simple hinges with only two antagonistic muscles moving the joint. Furthermore, the detection of changes in joint angle is performed either by hair fields deformed by the soft cuticle, near the joint, or by internal chordotonal organs which are positioned in parallel with one of the muscles. Stress is measured independently by sense organs in the hard cuticle close to the points of articulation. The discrete properties of these specialised receptors means that the importance of a given type of sense organ can be established for the control loop of each joint or group of joints.

In this paper I consider the stick insect as a walking machine and examine the properties that would be required in a model of this walking system. This model will be more complex than those that have so far contributed to our understanding of this system because it attempts to include support, stabilisation and the capability of walking on surfaces at any angle to the gravity vector. I wish to consider the essential features of such a model and what the most recent experimental work on insects has to tell us about the way in which these objectives are achieved.

The argument over whether leg movements are generated by a central motor score without recourse to peripheral sensory input or require sensory feedback for generating the cycle of movement now appears to have been resolved in insect locomotion (Graham 1985). A distinction must be made between step frequencies above and below 4 Hz. Below the critical cycle duration of 250 ms the time required for proprioceptive confirmation of placement via sensory feedback is greater than the time available (Delcomyn and Usherwood 1973; Graham 1978b, 1979). It is probable that central elements of coupling exist which prevent simultaneous recovery strokes in adjacent legs (Pearson and Iles 1973; Graham 1978a,b). At the higher step frequencies there may be no modification of gait after amputation (Delcomyn 1971).

At step frequencies below 4 Hz the normal leg cycle requires the completion of recovery or stance phases before the next part of the cycle is carried out (Baessler 1983). If either part of the cycle is blocked or interrupted, then adjacent legs in front and on the other side have their behaviour

Experimental Brain Research Series 15

modified (Cruse and Epstein 1982; Dean and Wendler 1982). At step frequencies below 1 Hz leg interactions which require 100–200 ms for completion such as the "tread on toe" reflex appear and the four phases of leg movement are separated by pauses of 50 ms (Graham 1979). These are recovery, depression, stance or power stroke and levation. The patterns of recovery movements at low speed are characterised by a relatively constant latency between the recovery of adjacent legs. This gives a tripod gait at high speed which is transformed into an asymmetric recovery pattern as the step period increases. In the slow walking gait, legs are recovered in diagonal pairs rather than three at a time.

What do these movements involve in terms of muscles, sense organs and the leg segments? I shall restrict myself to a description of stick insects. These animals were selected for the experimental investigation of locomotion because they walk slowly and are conveniently organised anatomically. They range in body length from 1 cm to as much as 30 cm in some species. The legs are only slightly specialised and this is primarily for camouflage (during twig mimesis). The leg articulations are defined by simple hinges rather than differential muscular support as in the cockroach and locust. There is no flight musculature and the thoracic musculature and nervous system are easily exposed for experimental recording or manipulation.

The half ganglion controlling a leg has a few 1000 neurones of which about 200 are used to activate 70 muscles. Some of the remaining interneurones transfer information within the half ganglion (controlling one leg), between half ganglia (coordinating pairs of legs within one body segment) and along the chain of ganglia (organising the movement of all six legs).

Nine muscles move each leg. Only eight are shown in Fig. 1. The missing thoracic (body) muscle passes through the coxa and is attached to the tendon of the coxal muscle depressing the femur. While the action of the other muscles is clear from the simple geometry of the leg segments, this intersegmental muscle has a complex action. It can contribute to either power stroke or recovery depending upon the orientation of the coxa relative to the body and always provides a component which assists in depressing the leg.

Sense organs are of three types: (a) position detecting hairs, (b) stress detecting organs (campaniform sensilla or CS) buried in the cuticle and often concentrated close to joints, and (c) chordotonal (CO) and myo-chordotonal (MCO) organs which span a joint and measure the angular relationship between segments. No MCOs have been reported within leg muscles. Hair fields and CS organs situated close to the joint are stimulated directly by joint movement and stress, respectively. In insects, it is possible to distinguish between stress and position or higher order derivatives of these parameters as the inputs controlling reflex loops.

Upright walking, with the body supported well clear to the ground, requires that the individual legs press down upon the walk-surface during a power stroke or stance phase to support the body. This is only possible for a limited time because, as the body is propelled forward, the leg moves to the rear and can no longer provide either support or propulsion. It is essential that the leg be lifted from the ground and swing forward into a new position to provide the next phase of support and propulsion. The leg has at least four functions. These are to recover itself, to drive the body forward (power stroke), to support the body weight (stance phase) and stabilise the body during the stance phase. In the supporting phase the leg may be required to perform three of these functions at the same time.

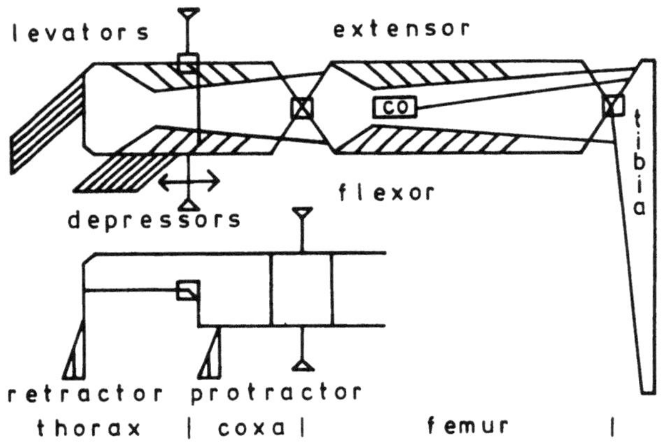

Fig. 1. Schematic drawing showing the three segments of the leg of a stick insect looking along the body axis. The lower view shows the coxa from above. Eight muscles or their tendons are shown of the nine muscles that are responsible for controlling the leg movement. The chordotonal organ which controls the reflex movement of the femur–tibia joint is also shown. Hair fields and stress–detecting organs are not shown, but are found close to the segment articulations. Rotation about an axis in the plane of the figure is shown by a line with a triangle at each end. An axis perpendicular to the figure is shown by a small square. The lower articulation of the body–coxa axis is constrained by a strut (not shown) to move towards and away from the body as shown by the arrow

Models have so far been restricted to the consideration of certain elements of the walking process and at the present time none are in any sense complete. Modelling has so far concentrated on the easiest part which is the temporal organisation of the recovery strokes.

The most recent models have begun to incorporate the control of power stroke and the responses of the system to load parallel to the body axis. (Cruse 1980 a,b; Cruse and Graham 1984). However, no model yet considers support, stabilisation or the changes in motor control of the muscles which must take place under conditions of varying orientation to the gravity vector. These appears to be very daunting problems, but I would like to look at them in a simplified system and see the extent to which recent experimental work is beginning to show how these various problems have been solved during the evolution of the stick insect Carausius morosus.

One might suppose that in supporting the body above the ground, in an upright position, the animal must activate the muscles depressing the femur. However, this is not essential for the body could be supported using the femur and tibia as a rigid strut. In this insect there are several joints in each leg so that appropriate torques must be applied to the more distal joints if a leg is to be used as a simple strut. The simple–stick model (SS) of Fig. 2 is a simplified version of the real animal. It has lost its coxal segments, i.e. the femurs are directly articulated onto the body and have two simple perpendicular hinge axes. The upper figure shows SS in a position that could solve the stability problem by letting the legs extend fully, but has two alternative extreme solutions to the task of holding the body up close to the surface above. If the femur is levated, then the body–to–surface distance can be maintained by a knowledge of the typical body–femur angle

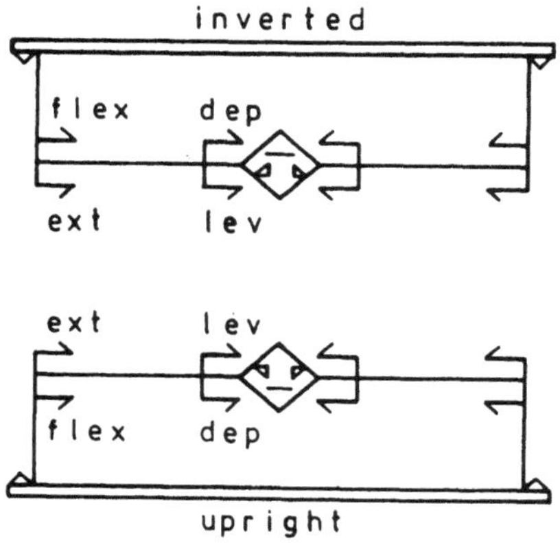

Fig. 2. A simplified model of the stick insect consisting of legs with only two segments (femur-tibia). The upper figure shows the model hanging below a substrate and the lower is for the model standing upright

required and direct feedback of this information onto the levator muscle for right and left sides. With only these muscles active SS would be in a position of stable equilibrium (this will be referred to as Mode I). Alternatively, it could use its femur and tibia as a strut giving both lateral stability and the correct body height by maintaining a fixed angle between femur and tibia. In this case only the flexor muscle would be active under the control of the CO that measures femur-tibia angle. Here the body is rigidly stabilised against horizontal movement of the body (Mode II). We do not have any data on stick insects hanging below a surface, but if we look at SS in the upright configuration we would expect that the depressors would be inactive if the strut method (Mode II) is in use and the extensors would be strongly active.

An early experiment indicates the mechanism used in the standing stick insect. Wendler (1964) was able to show that the body height of the stick insect was controlled by afference from hair fields measuring the coxa-femur angle. If these organs were shaved or immobilised by water soluble glue, then a load of only 2-3 times body weight was sufficient to depress the body to the ground. During the unoperated condition or if the glue was dissolved away, the animal used an active muscular output to resist loads of up to 6-7 times the body weight.

Because the depressors and levators directly influence this joint, it would appear that in the standing stick insect Mode I is used. In the cockroach, the passive forces in the femur-tibia joint are large but not sufficient to make a rigid strut in the absence of activity and it is reported that there is little activity at this joint in standing animals (Yox et al. 1982). In the absence of actual recordings of depressor activity in middle or front legs this suggests a similar solution to that proposed for the stick insect with passive forces used to provide lateral stability.

To a first approximation we might argue for two independent systems. The first keeps the angle body-femur constant using hair-field feedback as proposed by Wendler (1964), while a second system performs the same function for the femur-tibia joint, the latter being achieved partly by passive tension in the upright and inverted situations. In standing with one side down on a vertical

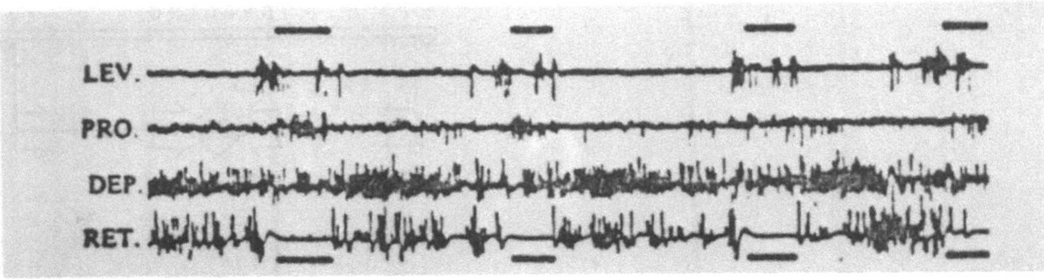

Fig. 3. Extracellular muscle activity (myogram) from LEVator, PROtractor (recovery), DEPressor and RETractor (propulsion) muscles of an adult stick insect during a walk on two light wheels exerting an upthrust against the legs. The protractor or recovery phase of the leg (determined from a film record) is marked by a horizontal bar above and below the records (recovery duration approximately 120 ms, cycle time 500 ms)

wall, the depressor and levator tensions could be reduced while the flexors are active on the upside and extensors on the downside (Fig. 3). To what extent do these control systems remain active during the walking state?

In the stick insect one unit of the depressor is active at 50–100 Hz during walking (Graham and Baessler 1981). This strong activity in the depressors supports Mode I. Records of flexor and extensor activity in a walking locust (Burns 1973) show little activity during the middle part of the power stroke and provide further support for this hypothesis.

Measurements of the vertical forces for a stick insect which stepped onto a force transducer in the walk-surface (Cruse 1976) show that the legs in support act like independent passive springs and modelling would begin with the assumption of independent leg support. This is not possible in the lateral stabilising system because the legs on the right and left sides of the body must be coupled in order to minimize energy requirements by differential activation of those muscles which are most appropriate (best mechanical advantage) for maintaining the femur-tibia angles. This depends upon the orientation of the body to the gravity vector.

So far we have only considered the lateral components of force in the stability plane where the problem may be considered as a static one during walking movements.

The problems of stability in the direction parallel to the body axis are more difficult because here the stituation really is a dynamic one. Let us consider an animal walking uphill so that the body mass acts as a stabilising influence and only the power-stroke muscles contribute to forward movement of the body.

In uphill walking, the problem of leg dynamics and control is analogous to horizontal walking under a load parallel to the body axis. This has been examined on a double treadwheel system in which friction loads could be applied to the wheel hubs (Foth and Graham 1983a,b). It has been possible to automate the process of leg position measurement so that one can derive position displays on a computer within a few minutes, analyse movement on-line and use the timing of these movements to disturb the system phasically.

A constant load or resistance to rearward movement of the legs produces two important changes in the timing of leg movements. The recovery duration is shortened to 100 ms under all conditions of load > 0.2 times the body weight and is independent of step period (in the absence of load, recovery duration varies between 100 ms and 400 ms and is directly proportional to step period). The lag between the recovery of adjacent legs remains relatively constant while the step period increases with increasing load.

There are at least two ways to model this behaviour. One proposes load or stress detecting organs in the cuticle of the legs as the origin of these changes. An alternative mechanism requiring only position-sensitive elements measuring the changing position of the legs during the support phase can be based upon a comparison of the "expected" rate of movement for a given level of motor output. If the actual movement of the leg is less than "expected" then this is considered to be equivalent to an external load, muscular fatique or the voluntary use of reduced motor output. Thus, the gait can be changed by higher level behavioural requirements as found in first instars under feeding conditions (Graham 1972). If fatigue is induced by prolonged walking the pattern of stepping is identical to that produced by an externally imposed load.

During a walk, the legs are recovered at regular intervals with reference to their extreme posterior position and the relative position of the other legs. The power-stroke muscles are activated to maintain a given joint angle in standing or increase it at the required rate in an uphill walk. When the leg reaches its posterior extreme position, provided neither a posterior nor a contralateral adjacent leg is in the recovery phase, the set point for support is altered to levate the leg and the recovery musculature is activated to move the leg forward for its next power stroke. At the anterior extreme position, the support angle is re-established and the rate of change of position along the body axis is compared with the rate of change in position for the required speed over the ground. This difference alters the activity in the retractor muscles. Such a mechanism responds to an external load, fatigue and volitional changes in exactly the same manner giving a tripod gait under low loading and the slow gait under conditions of muscle fatigue or high loading.

Downhill walking presents a number of interesting problems. Some current models based upon a two-state cycle of propulsion and recovery tend to run away catastrophically when under the equivalent of negative loading (force applied in the same direction as leg movement in the stance phase) or imply the existence of a controlled rate of movement. A model based on a velocity feedback servo of the kind proposed would use either power-stroke or recovery muscles depending upon the sign of the applied load. If during the power stroke the legs move faster relative to the body than is required, then muscle forces are reduced or antagonists are brought into action (Fig. 4).

Let us consider some of the strategies that various walking systems adopt when descending steep slopes. A man stands and moves forward and downward a step at a time keeping the body vertical. Horses and similar quadrupeds tend to hunker down at the back so that they can stand in equilibrium often at an angle to the line of steepest slope and again they step forward and down the slope. Only where the slope is very short will the horse or man run down it. If too long a down slope is attempted, the body may outrun the legs leading to a collapse forward. There appears to be no mechanism for providing a controlled reversal of power output except in a bipedal climber descending while facing up the slope. Hexapods normally have relatively short legs in comparison to their body length and must move down with the body axis approximately parallel to the slope. A

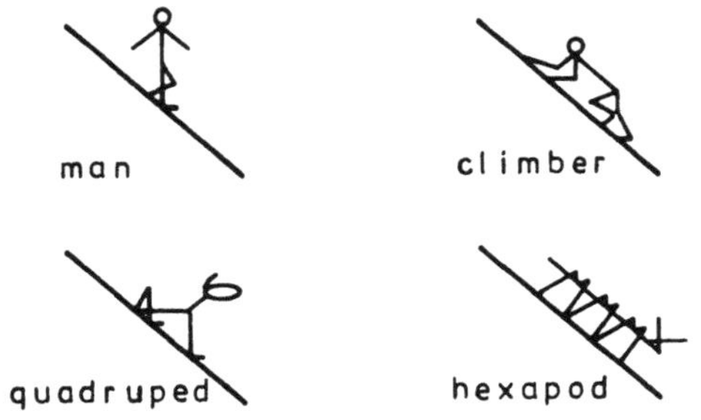

Fig. 4. Sketches of a man, horse, and climber and stick insect or robot hexapod descending a steep slope. See text for details

similar problem must be solved in robot vehicles. Both these systems require the application of forces opposing the direction of movement rather than balancing the body above the legs in a position of equilibrium.

There are several alternative strategies by which an insect might solve the problem.

1. Walk down backwards like a human climber descending a slope. This may be possible in scorpions and spiders (not strictly insects, but similar in structure), but has not been reported for these animals or any insect. In fact, a very precarious turn is made when an insect climbs to the top of a grass blade and finds it leads nowhere.

2. Walk forwards using some legs as props or anchors to oppose the pull of gravity and the propulsive forces of the other legs. This has been proposed in the cockroach to explain the significant changes in the coordination of the leading and following legs in down- and uphill walking, respectively (Spirito and Mushrush 1979). In locust, the long hind legs may be used as anchors, while the four anterior legs pull the body forward down the slope. The hind legs step much more slowly under this condition (Hustert, private communication).

3. The first part of the cycle may be used for braking purposes (Graham 1983). The following part of the stance phase then produces rapid forward movement. Stick insects use this method for 20% of their steps in downhill walking (Graham and Godden, in preparation).

4. The power stroke is controlled to produce a constant required velocity during the support phase regardless of the sign of the imposed load.

In the stick insect, a normal controlled displacement is used to walk downhill in 80% of the steps examined. The leg trajectory is identical to that found in horizontal and uphill walking. The "braking" type of step, in which the animal stands still momentarily after the placement of each tripod of legs, is most often observed in situations where different loads are applied to right and left leg sets and may be associated with an attempt to turn towards the more heavily loaded wheel. Stick insects, at least, and perhaps all insects appear to have the capability of controlled movement during the recovery phase with either a moving set point or velocity as the controlled variable (Cruse and Dean, this volume).

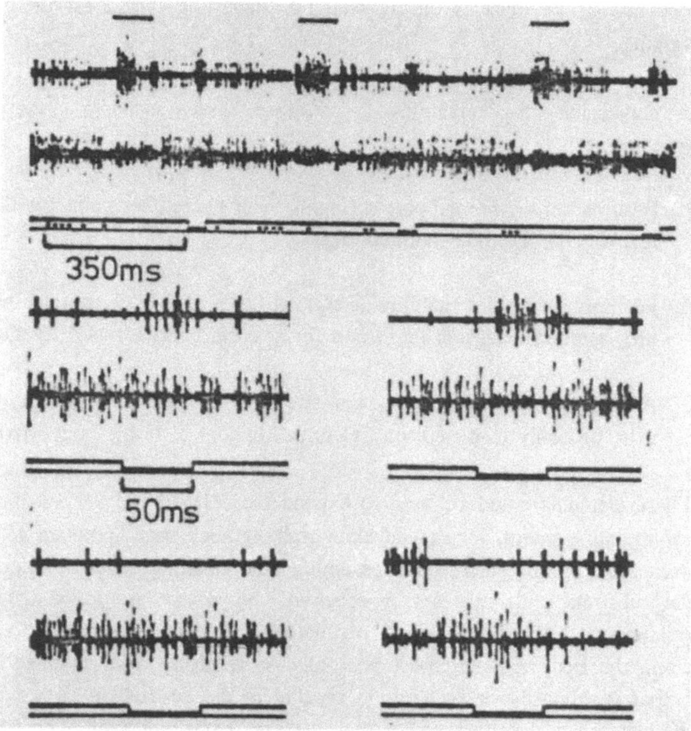

Fig. 5. Motor-nerve recordings for the protractor (recovery muscle, upper trace) and retractor (propulsion muscle, lower trace) of the middle leg of a stick insect are shown during the simulation of an uphill walk. In the upper record a pulse of downhill torque is applied three times for 50 ms (square pulse), 350 ms after the recovery stroke of the hind leg on the same side has been detected (a sequence of three or more dots on the lowest trace). The torque is applied in the same direction as the leg movement (momentary downhill condition). Before and after the pulse the torque opposes the direction of movement (uphill condition). The middle record shows the first and second pulses on an expanded scale. The lowest record on the same time scale shows a "control" situation in which the torque is zero before and after reversal (horizontal walking condition)

In order to test hypothesis (4) during the support phase, motor output to the propulsive and recovery (antagonist) muscles of a middle leg was recorded for a stick insect under simulated uphill conditions. These experiments were performed on a more sophisticated treadwheel in which a low inertia DC motor applied torque to the wheel, in the same direction or against the direction of motion, simulating downhill and uphill walking (Graham and Goddon, in preparation).

During an uphill walk, torque was applied against the direction of motion. A computer was used to detect the recovery movements of the hind leg. After a delay of 350 ms the voltage of the DC motor was inverted. This created a downhill torque for 50 ms. At low levels of torque, corresponding to a force of 0.2 times body weight, there was no detectable change in motor output. At higher levels where the legs were observed to be moved rearward by the pulse of downhill torque, the antagonist was activated and the power-stroke activity tended to be reduced (Fig. 5). This shows that induced movements in the direction of the normal power stroke are opposed by the recovery stroke muscles during walking suggesting that a servo- control system is active during walking.

In conclusion, for modelling a complete insect walking system the following elements are proposed:

1. Position servos for support springing with altered set point for the levation observed during the recovery stroke (hair-field control)

2. Position servos for stability lateral to body axis (CO control) with coupling between legs and programmed extension or flexion for rearward and forward oriented legs

3. A velocity servo for both power-stroke control and the recovery movement parallel to the body axis, probably mediated via CO or MRO control in the body-coxa joint.

These elements could be used to expand the delayed recovery-stroke coordinator already proposed for the stick insect, locust and slow walking cockroach (Graham 1977). The model would support itself clear of the surface and assuming that substrate contact confers grasping reflexes for holding the substrate until recovery is required, the system would be independent of the gravity vector orientation. This model would permit an appropriate response to either positive or negative load along the body axis imposed externally or resulting from fatigue. In addition, altering the motor output (voluntarily or by fatigue) relative to the desired or expected velocity would make possible the changes in gait observed in both the first instar and adult stick insect.

Can a meaningful model be based on these proposals? A detailed dimensional analysis of Carausius morosus has been made to determine the relative positions of muscle attachments, hinges and segment lengths and the masses of the individual components can easily be measured. Using a stick representation of the kind developed by Morasso (this volume) with the simple articulations of the insect leg it should be possible to incorporate the dynamics of locomotion and generate movements approximating a real time analysis.

It would appear that insect locomotion may provide an interesting test case for the new approach to the study of dynamic multisegmental systems advocated by Kelso and Saltzman (this volume). It should be possible to use both their methodology and the older reductionist or analytical approach in parallel. This would permit a direct comparison of these two perspectives to assess their respective heuristic and predictive merits.

REFERENCES

Baessler U (1983) Neural basis of elementary behaviour in insects. Springer. Berlin Heidelberg New York (Studies of brain function, vol 10)

Burns MA (1973) The control of walking in ortoptera. J Exp Biol 58: 45-58

Cruse H (1976) The control of body position in the stick insect Carausius morosus, when walking over uneven surfaces. Biol Cybern 24: 25-33

Cruse H (1980a) A quantitative model of walking incorporating central and peripheral influences.I. The control of the individual leg. Biol Cybern 37: 131-136

Cruse H (1980b) A quantitative model of walking incorporating central and peripheral influences. II. The connections between the different legs. Biol Cybern 37: 137-144

Cruse H, Epstein S (1982) Peripheral influences on the movement of the legs in a walking insect Carausius morosus. J Exp Biol 101: 161-170

Cruse H, Graham D (1984) Models for the analysis of walking in arthropods. In Bush BMH, Clarac F (eds) Co-ordination of motor behaviour. Cambridge University Press, Cambridge, pp 283-301

Dean J, Wendler G (1982) Stick insects walking on a wheel: perturbations induced by obstruction of leg protraction. J Comp Physiol 148: 195-207

Delcomyn F (1971) The effect of limb amputation on locomotion in the cockroach Periplaneta americana. J Exp Biol 54: 453-469

Delcomyn F, Usherwood PNR (1973) Motor activity during walking in the cockroach Periplaneta americana. I. Free walking. J Exp Biol 59: 629-642

Foth E, Graham D (1983a) Influence of loading parallel to the body axis on the walking coordination of an insect. I. Ipsilateral changes. Biol Cybern 47: 17-23

Foth E, Graham D (1983b) Influence of loading parallel to the body axis on the walking coordination of an insect. II. Contralateral changes. Biol Cybern 48: 149-157

Graham D (1972) A behavioural analysis of the temporal organisation of walking movements in the 1st instar and adult stick insect Carausius morosus. J Comp Physiol 81: 23-52

Graham D (1977) Simulation of a model for the coordination of leg movement in free walking insects. Biol Cybern 26: 187-198

Graham D (1978a) Unusual step patterns in the free walking grasshopper Neoconcephalus robustus. I. General features of the step pattern. J Exp Biol 73: 147-157

Graham D (1978b) Unusual step patterns in the free walking grasshopper Neoconcephalus robustus. II. A critical test of the leg interactions underlying different models of hexapod coordination. J Exp Biol 73: 159-172

Graham D (1979) The effects of circumoesophageal lesion on the behaviour of the stick insect Carausius morosus. II. Changes in walking coordination. Biol Cybern 32: 147-152

Graham D (1983) Insects are both impeded and propelled by their legs during walking. J Exp Biol 104: 129-137

Graham D (1985) Pattern and control of walking in insects. Adv Insect Physiol 18: 31–140

Graham D, Baessler U (1981) Effects of afference sign reversal on motor activity in walking stick insect (Carausius morosus). J Exp Biol 91: 179–193

Pearson KG, Iles JF (1973) Nervous mechanisms underlying intersegmental coordination of leg movements during walking in the cockroach. J Exp Biol 58: 725–744

Spirito CP, Mushrush, DL (1979) Interlimb coordination during slow walking in the cockroach. I. Effects of substrate alterations. J Exp Biol 78: 233–243

Stein RB (1982) What muscle variables does the nervous system control in limb movements? Behav Brain Sci 5: 535–578

Wendler G (1964) Laufen und Stehen der Stabheuschrecke Carausius morosus: Sinnesborstenfelder in den Beingelenken als Glieder von Regelkreisen. Z Vergl Physiol 48: 197–250

Yox DP, Dicaprio RA, Fourtner CR (1982) Resting tension and posture in arthropods. J Exp Biol 96: 421–425

Evidence for the Control of Velocity as Well as Position in Leg Protraction and Retraction by the Stick Insect*

J. Dean and H. Cruse

Fakultät für Biologie, Universität Bielefeld, Postfach 8640, 4800 Bielefeld 1, FRG

Observation alone does not reveal the nature of the control system underlying normal walking. In the past, theories about the control principles emphasized first, reflex chains and then later, central oscillators (reviewed by Delcomyn 1980). More recently, the very existence of a central pattern generator has been questioned for stick insect walking (Bässler and Wegner 1983). Nevertheless, the current consensus is that a basic pattern for locomotion may be generated centrally by the nervous system, but this pattern can be significantly shaped and altered by peripheral influences.

Not only the pattern, but also each individual movement, is subject to peripheral control. Precisely defining these peripheral influences has turned out to be a major challenge. Control theory derived from engineering and robotics offers a variety of standard parameters which might be controlled during movement. As summarized in a recent article by Stein (1982), these include position, velocity, force, muscle stiffness, viscosity, and impedance. The parameters actually represented in and controlled by the nervous system are not so easily equated with these terms. To give just one example, even tonic proprioceptors which signal position information usually have some degree of phasic response so that mixing of position and velocity information occurs at the very input to the nervous system. In the present paper, we would like to address this fundamental problem of motor control by examining the leg movements of stick insects during walking.

The rhythmic stepping of a leg can be divided into two phases. The stance or power stroke is the interval in which the leg is in contact with the substrate. During this time the leg can provide support for the body and contribute to propulsion by exerting force against the substrate. The swing or recovery stroke is the interval in which the leg is free of the substrate and moving to a position where it can begin a new stance. During this phase the leg contributes neither support nor propulsion. Two additional terms, retraction and protraction, are often used in the literature on insect locomotion. When a stick insect walks forward, the stance or power stroke typically involves retraction of the leg to the rear while the swing is primarily a protraction of the leg to the front.

The functional requirements placed on stance and swing differ in several important ways and these, in turn, affect the type of experiment that can be performed. Therefore, it is useful to consider these two movements separately in attempting to analyze the way in which they are controlled by the nervous system. (Experiments on the swing were performed by JD; those on the stance by HC. See also Dean 1984 and Cruse 1985a,b.)

* The research was supported by a grant from the Deutsche Forschungsgemeinschaft (Cr58/4).

Experimental Brain Research Series 15

SWING PHASE

By definition, the function of the swing or recovery phase is to return the leg from its position at the end of one power stroke to a position where it can begin a new one. Hence, the main program for the leg controller needs to perform two tasks: it needs, first, to find a substrate for the leg at the end of the swing and, second, to ensure continuous locomotion, it needs to complete the swing quickly enough for the leg to contribute propulsion and support before the remaining legs reach the limits of their stance positions.

With respect to the second task, swing duration, the adult stick insect is an exception. Most insects walk using a tripod gait in which both swing and stance durations are shortened proportionately with decreases in step period (reviewed in Delcomyn 1981). In contrast, for the adult stick insect the duration of the swing phase is almost constant for all walking speeds under normal conditions (Wendler 1964; Graham 1972). Recently, it has been shown that the load against which the animal is walking affects the speed of the swing phase (Graham 1981). High loading leads to faster swings. This change lengthens the duty cycle of the leg, increasing its contribution to overcoming the increased resistance. Apparently, the large abdomen of the adult stick insect creates a high load and the duration of the swing is minimized.

With respect to the first task, finding a substrate, the stick insect shows a unique adaptation to the irregular structure of its normal arboreal habitat. During the swing, the middle and rear legs are guided towards a suitable foothold by aiming at the location of the ipsilateral front and middle legs, respectively (Cruse 1979, Dean and Wendler 1983). The coordination is such that the target leg is nearing the end of its stance on a load-bearing substrate. Hence, each foothold located by a front leg is used in turn by the ipsilateral middle and rear legs. For their part, the front legs are guided by information from the antennae (Cruse and Krieger, in preparation).

These two features of the main program, short duration and aiming at a target, make the swing of the stick insect similar to many of the movements considered in studies of motor control in humans and other vertebrates. These similarities make the stick insect movement an attractive model for asking the same basic questions concerning control of movement. In essence, these questions are: first, whether feedback is active in controlling the movement, and if so, what parameters of the movement are controlled?

Unless a movement is so fast as to preclude use of feedback, it does not directly reveal any characteristics of the underlying control. In order to obtain such information, it is usual to perturb the movement in some way and to extrapolate from the elicited changes to the nature of the control system. One common perturbation is to change the load. This was the approach taken in studying the stick insect swing (Fig 1a).

Several possible control mechanisms were tested. The simplest hypothesis is that the leg behaves like a position servomechanism in which the force is simply proportional to the distance between the target position and the current position of the leg (Fig. 2a). If this were the case, then the actual end point of the protraction (AEP: anterior extreme position) should shift caudally when the load is increased. This was tested in the first experiment. Placing the target leg on a fixed platform, as shown in Fig. 1a, ensured that the target position did not change during the movement of the rear leg.

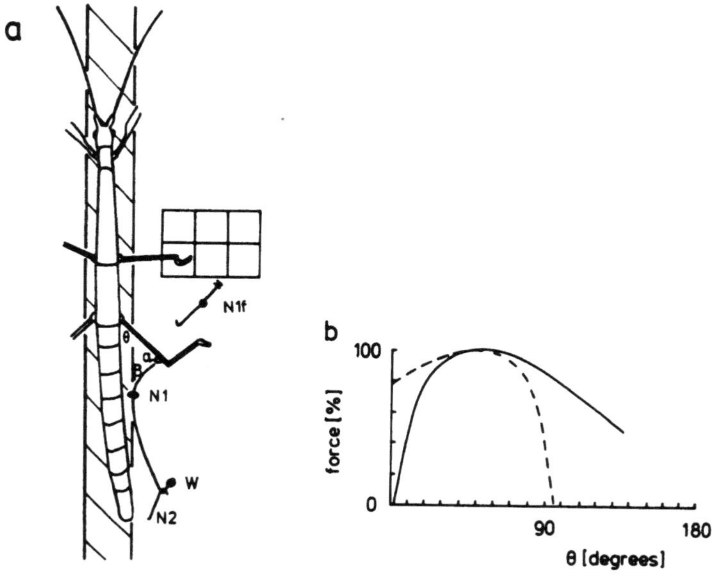

Fig. 1. Method for changing the load on leg protraction. (a) The insect is fixed along its dorsal surface to a holder (not shown) and walks by turning a styrofoam treadwheel (cross–hatched band). Weights (W) are attached to a fine thread passing through two guides (N1, N2) and fixed to the distal femur of the right rear leg. The position of N1 is shown for the condition in which the applied force opposed leg protraction. When the applied force was to aid protraction, N1 was located at N1f, slightly rostral to the metathoracic coxa and just beyond the reach of the femur. In the first series of experiments, the right middle leg, which serves as a target for the rear leg protraction, was placed upon a fixed platform beside the treadwheel. (b) The component of the applied force acting perpendicular to the femur is graphed as a function of leg angle. This component is the effective force opposing (solid line) or aiding (dashed line) leg protraction. (Fig. adapted from Dean 1984)

The end point of the rear leg step was measured first with no added load and then with weights of 110 mg and 250 mg opposing leg protraction. In general, the end point did show a small caudal shift with increased load, but this was not significant for most animals (Dean 1984). In particular, those animals for which the aiming was better in the control condition showed very little change with increased loading. The weights tested applied forces 14x and 32x the calculated maximum force needed by the leg during an unobstructed swing. Therefore, it is evident, first, that the force is not simply proportional to the distance from the target and, second, that feedback is involved in adjusting the force to meet the altered load.

With the platform, only the first step at the start of walking could be measured. Dynamic effects of loading could be better studied when all six legs were allowed to walk continuously on the treadwheel. Loads were again varied by adding weights to resist the protraction of the right rear leg. Weights of 50 to 750 mg were tested. Increasing the load elicited systematic changes in the leg's movement (Fig. 3). The AEP shifted caudally, the amplitude of the swing decreased, the duration increased, and overall velocity decreased. The decrease in velocity probably reflects basic

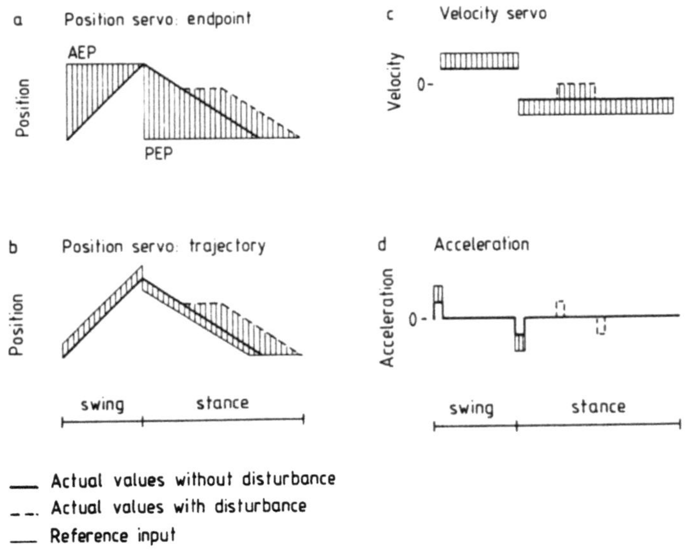

Fig. 2. Four different models for control of leg movement during stepping. For each model the thick solid line represents the actual value and the thin line represents the reference value of the controlled parameter as a function of time. The dashed lines deviating from the normal course of retraction illustrate changes induced when the retraction movement of the leg is temporarily prevented. Muscle activation is considered to be proportional to the difference between actual and reference values (vertical hatched area). The actual force is modified by the low−pass characteristics of the muscles. For position models, the anterior and the posterior extreme points (AEP and PEP) define the end points of swing and stance, respectively

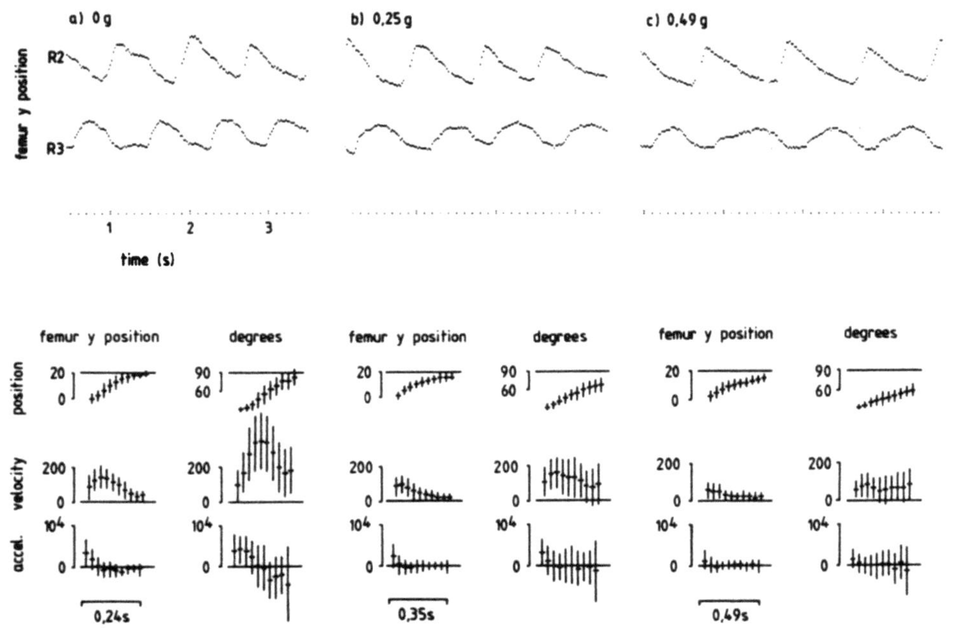

muscle properties: speed of contraction decreases with increasing load. The shift in the AEP might suggest that the targeting reaction no longer was maintained, but this was not entirely the case. The target middle leg continued its retraction during the prolonged swing of the rear leg. Hence the position of the middle leg at the time the rear leg completed its swing was also farther to the rear. This adjustment in the targeting behavior indicates that the target position is not fixed at the start of the movement.

The speed with which the feedback systems increase the force could be determined only approximately. For these measurements, a stop was positioned to support the weight during the initial part of the swing. For the lighter weights of 110 and 250 mg, the leg often continued moving forward, albeit with reduced velocity, through the onset of the opposing force (Fig. 4a). Overcoming the increased load required forces considerably greater than the maximum calculated for unobstructed protractions. Therefore, this observation indicates a response time of less than 20 ms, the limit of resolution for the video recording system.

As a further test of the "position servo" hypothesis, the rear leg was subjected to forces which aided rather than opposed protraction. If the leg behaves as a position servo with its reference position set to the end point, then the leg should move forward more rapidly under the influence of the applied weight. This was not the case: the velocity of the protraction generally did not increase significantly.

A more complicated version of a position servomechanism is one in which the reference position for the servo is not set at the end point but rather moves along a trajectory from the start to the end point (shown in Fig. 2b, for stance). This model could also be tested with external forces aiding rather than opposing protraction. In this case it was assumed that the applied weights would move the leg forward ahead of the intended trajectory, leading to a pause or to small retractions when the weight was suddenly removed. Such aberrations were not usually present (Fig. 4b).

The results concerning the swing phase can be summarized as follows. Position parameters are used to determine the end point of the movement and this information is updated during the swing. Position parameters do not appear to influence the force. Instead it seems that the velocity of the movement is controlled and that muscle force can be increased or decreased or antagonists activated when the velocity varies from that intended.

◄ Fig. 3. Effect of force opposing rear leg protraction during continuous walking. Three different load conditions are illustrated in (a–c). Typical steps are shown in the upper row. The two traces represent movement of the rear leg subjected to load (R3) and the ipsilateral middle leg (R2), with protraction corresponding to upward change in the trace. With increasing loading, the protraction movement of the rear leg is prolonged and slower, and the AEP is shifted caudally. The lower row shows averaged profiles for leg position, velocity, and acceleration measured both as femur y position and as leg angle. The durations of the swings were normalized and divided into 10 classes for computing means and standard deviations. The time bars beneath each abscissa show the mean swing duration for each condition. Units on the ordinate are either femur y position, at the left, or degrees, at the right; the corresponding time derivatives are in sec^{-1}. Femur y position is proportional to tan $(90-\theta)$: it magnifies changes in leg position for small θ but is more directly related to tarsus position on the wheel.

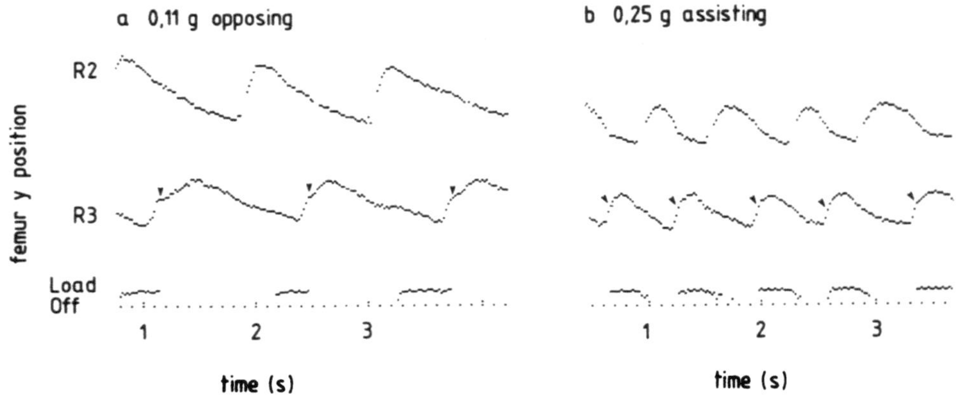

Fig. 4. Protractions influenced by sudden changes in external force. Representation of leg movement is similar to that in Fig. 3. The line of points above the time axis indicates when the load was being carried by a stop and not acting on the leg. (a) The load opposing rear leg protraction takes effect at the arrows. (b) The force assisting protraction ceases at the arrows. (Figure from Dean 1984)

STANCE PHASE

The stance is the working part of the leg's cycle. In this phase the leg is contacting the substrate and can contribute to supporting and propelling the body. Because the leg is mechanically coupled through the substrate to all the other supporting legs, the analysis of movement in an individual leg is less straightforward than for the swing. Nevertheless, several features of stance can be characterized by observation. First, the speed of forward progression is directly determined by the speed of leg movement during stance. Since the amplitude of leg movement usually does not change during straight walking, stance duration also depends upon walking speed (Graham 1972). Second, the end point of retraction (PEP: posterior extreme position) is determined by a combination of influences both from the leg itself and from intersegmental or central sources. These include the position of the leg as measured by proprioceptors (Bässler 1977, Dean and Wendler 1983), the load on the leg either in the form of cuticle deformation (Bässler 1977) or as the weight supported by the leg (Jander 1985), inhibition arising from an adjacent posterior leg during its swing (Cruse and Epstein 1982), and undefined central oscillations (Cruse and Saxler 1980). Third, although leg retraction in arthropods is often represented as having a constant velocity, for the stick insect, at least, it proceeds in a characteristic "lurching" manner (Graham 1983). Part of the decrease in velocity of leg retraction is thought to arise through mechanical coupling as other legs oppose forward motion immediately after beginning their stances. Fourth, retraction generally proceeds with a higher speed when the longitudinal resistance is decreased (Graham and Cruse 1981, Epstein and Graham 1983).

More information on the control of individual legs can be gained by adapting a finding of Wendler (1964). When an animal walks on a treadwheel and one leg is placed on a fixed platform beside the wheel, that leg generally remains in stance. The force exerted by the leg in the longitudinal

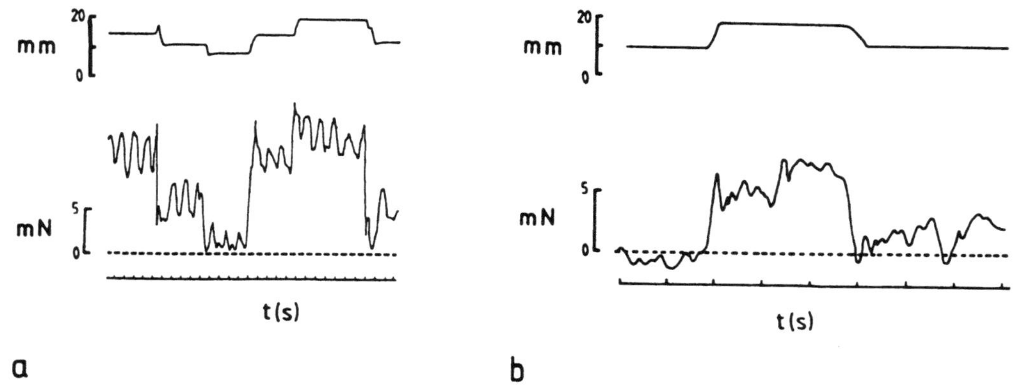

Fig. 5. Measurements of retractor force from a front leg standing on a platform. The leg stood on a force transducer which was moved forward and backward by hand. The upper trace shows the position of the platform with anterior being upward; the lower trace shows the force in the anterior–posterior direction with force which would retract the leg shown as positive. The force trace shows large DC changes which parallel the shifts in platform position. The dynamic oscillations around these static levels are coordinated with the leg movements of the walking legs. The measurements were made before (a) and after (b) severing the tendon of the chordotonal organ and thereby removing the major proprioceptive input monitoring position in the anterior–posterior direction. The leg continues to exert more force at anterior positions, which points to properties of muscles and joints rather than proprioceptive loops as the origin of this effect

direction typically can be divided into static and dynamic components (Fig. 5a). The static component can be measured to test hypotheses for the control of stance which are analogous to those already discussed for the swing.

The simplest hypothesis is again that the force exerted by the leg is proportional to the distance from the intended end point (Fig. 2a). According to this model, the mean force exerted by a leg should increase for more rostral platform positions: this is the case, at least qualitatively (Fig. 6). However, it is known that the isometric force which muscles can exert varies with their length. Control measurements made with the major proprioceptive input eliminated showed the same pattern of force changes (Fig. 5b). Therefore, this pattern can be attributed to muscle properties or to lever arm changes.

Hence this finding does not answer the original question of whether the end point position is used to control the force exerted by the retractors. This question can be tested directly by comparing the force curve presented above with one measured while the swing of the adjacent posterior leg is obstructed. The obstruction keeps the posterior leg in continuous swing phase. During its swing the posterior leg inhibits a swing by the adjacent leg (Dean and Wendler 1982). When this inhibition is prolonged, the mean retraction end point of the forward leg shifts caudally (Cruse and Epstein 1982). This inhibition can be demonstrated for leg positions throughout the step (Cruse 1985b). If it is assumed that this actual end point also serves as the target during the movement, then a position servomechanism using this end point as its reference point should produce a parallel shift

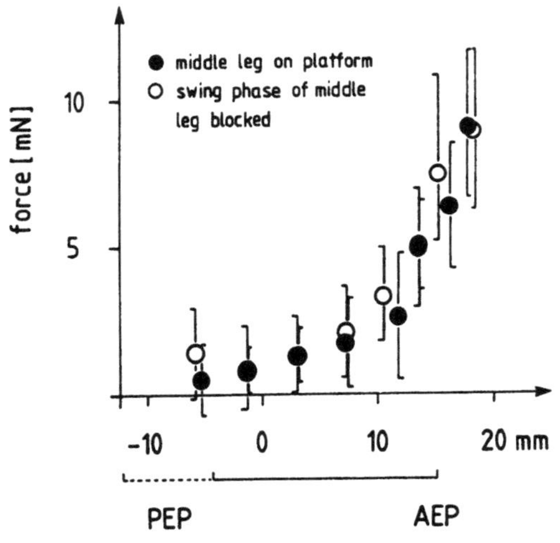

Fig. 6. Lack of effect of changed end point on force during retraction. Measurements of the retraction force of the front leg were made with the ipsilateral middle leg either on a platform (•) or with its swing phase blocked (○). Obstructing the swing phase causes the retraction end point (PEP) of the front leg to shift caudally, as indicated by the dashed line on the abscissa. Nevertheless, the force exerted by the leg as a function of position does not change, indicating that the position of the end point is not used to determine the force

in the relationship between force and position. In fact, the force curve is unchanged (Fig. 6). Therefore, the altered end point is not involved in determining the retractor force.

As was the case for the swing, a further variant of the position servo hypothesis is one in which the reference input is moved along a trajectory. If this trajectory is independent of peripheral feedback, then several predictions can be made about how the force exerted by the leg should change when the actual movement is speeded up or slowed down. First, however, it is necessary to measure the force produced by the leg as it moves through a normal retraction (Fig. 7a). This force reaches a maximum midway in the retraction.

Using this data, the responses predicted for different control models can be formulated (Fig. 2). For a position servomechanism based on a planned trajectory, stopping the movement of the leg should cause the difference between the actual position and the intended trajectory to increase for the duration of the halt and this should be reflected in the force (Fig. 2b). Figure 7b shows that this was not the case: after an initial increase, the force actually declines during the halt. (Some initial increase might be expected simply on the basis of muscle properties, but several observations indicate that a large active response is present as well. First, the measured front leg sometimes resists the imposed movement, indicating that the protractors and/or extensors are active. However, if the imposed movement then is stopped, the force reverses direction, indicating that the normal power stroke muscles are turned on. Second, Graham (this volume) presents analogous electrophysiological evidence for activation of protractors or retractors as appropriate to counter torque pulses perturbing the power stroke.)

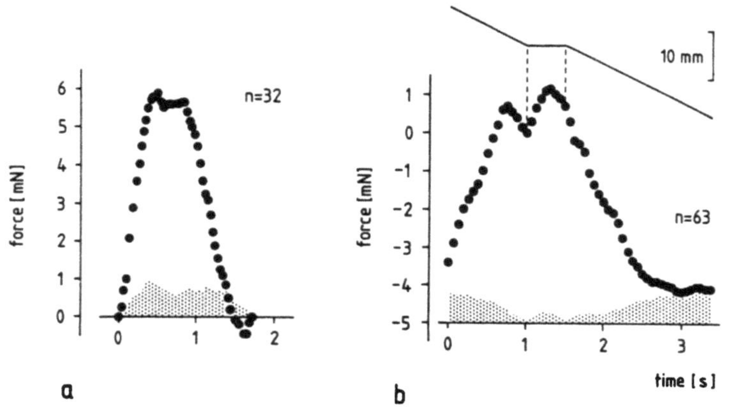

Fig. 7. Measurements of retraction force with and without an interruption in leg movement. Measurements were made by moving a force transducer beside the treadwheel at about the walking speed and recording the force when a leg stepped on the transducer. The trace at the top illustrates the movement of the transducer and leg. (a) The transducer was moved smoothly throughout the retraction. (b) The transducer was halted briefly during the retraction. When leg movement is prevented, the force record shows a maximum. This pattern best fits a model in which force is derived from a combination of velocity and acceleration signals (models c and d in Fig. 2). The sizes of the standard deviations are shown by the dotted area above the abscissa

If position is not the parameter controlling the force exerted by the leg, what other possibilities are there? One is movement velocity, the first derivative of position. If the leg is attempting to maintain a constant velocity, then stopping it should result in a constant deviation between the desired velocity and the actual velocity (Fig. 2c). If this deviation is the input to the force controller, then force should remain constant for the duration of the halt. This is a better approximation to the observed results but still not completely accurate. The decline in the force as the halt is maintained suggests that acceleration, the second derivative of position, also affects the force (Fig. 2d). Our measurements are not sufficient to determine the exact relationship of output force to either velocity or acceleration and thus do not meet a strict definition of "control". Therefore we will use the expression "velocity sensitive control".

CONCLUSION

The preceding discussion has treated the swing and stance movements separately, but it should be clear that there are many similarities in their control. First, as many previous studies have demonstrated, peripheral feedback greatly influences the motor pattern during walking and this is true both for swing and for stance. These peripheral influences include both the position of the leg and the load upon it. However, the results presented here show that different movement parameters influence different aspects of the movement. Position information is involved in the decision to terminate both swing and stance. The position of the target leg provides the most important criterion for ending the swing phase and the position of the protracting leg determines when this criterion is met. Proprioceptive information from both the target leg and the moving leg is involved (Cruse et al. 1984; Dean and Wendler 1983). Position is one of several factors contributing to ending the stance phase. A second factor is the load on the leg both in the axial and in the vertical directions. A third factor is coordination from intersegmental or central sources. Inhibition during the swing of the adjacent caudal leg causes the PEP to move caudally. Undefined central sources cause the force exerted by a leg held at a constant position to vary in rhythm with the walking legs. This influence could arise either from a central oscillator for the leg or from coordination signals transmitted from the other legs. In either case, it could affect the stance end point either indirectly by way of the self-induced loading on the leg or directly through the relevant interneurons.

Position information does not appear to be important in determining the force developed by the muscles during the movement. Neither the position of the end point to be reached nor proprioceptive input signalling leg position affect the output force. Force does vary as a function of position due to the geometry of the joints and the mechanical properties of the muscles involved, but these factors are not under the control of the nervous system.

The nervous system does modulate force in response both to deviations from the normal velocity and, to some extent, to the accelerations accompanying these deviations. This amounts to a velocity sensitive control during movement. For both stance and swing, the velocity of the movement has a direct bearing on the overall performance of locomotion. During stance, retraction velocity governs the speed of forward locomotion; during swing, protraction velocity enables the leg to assume its stance position at the proper time. Therefore, it is appropriate that the velocity of the leg movement be controlled in the longitudinal plane. In other planes, other parameters may be more appropriate. For example, position information is surely important for controlling the height of the body during stance.

The swing movement, and the stance to a lesser extent, resemble what have been called fast ramp movements in vertebrates (Desmedt and Godaux 1978). These are rapid movements in which muscle activity continues throughout the movement. As in the stick insect, rapid reflex pathways are active to meet variations in load. The normal load for the swing is simply the mass of the moving limb. This load is predictable and load compensation probably plays a negligible role in normal walking. For the stance, the load can vary considerably depending upon the insect's orientation with respect to gravity.

The findings presented above and those from the literature can be used to construct the diagram of the leg controller shown in Fig. 8. This figure separates factors affecting the termination of a step phase from factors which influence the output force during a phase. The first group, which includes position information, contains factors which can be considered part of the oscillator. The second group, which includes velocity, contains those which participate in a servomechanism. The separation made here does not necessarily correspond to structural separation in the nervous system. Force compensation could occur either through feedback at the motoneuron or at the interneuron level. Nevertheless, it is interesting to compare the scheme presented here with recent physiological findings in the locust flight system. In the steering pathways from the ocelli, the sensory input has a weak direct effect on motoneurons but the stronger effect is on premotor interneurons which are not part of the flight oscillator (Reichert et al. 1985).

Acknowledgements. We are grateful to D. Forsythe for helpful criticism of the manuscript.

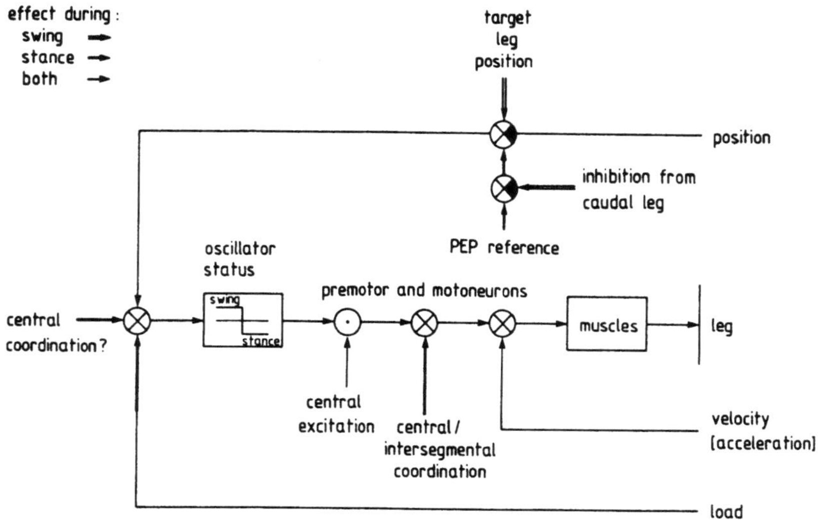

Fig. 8. A schematic summary of control influences during stepping. The leg and various parameters of its motion are represented on the right margin. Factors considered part of the oscillator act on the oscillator status box at the left and control switching between stance and swing. Factors affecting movement during a phase enter the control path between oscillator and muscles

REFERENCES

Bässler U (1977) Sensory control of leg movement in the stick insect Carausius morosus. Biol Cybern 25: 61–72

Bässler U, Wegner U (1983) Motor output of the denervated thoracic ventral nerve cord in the stick insect Carausius morosus. J Exp Biol 105: 127–145

Cruse H (1979) The control of the anterior extreme position of the hindleg of a walking stick insect, Carausius morosus. Physiol Entomol 4: 121–124

Cruse H (1985a) Which parameters control the leg movement of a walking insect? I. Velocity control during the stance phase. J Exp Biol 116: 343–355

Cruse H (1985b) Which parameters control the leg movement of a walking insect? II. The start of the swing phase. J Exp Biol 116: 357–362

Cruse H, Saxler G (1980) The coordination of force oscillations and of leg movement in a walking insect (Carausius morosus). Biol Cybern 36: 165–171

Cruse H, Epstein S (1982) Peripheral influences on the movement of the legs in a walking insect Carausius morosus. J Exp Biol 101: 161–170

Cruse H, Dean J, Suilmann M (1984) The contributions of diverse sense organs to the control of leg movement by a walking insect. J Comp Physiol A 154: 695–705

Dean J (1984) Control of leg protraction in the stick insect: a targeted movement showing compensation for externally applied forces. J Comp Physiol A 155: 771–781

Dean J, Wendler G (1982) Stick insects walking on a wheel: perturbations induced by obstruction of leg protraction. J Comp Physiol 148: 195–207

Dean J, Wendler G (1983) Stick insect locomotion on a walking wheel: interleg coordination of leg position. J Exp Biol 103: 75–94

Delcomyn F (1980) Neural basis of rhythmic behavior in animals. Science 210: 492–498

Delcomyn F (1981) Insect locomotion on land. In: Herreid CF II, Fourtner CR (eds) Locomotion and energetics in arthropods. Plenum, NY, pp 103–125

Desmedt JE, Godaux E (1978) Ballistic skilled movements: load compensation and patterning of the motor commands. In: Desmedt JE (ed) Cerebral motor control in man: long loop mechanisms. Karger, Basel, pp 21–55 (Progress in clinical neurophysiology, vol 4)

Epstein S, Graham D (1983) Behaviour and motor output of stick insects walking on a slippery surface. I. Forward walking. J Exp Biol 105: 215–229

Graham D (1972) An analysis of walking movements in the first instar and adult stick insect. J Comp Physiol 81: 23–52

Graham D (1981) Walking kinetics of the stick insect using a low–inertia counter–balanced pair of independent tread–wheels. Biol Cybern 40: 49–57

Graham D (1983) Insects are both impeded and propelled by their legs during walking. J Exp Biol 104: 129–137

Graham D, Cruse H (1981) Coordinated walking of stick insects on a mercury surface. J Exp Biol 92: 229–241

Jander JP (1985) Mechanical stability in stick insects when walking straight and around curves. In: Gewecke M, Wendler G (eds) Insect locomotion. Parey, Hamburg, pp 33–42

Reichert H, Rowell CHF, Griss C (1985) Course correction circuitry translates feature detection into behavioural action in locusts. Nature 315: 142–144

Stein RB (1982) What muscle variable(s) does the nervous system control in limb movements? Behav Brain Sci 5: 535–577

Wendler G (1964) Laufen und Stehen der Stabheuschrecke Carausius morosus: Sinnesborstenfelder in den Beingelenken als Glieder von Regelkreisen. Z Vergl Physiol 48: 198–250

Subject Index

MIX
Papier aus verantwortungsvollen Quellen
Paper from responsible sources
FSC® C105338

If you have any concerns about our products,
you can contact us on
ProductSafety@springernature.com

In case Publisher is established outside the EU,
the EU authorized representative is:
Springer Nature Customer Service Center GmbH
Europaplatz 3, 69115 Heidelberg, Germany

Printed by Libri Plureos GmbH
in Hamburg, Germany